IET TRANSPORTATION SERIES 23

Behavioral Modeling and Simulation of Bicycle Traffic

Other related titles

Volume 1	**Clean Mobility and Intelligent Transport Systems** M. Fiorini and J.-C. Lin (Editors)
Volume 2	**Energy Systems for Electric and Hybrid Vehicles** K.T. Chau (Editor)
Volume 5	**Sliding Mode Control of Vehicle Dynamics** A. Ferrara (Editor)
Volume 6	**Low Carbon Mobility for Future Cities: Principles and applications** H. Dia (Editor)
Volume 7	**Evaluation of Intelligent Road Transportation Systems: Methods and results** M. Lu (Editor)
Volume 8	**Road Pricing: Technologies, economics and acceptability** J. Walker (Editor)
Volume 9	**Autonomous Decentralized Systems and Their Applications in Transport and Infrastructure** K. Mori (Editor)
Volume 11	**Navigation and Control of Autonomous Marine Vehicles** S. Sharma and B. Subudhi (Editors)
Volume 12	**EMC and Functional Safety of Automotive Electronics** K. Borgeest
Volume 16	**ICT for Electric Vehicle Integration with the Smart Grid** N. Kishor and J. Fraile-Ardanuy (Editors)
Volume 17	**Smart Sensing for Traffic Monitoring** N. Ozaki (Editor)
Volume 18	**Collection and Delivery of Traffic and Travel Information** P. Burton and A. Stevens (Editors)
Volume 25	**Cooperative Intelligent Transport Systems: Towards high-level automated driving** M. Lu (Editor)
Volume 26	**Traffic Information and Control** R. Li and Z. He (Editors)
Volume 30	**ICT Solutions and Digitalisation in Ports and Shipping** M. Fiorini and N. Gupta
Volume 38	**The Electric Car** M.H. Westbrook
Volume 45	**Propulsion Systems for Hybrid Vehicles** J. Miller
Volume 79	**Vehicle-to-Grid: Linking electric vehicles to the smart grid** J. Lu and J. Hossain (Editors)

Behavioral Modeling and Simulation of Bicycle Traffic

Ling Huang and Jianping Wu

The Institution of Engineering and Technology

Published by The Institution of Engineering and Technology, London, United Kingdom

The Institution of Engineering and Technology is registered as a Charity in England & Wales (no. 211014) and Scotland (no. SC038698).

First published 2021

The Institution of Engineering and Technology
Michael Faraday House
Six Hills Way, Stevenage
Herts, SG1 2AY, United Kingdom

www.theiet.org

British Library Cataloguing in Publication Data
A catalogue record for this product is available from the British Library

ISBN 978-1-78561-951-9 (hardback)
ISBN 978-1-78561-952-6 (PDF)

Typeset in India by MPS Limited

Contents

About the authors

Ling Huang, PhD is an associate professor in the School of Civil Engineering and Transportation at South China University of Technology, China. She is also a member of Traffic Modeling and Simulation Professional Committee of China. Her research interests include cyclist behavior, bicycle sharing system, traffic simulation, behavior analysis, driving behavior, pedestrian, computer vision and traffic simulation applications.

Jianping Wu, PhD is a professor in the School of Civil Engineering at Tsinghua University, China, and director of the Future Transport Research Centre of Tsinghua University, the University of Cambridge, Massachusetts Institute of Technology. He is associate editor of IET ITS, a fellow of the IET, member of the Environment and Engineering Committee (CEE) of the WFEO, executive director of the China Association of Simulation, and "Cheung Kong Scholar" Professor of the Ministry Education of China.

Chapter 1
Section bicycle individual riding characteristics

1.1 Concepts of bicycle traffic

Bicycles, also known as a bike or a cycle, are usually small two-wheeled land vehicles. After people get on the bicycle, they pedal as the power. Thus, a bicycle is a green, healthy and environment-friendly means of transportation. Figure 1.1 is the typical bicycle. There are many kinds of bicycles, including single bicycle, double bicycle and multi-person bicycle.

Bicycles can be used as environment-friendly vehicles for walking and traveling.

More and more people use bicycles as fitness equipment for cycling exercise and bicycle travel. Cycling itself is also a sport. There are road bicycle race, mountain bicycle race, field bicycle race, acrobatic bicycle race and so on.

Although kinds of bicycles are mentioned earlier, the bicycle traffic discussed in this book is the ordinary bicycles for commute trips. Cycling is a kind of flexible, portable and relatively low-speed traffic mode. The power of bicycle is generated by human body, so it is not suitable for long-distance travel.

The advantages of bicycle traffic include mobility, flexibility, economy and environmental friendliness (occupying small road area, no air pollution, etc.). Bicycle traffic plays an important role in urban transportation in developing countries, such as China.

In recent years, with the increasingly prominent environmental problems, bicycle traffic has become more and more valued as a green, flexible and efficient alternative traffic mode in developed countries with dispersed residence and developed motor vehicles.

From the road area occupied by a vehicle, a running bicycle with a speed of 15 km/h occupies a road area of 8 m^2, while a car at the same speed occupies a road area of 50 m^2, more than six times that of a bicycle. In fact, the average number of people in a car is far less than six.

Therefore, compared with motor vehicles, bicycle traffic has higher road efficiency.

As the microscopic dynamic characteristics of bicycles are quite different from those of motor vehicles, following we will introduce some unique bicycle microscopic riding characteristics.

Figure 1.1 Typical bicycle (GIANT bicycle)

1.2 Overview of bicycle microscopic riding characteristics

Bicycle microscopic riding characteristics are produced by the two wheel structural characteristics of bicycles.

These microscopic dynamic characteristics are not directly related to the macroscopic state of traffic flow, such as the traffic volume. They embody the individual motion characteristics of bicycle riding.

Following are basic motion characteristics in cycling.

1.2.1 Instability

The contact area between a bicycle tire and ground is small. When the tire pressure is 35, 28, 21 and 14 N, the contact area between the tire and the ground is 23, 26, 30 and 39 cm^2, respectively [1].

When riding the bicycle, the cyclist controls the direction by moving the handlebars and the center of his/her own body to the left and right.

So the motion of bicycle riding is dynamic equilibrium. This can be seen in the serpentine tracks that bicycles leave on the road.

1.2.2 Lateral pressure

When bicycles and motor vehicles are mixed together, cyclists are subjected to psychological lateral pressure from motor vehicles.

This psychological lateral pressure is inversely proportional to the lateral distance between the bicycle and the motor vehicle. The smaller the lateral distance between bicycle and motor vehicle, the greater the pressure and the more threatened and unsafe the cyclists feel. The farther apart from the motor vehicle, the cyclist feels less stressful and safer.

In the case of the same lateral distance, the higher speed or the larger size of the peer motor vehicle, the greater the psychological lateral pressure on the peer cyclist [2].

1.2.3 Short travel distance

The driving force of bicycles is generated by the human body.

If continuous pedaling for more than 30 min, an adult man can only produce an average power of 0.15 kW, and an adult woman may only produce 70% of that. Therefore, bicycles are not suitable for long-distance trips. Most bicycle travels are no more than 50 min, and generally bicycle travels are about 30 min.

1.2.4 Quick start at intersection

Compared with motor vehicles, bicycles start more quickly at intersection stop lines. According to experiments conducted in China city, bicycles start at intersections more than 1 s earlier than motor vehicles [3].

1.2.5 Clustering

Bicycles, unlike motor vehicles, do not keep a certain distance and move in a straight line according to the lanes. Instead, they move in cluster, unwilling to slow down and stop [2].

1.2.6 Agility

Compared with a motor vehicle, a bicycle is a tool with strong mobility and agility.

Bicycles are easy to steer, accelerate or decelerate. At mixed traffic flow intersections, if the speeds of the motor vehicles are low, the bicycles can "pass" through the interval between the motor vehicles (Figure 1.2) [2].

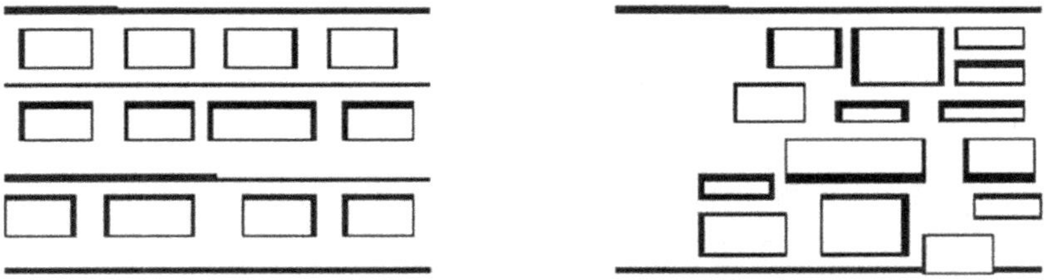

(a) (b)

Figure 1.2 (a) Lane-based motor vehicles and (b) non-lane-based mixed traffic flow

1.3 Overview of bicycle traffic in various countries

In some countries, a bicycle is an important traffic mode, such as China and the Netherlands. They are called "bicycles kingdoms."

In recent years, the rapid expansion of motor vehicle traffic has brought serious social and environmental problems.

Environmentalists, social activists and urban planners have come to realize that motorcar-oriented urban transport system may be the main source of urban economic, social and environmental problems.

A bicycle is a kind of green transportation in city. It not only saves energy and reduces pollution but also is more flexible and convenient. Therefore, governments around the world have begun to actively encourage the use and development of bicycle transportation in cities.

1.3.1 China

Bicycles are the most common and widespread vehicles in Chinese cities.

By survey, in the 1980s and 1990s, the level of ownership of urban cyclists (aged 14–60) in China has exceeded that of one bicycle per capita.

Xu and Chen [4] analyzed the survey data of origin-destination (OD) spatial distribution of bicycle trips in several cities. The analysis shows the following: the bicycle traffic in downtown or central business district (CBD) is densely distributed and has a large flow. It also shows the phenomenon that the bicycle traffic flow from all directions to the center increases with the approach of the center. This often results in bicycle traffic congestion and road overload in urban central areas.

As the power is generated by human body, bicycles are not suitable for long-distance trips.

Most bicycle travels are no more than 50 min, and generally bicycle travels are about 30 min.

Table 1.1 shows the percentages of walking, bicycle and transit traffic modes for different travel time groups in China's cities.

1.3.2 The Netherlands

The Netherlands can be called the "kingdom" of bicycles. According to the average population ownership, in the Netherlands, every 1.6 people have a bicycle.

Early in 1937, the world's first bicycle tunnel was built in Rotterdam, the second largest city in the Netherlands. The tunnel was built at the bottom of the Maas River 30 m below the ground. It connects the new and old districts of Rotterdam. At the entrances and exits of the tunnel, five elevators were set up for bicycles to enter and exit the tunnel automatically.

Then, in 1941, a road tunnel was built at the bottom of the Mass River in Rotterdam. There are four lanes in the tunnel with dedicated lanes for bicycles and pedestrians. At present, there are 10,000 km of bicycle lanes/roads in the Netherlands.

In order to improve the environment for bicycle, the Netherlands Ministry of Transport has promoted the transformation from the use of automobiles to bicycles

Table 1.1 Percentage of various travel modes grouped by travel time in cities of China [4]

Travel time (min)	Traffic mode		
	Walk (%)	**Bicycle (%)**	**Transit (%)**
<10	52.6	44.0	2.0
11–15	39.0	55.7	3.7
16–20	35.0	52.8	9.2
21–25	27.5	52.5	17.3
26–30	21.9	50.9	23.7
31–35	13.6	36.0	43.7
36–40	11.6	35.6	45.4
41–45	8.0	32.7	50.6
46–50	8.2	24.9	58.2
51–55	6.0	19.9	64.4
56–60	7.0	6.2	62.9
<60	7.6	14.4	15

and bicycles and public transit. The measures include rectifying and add bicycle parking lots and strengthening anti-theft measures and the "bicycle expert program," which has have completed 112 designs and achievements in 1990–1999.

1.3.3 The United States

The West Coast of the United States is considered the golden zone for cycling, where bicycles can be used as transportation throughout the year. Thus, California has the largest number of cyclists. The Davis City is called "University town," where most students like to ride bicycles, so it is also called "Bicycle town."

The city has a population of 38,000 people, 30,000 bicycles and more than 60 km of bicycle roads, underground bicycle tunnels and underground garages.

Americans admire the convenience of cycling. Recently, the purchase of bicycles has exceeded that of automobiles. There are nearly 100 million bicycles in the country. There are 155,000 km of bicycle roads. Every day, more than 14 million people ride bicycles to go to working places and schools.

1.3.4 Germany

In recent years, the German government has begun to actively encourage the use of non-motor vehicles in cities, vigorously advocating pedestrian and bicycle travel as energy conservation and environmental protection. This measure has effectively alleviated the traffic pressure and improved the traffic conditions, the urban economic, environmental and living conditions.

According to the data from the Federal Bureau of Statistics, in Germany 814 bicycles per 1,000 inhabitants, 38% of whom ride to work because they think cycling is not only simple, convenient and safe, but also a good way to exercise. The German Ministry of Transport spared no expense in this regard. In 2002 alone, it spent 200 million euros on the construction of bicycle lanes, which gradually formed a dense network of bicycle lanes in Germany. Well-planned bicycle lanes have greatly reduced the incidence of traffic accidents.

In addition, more advanced bicycle lanes are being developed all over Germany. For example, in 2010, bicycle expressways have been built in major cities for bicycle use, with a maximum speed of 50 km/h. The "Bicycle Highway" refers to the intercity traffic lines with flat and wide roads, which are specially used for cycling.

Although the two sides of these dedicated "Bicycle Highways" are not completely closed, there are no intersections throughout the journey, so there is no need to set up traffic lights. Pedestrians and automobiles are prohibited on the road, which enables cyclists to travel between the two cities at a very high speed. In addition, a large "aerodynamic" system is installed beside the bicycle highway, so that cyclists can move forward quickly with the help of the wind, which is both fast and labor-saving.

It can be seen from the previous text that bicycles will not be reduced or disappeared as an alternative traffic mode in the future. In large modern cities, bicycle traffic will be connected with the transit system, such as buses, metro, urban rail and so on, to form an integrated urban transportation system. While in small- and medium-sized cities, bicycle traffic will still be one of the most important commute modes because of the relatively short trip distance.

In addition, bicycle vehicles and facilities will also have new developments, such as the rapid growth of electric bicycles in many cities, many regions began to plan special bicycle network, develop more advanced bicycle-related facilities, such as Germany's "Bicycle Highway," in order to promote and encourage the development of bicycle traffic.

1.4 Bicycle traffic features in comparison to other transport means

The development of bicycle traffic can not only effectively solve the problems of urban resources, urban environment, urban traffic but also improve the health level of citizens. Therefore, it is a kind of energy-saving, environmental protection, convenient, economical and healthy low-carbon transport, especially suitable for short-distance travel.

1.4.1 Economize on energy consumption

The energy used by modern vehicles is nonrenewable and limited on the earth, but any motorized vehicle relies on energy. The rapid increase of motor vehicles will aggravate the contradiction between supply and demand between energy and motor vehicles.

Relevant data analysis shows that the proportion of oil consumption in transportation industry in the whole country is 3% in Britain, 27% in Japan, 28% in France and 35% in the United States (in 1992) and about 30% in China (in 2007).

Table 1.2 Fuel consumption of various traffic modes (1 for standard bus) [5]

Traffic mode	Fuel consumption per passenger-kilometer
Walking	0
Cycling	0
Motorcycle	5.5
Private car	8.2
Standard bus	1
Articulated bus	0.9
Bus (with dedicated lane)	0.8

Table 1.3 Energy intensity of various traffic modes [6]

Traffic modes	Assume passengers (person)	Energy intensity (kJ/person-kilometer)
Walking	1	418
Cycling	1	148
1990, Chrysler LeBaron, General car	2	1,460
1991, Honda for city	2	650
Volvo, LCP2000 Trial car	2	576
Toyota, AXV Test Trial car	2	420
Light rail vehicle	55	676
Intercity Bus	40	505
Intercity Railway Car	80	466

With the development of economy and society, the number of automobiles is increasing, the demand for oil will increase rapidly. If the growth of motor vehicles usage is not contained, the situation of oil shortage will become more and more serious.

However, because the amount of energy consumed by different traffic modes varies greatly, it can be effectively saved by choosing proper traffic modes. Tables 1.2 and 1.3 compare the fuel consumption and the energy intensity of various traffic modes.

Comparing Table 1.2 with Table 1.3, bicycle traffic does not consume fuel, and cars consume the most fuel per person-kilometer. In terms of energy intensity, bicycle traffic is also the least, and the private car is the highest, which is ten times as much as bicycle traffic. Therefore, encouraging the development of bicycle traffic will greatly save our fuel resources, which is of great significance for those energy scarce countries.

1.4.2 Saving traffic land consumption

With the gradual progress of global urbanization, land has become a very scarce and expensive resource in some cities. Transportation is one of the main users of land, especially with the popularity of private cars, which occupies more and more land in the city and becomes the greediest user of urban space. In many cities of developed countries, the proportion of land occupied by transportation has reached 30%. Even in some cities, the land occupied by urban roads and parking lots accounts for 60%–70% of the area [7]. While in a city, the land available for transportation is limited. Overexploitation and utilization of urban land resources for transportation will restrict the overall construction of the city and the development economy.

The consumption of urban land resources by urban traffic is mainly reflected in two aspects: one is the demand for roads and the other is the demand for parking lots.

1.4.2.1 Demand for roads

In the same road section, the larger the passenger volume per unit time, the larger the effective traffic volume, the lower the road area occupied per capita, the higher the utilization rate of road resources and the more economical the utilization of urban land resources.

Table 1.4 shows the per capita road area occupied by various traffic modes.

From Table 1.4, we can see that the per capita road area occupied by bicycle traffic is less than one-third of the road area occupied by cars. Therefore, the use of bicycle will save urban road resources and urban land development. And the increase of the proportion of urban car travel will increase the demand for the entire urban traffic road.

1.4.2.2 Demand for parking lots

Due to the increasing number of urban motor vehicles, especially private cars, the problem of parking difficulty becomes more and more prominent in the case of limited urban road resources. The demand for parking lots varies greatly with the different modes of urban transportation. Table 1.5 compares the static occupancy of road space by several modes of transportation.

Table 1.4 Per capita road area occupied by various traffic modes [8]

Traffic modes	Dynamic per capita road area occupation	
	Range (m^2/p)	Value (m^2/p)
Walking	0.72	0.75
Cycling	5–10	6.5
Motorcycle	7–25	16
Private car	10–30	20
Mini bus	2.5–5	4
Standard bus	1.5–4.2	2.8
Articulated bus	1.5–4.2	2.8
Light rail vehicle	2.5–5	3.8
Metro	2.5–5	3.8

Table 1.5 Static parking area of main urban traffic modes

Traffic modes	Provisions of the parking lot area		Parking garage and building area (m^2)	Parking lot/garage area per passenger (m^2)
	Range (m^2)	Value (m^2)		
Bicycle	1.5–1.8	1.6	2.0	1.6/2.0
motorcycle	2.4–2.6	2.5	3.0	2.2/2.5
Private car	25–35	30	40	18/26.5
Standard bus	50–70	55	65	1.4/1.6
Articulated bus	60–90	65	80	1.1/1.3

Table 1.6 Emissions of hazardous gases from vehicle kilometers in different traffic modes [9]

Traffic modes	CO	NO_X	HC	Σ
Bicycle	0	0	0	0
Private car	36.1	0.91	3.16	40.17
Bus transit	37.2	16.85	15.99	70.04
Rail transit	0	0	0	0

According to the analysis of Table 1.5, the parking area occupied by bicycle users is only about 1/12–1/15 of that of private cars and is similar to that occupied by standard buses. Therefore, compared with the development of cars, the vigorous development of bicycles will greatly save urban static land resources.

1.4.3 Effective reduction of urban traffic emission pollution

With the rapid growth of urban motor vehicles, more and more exhaust gases are emitted. The concentrations of carbon dioxide (CO_2), nitrogen oxides (NO_X), carbon monoxide (CO) and hydrocarbons (HC) in the atmosphere are increasing, which aggravates the speed of global temperature rise and environmental deterioration.

Different types of vehicles emit different amounts of exhaust gas, so in order to achieve low-carbon traffic, it is necessary to rationally choose the right vehicles and optimize the travel structure. The bicycle is driven by manpower, so it does not emit greenhouse gases and toxic gases. It is a green and environmental-friendly mode of transportation. Developing bicycle traffic can not only reduce the emission of toxic and harmful gases, control the greenhouse effect, but also improve the urban environment and air quality. Table 1.6 compares the main harmful exhaust emissions from different traffic modes.

As can be seen from Table 1.6, bicycles are clean and pollution-free green vehicles. Driving at the same distance, the exhaust emission of cars is the highest, and the popularity of cars will undoubtedly lead to the increase of total pollution emissions, resulting in the deterioration of urban atmospheric environment and ecological environment. Bicycle traffic is the most environment-friendly way of transportation. The government should vigorously advocate the development of bicycle traffic and low-carbon lifestyle.

1.4.4 A healthy traffic mode

With a continuous improvement of people's living standards, nutrition is becoming more and more abundant, and due to the general pressure of modern people, less exercise time, obesity is becoming increasingly prominent, which brings a series of health problems. And the use of bicycle travel can play a very good fitness role. The main manifestations are as follows:

1. Riding a bicycle is like running exercise, which can better exercise the endurance of visceral organs, improve cardiopulmonary function and thus enhance the body's immunity.
2. Choosing bicycle traffic can achieve the effect of fitness and weight loss. Biking is an aerobic exercise, which can consume more calories and effectively solve the problem of obesity.
3. Choosing bicycle traffic can prevent obesity, thereby preventing a series of diseases caused by obesity, such as cardiovascular diseases, diabetes, hypertension, hyperlipidemia and so on.
4. Choosing bicycle traffic for outing can also integrate into nature, which can make people feel comfortable and eliminate worries and pressures.

Therefore, nowadays, people pay more and more attention to healthy life and green travel, bicycle as a healthy, environment-friendly and leisure way of travel should be vigorously promoted.

1.4.5 Low travel cost

For the low-income group, using bicycle travel can save a lot of car purchase, maintenance, parking and other costs. For the low-income group, bicycle is the most economical and practical traffic mode. The bicycle traffic makes up for the economic deficiencies of cars and buses with its own characteristics. The remarkable economy of bicycle travel plays an important role in the development of bicycle traffic. Therefore, in many cities of developing countries, bicycle traffic accounts for a very important role.

1.4.6 Lower internal and external transport costs

Transportation cost includes two parts: internal and external. The internal costs are direct traffic costs, including infrastructure investment costs, user costs and travel time costs. The external cost refers to the cost caused by the external effect of traffic (mainly negative external effect), including traffic congestion cost, environmental pollution cost and traffic accident cost.

1.4.6.1 Internal costs

The cost of infrastructure construction in internal cost mainly includes the cost of dynamic infrastructure construction, that is, the cost of road infrastructure construction and the cost of static infrastructure construction, that is, the cost of parking facilities construction.

User cost mainly includes vehicle purchase fee, maintenance fee, insurance fee, tax fee, fuel fee and surcharge.

The time cost mainly includes the cost of travel time, waiting time and changing time.

Table 1.7 shows the average number of passengers per vehicle and the average cost per passenger per vehicle.

From Table 1.7, we can see that the order of road infrastructure construction cost shared by each passenger is car>motorcycle>bicycle>bus. The average cost of bicycle infrastructure is lower than that of automobiles and motorcycles.

From the user cost point of view, the purchase, maintenance and storage of motor vehicles take up a large amount of funds. With the rapid increase of private cars and the increasingly tense contradiction of garage resources, the price of parking spaces has been rising year by year. For example, the price of car parking spaces in Hong Kong has been equivalent to 20 or even 30 times the price of cars. The development of car transportation will consume a lot of funds from individuals and the state. If these funds are used for the construction of green transportation facilities and the development of new technologies of green transportation, it will be more conducive to the development of urban traffic. If a bicycle is used to travel, its purchase, maintenance and storage costs are lower, which will greatly save personal expenses.

In terms of travel time, different modes of transportation and the time required to travel the same distance are also different. Table 1.8 shows the travel time required for several different modes of transportation.

From Table 1.8, it can be seen that when the travel distance is less than 6 km, the advantages of a bicycle are particularly obvious. Choosing bicycle traffic trip greatly saves travel time and accordingly saves travel time cost. Therefore, bicycles are the best traffic mode for short-distance travelers.

1.4.6.2 External costs

Not every traffic mode has all the external costs. There is no external cost for walking, the external cost of rail transit mode is almost zero, and there is only congestion cost for bicycle traffic. While there are three external costs for car, motorcycle and bus modes of travel: traffic congestion cost, environmental

Table 1.7 Infrastructure costs shared per passengers of different traffic modes in China [10]

	Traffic modes			
	Bicycle	**Motorcycle**	**Private car**	**Bus**
Average number of passengers (person)	1.0	1.2	1.5	40
Average cost per passenger (RMB)	0.5	2.5	5.5	0.2

Table 1.8 Travel time cost for different traffic modes [11]

Traffic modes	2 km	4 km	6 km	8 km	10 km
Bicycle (no transfer)	9	19	30	42	55
Bus (no transfer)	16	24(20)*	32(30)	40(35)	48(41)
Bus (transfer once)	20	30(24)	38(30)	45(37)	54(45)
Metro (no transfer)	32(23)	36(27)	39(30)	43(34)	46(37)
Metro (transfer once)	39(29)	43(33)	45(35)	49(38)	52(42)

*Note: The data in brackets refer to the time required to travel by bicycle instead of walking to and from the station.

Table 1.9 External costs of different traffic modes [12]

	External costs (euro/1,000 km)				
	Walking	**Cycling**	**Private car**	**Bus**	**Rail transit**
Air pollution	0	0	8–12	<1	<1
Noise	0	0	12–25	2–3	2–3
Traffic accident	0	0	5	<1	<1
Traffic congestion	0	<1	50	<10	0

pollution cost and traffic accident cost. Table 1.9 shows the external costs of different traffic modes.

From Table 1.9, we can see that besides walking, the external cost of a bicycle is relatively low and that of car is the highest. In order to reduce the external cost of transportation, bicycle travel should be advocated and car travel should be restricted.

To sum up, bicycle traffic is a green and healthy traffic mode, which supports the sustainable development of society. It should be vigorously promoted and encourage people to use it.

However, at present, most urban traffic facilities, including road segments and intersections, are planned and designed from the perspective of motor vehicles. Especially the intersections, bicycle traffic is even less considered, which makes the safety situation of bicycle traffic at intersections worrying.

Statistics show that most vehicle–bicycle collisions occur at intersections. Klassen *et al.* found 424 vehicle–bicycle collisions at Edmonton, while only 147 vehicle–bicycle collisions occur on road segments [13]. That means about 65% of the vehicle–bicycle collisions occur at intersections at Edmonton.

The low level of bicycle safety at intersections is a common phenomenon in many countries, which greatly restricts the application and promotion of bicycle traffic.

Therefore, it is necessary to conduct a comprehensive analysis and model the cyclists' microscopic behaviors at intersections, so as to provide theoretical basis for improving bicycle safety at intersections.

1.5 The main content of this book

The purpose of this book is to analyze and model the microcosmic behavior of bicycle traffic on roads, signalized intersections and un-signalized intersections, so as to lay a theoretical and data foundation for bicycle facilities planning, design and the relevant traffic management measures. At the same time, we also analyzed and modeled the cyclists' microscopic behaviors on various facilities, which can lay a foundation for the microscopic simulation model for urban mixed traffic flow.

The main contents and structure of this book include the first chapter introduces the concept and development history of bicycle traffic, the overview of bicycle traffic development in various countries and the bicycle traffic features in comparison to other transport means.

Chapter 2 introduces the bicycle riding characteristics on the road section, including the psychological characteristics of bicycle riders, static and dynamic dimensions, snake shape track, speed, braking and turning characteristics.

Chapter 3 introduces the micro-behavioral characteristics of bicycles at signalized intersections, and the data of speed, acceleration, time and density of bicycles are analyzed. Chapter 4 introduces the model of bicycle crossing conflict vehicle flow at signalized intersections, includes analyses and modeling of the selection behavior of acceptance gap and lag when bicycle crossing conflicts vehicle flow.

Chapter 5 discusses the bicycle microscopic behavior analysis paradigm, including the behavior analysis paradigm and human behavior characteristics. And from the fields of psychology, economics and traffic engineering, we discussed the microscopic behavior analysis paradigm.

Chapters 6–10 analyze the microscopic behavior of bicycles at mixed traffic flow un-signalized intersections. Chapter 6 presents the framework of bicycle microscopic behavior analysis at un-signalized intersections. Chapter 7 builds the micro-behavior model of bicycles at un-signalized intersections. Chapter 8 introduces an empirical analysis of the micro-behavioral model of bicycles using field data. Chapter 9 validates the bicycle micro-behavior model from the perspective of micro-simulation model. Chapter 10 proposes a model of bicycle collision avoidance behavior at un-signalized intersections based on NN.

References

[1] American Association of State Highway & Transportation Officials (AASHTO). Guide for the Development of Bicycle Facilities. Washington, DC, AASHTO, Aug. 1994.

[2] J. Forester. Bicycle Transportation (1st edition). Cambridge, MA, MIT Press, 1983.

[3] X.M. Liu, L.D. Shen, and F. Ren. Operational Analysis of Bicycle Interchanges in Beijing, China. In Transportation Research Record 1396, Washington, DC, TRB, National Research Council, 1993, pp. 18–21.

[4] J. Xu and X. Chen. Chapter 10 Pedestrian and Bicycle Traffic. In Traffic Engineering Manual, Beijing, China Communications Press, 1997.
[5] X. Zhou. Development Direction of Passenger Transport in Big Cities. China Transportation Review, 1996, 12:34–38.
[6] S. Yang. Study on the Sustainable Development Indicators for Urban Public Transit System. Master Dissertation. Chang'an University, May 2004.
[7] M. Gao. Study on Bicycle Traffic in the Sustainable Development of Urban Traffic. Master Dissertation. Chang'an University, May 2005.
[8] F. Wang. Decision-Making for the Future Development of Bicycle Transport in China. Planners. 2001, 17(6).
[9] Y. Quan and X. Liu. Where Is the Way – Talking About Urban Transportation. China City Press. 2002.
[10] T. Gao. The Quantization Study of Trip Cost About Urban Different Trip Ways. Master Dissertation. Northeast Forestry University, Jun. 2006.
[11] A.G. Verman, C.J. Louis, and D.M. Litment. Urban Bicycle. China Urban Transport Development Strategy Seminar, 1997.
[12] J. Cai. World Urban Transportation. Shanghai, Tongji University Press, Aug. 2001.
[13] J. Klassen, K. El-Basyouny, and M.T. Islam. Analyzing the Severity of Bicycle-Motor Vehicle Collision Using Spatial Mixed Logit Models: A City of Edmonton Case Study. Safety Science, 2014, 62:295–304.

Chapter 2
Microscopic bicycle microscopic riding characteristics

Bicycle traffic is not only very different from motor vehicle traffic in terms of trip length, resource occupation and environmental conservations, but also it has its own riding characteristics in terms of microscopic motion.

Compared with other motor vehicles, a bicycle has its own structural characteristics, such as the unstable two-wheel structure, relatively small size, flexibility, powered by human, etc. These characteristics make its microscopic riding motion, quite different from the dynamic motions of motor vehicles.

This chapter focuses on the psychological characteristics of cyclists and the microscopic riding characteristics of cyclist-bicycle unit (also known as bicycle individual). We first analyze the psychological process and main psychological characteristics of bicycle riders are elaborated, and then the riding characteristics of individual bicycles, such as dynamic and static traffic characteristics, serpentine trajectory, speeds distribution, factors affecting speed, braking and turning characteristics.

2.1 Cyclists' psychological characteristics

2.1.1 Cyclists' psychological process

In his book *Traffic Psychology*, Lai puts forward the cyclists' psychological process as shown in Figure 2.1 [1]. First of all, cyclists use perceptual organs, mainly eyes and ears, to penetrate the external environmental information into the central nervous system of the brain. The central nervous system of the brain, through a certain information processing process, controls the bicycle's movement (acceleration, deceleration and turning) by controlling the handlebars, pedals and saddles of the bicycle through the body's locomotive organs, such as hands, feet, buttock. And at the same time, the cycling also changes and influences the external environmental information.

In their paper "Young cyclists' psychological behavior and traffic safety," Wang Zhaoxian and Nie Lei pointed out that the road traffic system is a dynamic system composed of people, cars, roads and environment [2]. In this system, human factor is the most important factor in the safety analysis. From the viewpoint of human–machine system, man and bicycle constitute a human–machine system. The system is an organic whole that includes the combination of human, machine (bicycle) and human–machine (bicycle). It can be represented by Figure 2.2.

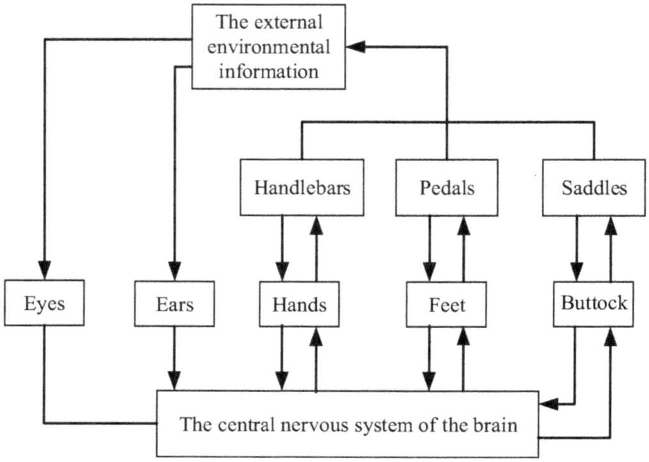

Figure 2.1 The diagram of a cyclist's psychological process

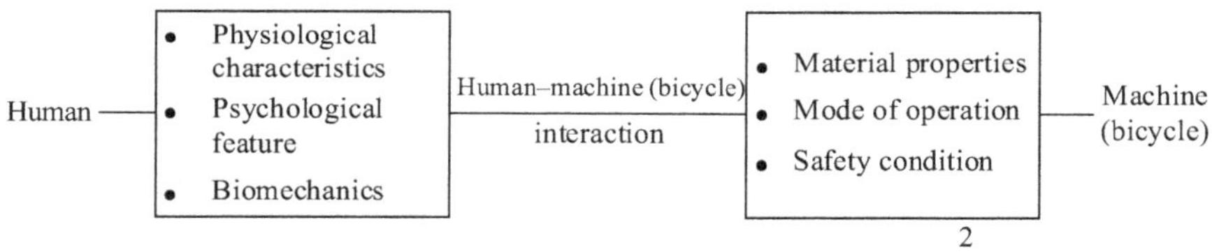

Figure 2.2 Diagram of human–machine (bicycle) relation [2]

In the human–machine system, the role of human in the manipulation process is mainly described by the following psychological formulas:

$$\text{External stimulus}(S) \rightarrow \text{consciousness}(O) \rightarrow \text{response}(R)$$

That is, the outside world through the input of information, and then through the brain for information processing, forming instructions, commanding hand and foot operations, thus producing the output of behavior.

In the man–machine system of bicycle riding process, bicycle riders and roads constitute a whole system. The main factors that constitute the system are riders, bicycles (including motor vehicles and other non-motor vehicles), roads and environment.

In this system, the rider receives information from road, vehicle and pedestrian environment through sensory system, such as hearing and vision; then, the information enters the central nervous system of human brain for analysis, processing and processing to form decision-making and instructions; and then, through the combination of motor system, such as hands and feet, it is transmitted to bicycles to carry out decision-making and instructions. Bicycles run according to people's wishes and then show various driving characteristics of bicycles. The process of information acceptance and processing of cyclists is actually the process of their psychological activities.

This process can be regarded as a closed-loop mental activity process, as shown in Figure 2.3.

2.1.2 Cyclists' psychological characteristics

According to Lai's summary in his book "Traffic Psychology," the main cyclists' psychological characteristics during cycling are as follows.

2.1.2.1 Fearful to the motor vehicles

Cyclists are afraid of motor vehicles, especially large ones. One of the reasons is that the rider has neither a cab nor a cockpit. Whether in physical or psychological aspects, he is directly affected and disturbed by the external environment, and his own safety is not protected. Bikers are weak in road traffic. The second reason is that the bicycle is unstable, the motor vehicle can stop at will, and the bicycle will fall when it stops.

These two reasons make bicycle riders suffer from psychological pressure from motor vehicles when they participate in traffic. When a motor vehicle and a bicycle riding together, the closer the distance between the two, the greater the psychological pressure and fear of the rider. Especially in the middle of two motor vehicles, the psychological fear of cyclists is particularly serious. Under the influence of fear psychology, cyclists will have a moment of high tension and panic, so as to increase the serpentine width and fall, resulting in traffic accidents.

2.1.2.2 Haste in riding

Whether a cyclist travels for any purpose, it is their common psychological need to arrive at his destination quickly and in a time-saving manner.

The bicycle is light and flexible. It does not delay time due to long waiting in traffic jams. It can also avoid the suffering of bus congestion. Some cyclists scramble for time. Sometimes they know that the motor vehicle has arrived but also dare to risk overtaking and even forced the driver to brake to avoid collision. Such rapid overtaking is very easy to induce accidents.

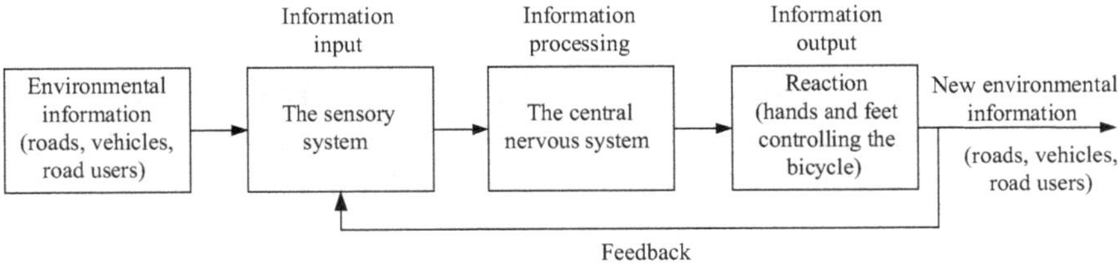

Figure 2.3 Mental activity processes of cyclists [2]

2.1.2.3 Casual and varied riding patterns

Some cyclists have some unsafe riding habits, such as casual riding and weak traffic awareness, which makes the distribution of bicycles on the road discrete.

Its main performance is some bicycles ride fast, others ride slowly, some ride in small groups, some ride along.

Especially in the absence of traffic command and management, some cyclists occupy the whole road, leading to traffic disorder, seriously affecting the traffic of pedestrians and motor vehicles, and endangering traffic safety.

2.2 Static and dynamic measurements

2.2.1 Static measurements

The so-called static measurements refer to the basic size and mass characteristics of the pedestrian–bicycle composed unit when the bicycle is in a static state. The center distance between common bicycles' front and rear wheels is 1,050–1,130 mm, and the weight of the whole bicycle is 16.0–20.5 kg [3].

According to the description of a bicycle's outlined in *Traffic Engineering Manual (TEM)* [4], the bicycle's length is generally 1.6–1.9 m, and the width of bicycle handlebars is about 0.36–0.60 m. The maximum size of the bicycle is 1.9 m in length, 0.6 m in width, 2.25 m in height and 1.2–1.8 m^2 in static parking area.

In the United States, the length of bicycles is 1.75 m and the width of handlebars is 0.6 m [5]. In the Netherlands, 95% of bicycles are less than 1.90 m in length, while all bicycles are less than 0.75 m in width [6].

2.2.2 Dynamic measurements

The so-called dynamic measurements of bicycle refer to the sizes of individual bicycle and cyclist in the process of movement.

According to the data in the China's TEM [4], it is considered that the running process of white bicycle is serpentine, and the width of one-sided serpentine track is about 0.20 m, so the lateral safety width of bicycle is 1.0 m (the width of handlebars as 0.6 m plus the width of serpentine track as 0.4 m); the running height space is 2.5 m, where 25 cm is the safe space. The longitudinal safety space of bicycle is 2.50 m.

According to AASHTO [7], it is considered that the bicycle rider's legs are the widest, 0.75 m, and the lateral moving distance of one side is 0.125 m. Therefore, the necessary lateral width is 1.0 m, and the safe distance of 0.2 m is needed. Therefore, it is considered that the width of bicycle lane is the smallest, 1.2 m. When there is no physical isolation between bicycle lane and motor lane and the traffic is large, the white bicycle lateral is recommended. The running space is 1.5 m. The running height is also 2.50 m. Other American studies show that the typical lateral width of a bicycle running is 0.75–1.40 m, which is also closely related to the effective width of the bicycle lane. Early studies have shown that when bicycles share the same road with motor vehicles, the minimum lateral width of bicycles is 1.28 m [8]. In the Dutch and German studies, the lateral clearance distance required for bicycles was 1.0 m [9].

2.2.3 Running area

The area occupied by bicycle unit when running is also known as dynamic-space-occupied area. According to China's TEM [4], the road area occupied by bicycle units at different speeds can be estimated as follows. Assuming that the width of a running bicycle is 1 m, the formula for calculating the road area occupied by running vehicle is generally as follows:

$$Sr = W \times h_v \tag{2.1}$$

where Sr represents the road area occupied by the running vehicle in m^2; W represents the width of the running vehicles; and h_v represents the time headway under speed v. According to the previous formula, Table 2.1 lists the road area occupied by bicycles at different speeds. It can be seen that with the increase of speed, the road area occupied by bicycles is also increasing.

Early studies on the campus of the University of California Davis in the United States showed that bicycles could ride freely when the occupied area of a single vehicle is larger than 20 m^2/bicycle, while congestion might occur when the occupied area is less than 3.7 m^2/bicycle [10].

Canadian studies show that bicycles can travel freely when occupancy area is greater than 9.3 m^2/bicycle, and that when occupancy area is less than 3.0 m^2/bicycle, cyclists' control of bicycle is greatly limited [11].

A research in China shows that when the control space of bicycle is greater than 10 m^2/bicycle, cyclists can drive bicycles comfortably, on the other hand when the occupied area is less than 2.2 m^2/bicycle, some cyclists need to get off and implement it [12].

From the data of the previous countries, it can be seen that the free driving space of bicycles in the United States is obviously larger than that of Canada and China, which is closely related to the free driving speed of bicycles. From later analysis, it can be seen that the free driving speed of bicyclists in the United States is the highest in the survey data of all countries, so the free driving space is also the highest.

Road area required by running bicycles can be converted into typical traffic flow density data. Therefore, according to the road occupancy data of running bicycles, we can think that when the density is below 0.05 bicycle/m^2, bicycles can travel freely; when the density is more than 0.45 bicycle/m^2, the bicycle traffic is seriously congested.

According to the description of congestion density in China's TEM, it is considered that when the density in the system is 0.54–0.67 bicycle/m^2, the bicycle

Table 2.1 Road area occupied by running bicycles at different speeds

Speeds (km/h)	5	10	15	20	25	30
Occupied area (m^2)	4.1	5.2	8.1	10	12	16

traffic system will be blocked. The related research of Jilin University also shows that the blocking density is 0.54–0.65 bicycle/m^2 [13].

2.3 S-shaped trajectory

Unlike motor vehicles, bicycles have S-shaped tracks on the road. A research on the width of bicycle S-shaped trajectory is of great significance to the safety design of bicycle facilities. According to the experiment of Japan Transportation Engineering Research Association on flat lane, the higher the speed of riding, the smaller the width of serpentine trajectory is for adults.

For example, when the average speed of adult riding is 17 km/h, the width of S-shaped trajectory is 40 cm; for middle school students, if the riding speed is in the range of 13–18 km/h, the width of S-shaped trajectory decreases with the increase of speed, but when the riding speed is in the range of 18–23 km/h, the width of S-shaped trajectory increases with the increase of speed.

For pupils, when the riding speed is in the range of 11–13 km/h, the width of the S-shaped trajectory decreases with the increase of the speed, but when the speed exceeds 13 km/h, the width of the S-shaped trajectory increases accordingly.

Figure 2.4 illustrates the S-shaped trajectory width of cyclists at different ages on an unrestricted lane.

They also found that the width of the S-shaped trajectory of a bicycle on a certain gradient road was larger than that on a flat road at the same speed. Table 2.2 shows the average bicycle speeds and corresponding S-shaped track widths of different cyclists on roads with a longitudinal slope of 4%.

In addition, British scholar John Forester pointed out that in the mixed traffic flow of bicycle and motor vehicle, there is a correlation between the width of the bicycle's S-shaped track and the speed and distance of the neighboring motor vehicle. That is to say, the width of the S-shaped track of bicycle riding is related to the psychological pressure of the cyclist. When the motor vehicle in mixed traffic flow is closer to the cyclist, the speed is higher. The higher the cyclist's psychological pressure, the wider the S-shaped trajectory is [15].

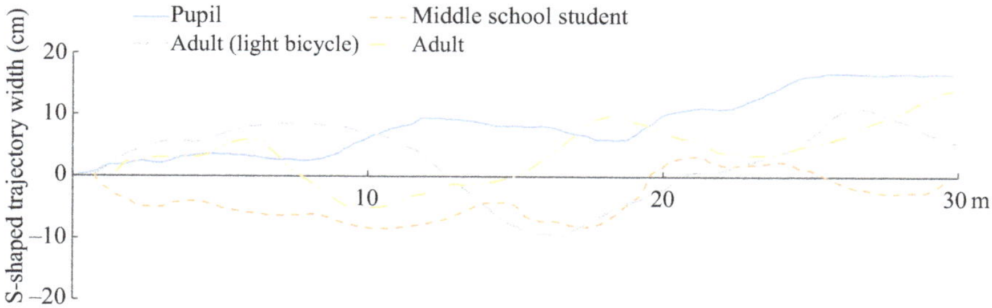

Figure 2.4 S-shaped trajectory range of bicycle driving without protection restriction [14]

Table 2.2 S-shaped trajectories in different age groups

Slope	Cyclists type	Mean speeds (km/h)	Average S-shaped trajectory width (cm)
+4%	Adults	8.3	36.0
	Middle school student	11.0	47.0
	Pupil	10	36.7
	Average	9.8	39.7
−4%	Adults	14.2	40.0

Source: Reference [14].

2.4 Speed characteristics

Cycling speed can be divided into two categories: one is the free riding speed of the bicycle when it is not affected by other obstacles, the other is the average speed of the bicycle when it is riding, that is, running speed or riding speed.

2.4.1 Free riding speed

Free riding speed is a key indicator of bicycle traffic flow, which has been studied by scholars in various countries. An early survey of bicycle traffic on campus of California University Davis found that the free speed of bicycles was greater than 17.7 km/h [10].

The report published by FHWA points out that the 85th percentile speed of a bicycle is about 24 km/h when it runs freely, and the designed speed of a bicycle on a flat area is 32 km/h, at which time each rider can travel at the desired speed [16].

Swedish research shows that 85% of bicycles travel freely at speeds ranging from 16 to 28 km/h [17].

Canadian research shows that bicycles can ride freely at a speed of 25 km/h [11].

The Dutch study shows that the average speed of bicycles is 18 km/h and the standard deviation is 3 km/h.

The study also shows that the observed average speed is not affected by the width of the road. Research in China shows that the free speed of a bicycle is 18.2 km/h on organic and non-isolated roads, while the free speed of a white bicycle is 13.9 km/h on non-mixed roads [18].

2.4.2 Riding speeds

As an important operation analysis parameter, there are many studies on bicycle riding speeds in some countries. As early as 1975, the Campus Survey of California University Davis found that the average speed of bicycle lanes was 19 km/h, while the data analysis of bicycle lanes showed that the speed of bicycles ranged from 17.7 to 20.1 km/h [8]. Another bicycle survey on Michigan Campus in the United States found that the average speed on the special bicycle road was 24.9 km/h, while the average speed on the special bicycle path was 20.3 km/h [26]. In addition, [16] Forester [15], [19,20] all measured the running speed of bicycles in different environments.

There are many studies on the speed of bicycles in China. As early as 1979–1981, Beijing conducted a field survey on the speed of bicycles on the road sections. Data analysis shows that the average speed of bicycles on isolated main roads is 16.28 km/h, while that on non-isolated roads is 14.21 km/h [4].

Liu *et al.* showed that the speed of a bicycle ranged from 12 to 16.3 km/h, and the overall average was 14 km/h [21]. Jilin University has studied the running speed of bicycle flow under different isolation conditions in peak period and obtained that when there is hard isolation between machines, the average speed is 4.16 m/s (14.98 km/h), and the main speed distribution is 4–5 m/s (14.40–18.00 km/h), which satisfies the normal distribution; when there is no isolation between machines, the average speed is 3.83 m/s (13.79 km/h), which also satisfies the normal distribution [22].

In addition, the researchers analyzed the distribution characteristics of bicycle riding speed in different environments. The general conclusion is that the speed of bicycle obeys normal distribution [23].

Some scholars have also studied the speed of bicycle running at intersections and found that the speed of left turn is the lowest, followed by right turn, and the speed of straight line is the highest. The overall analysis shows that the speed of bicycle running at intersections is less than the speed of road section [24,25].

It is worth noting that the earlier studies are measured in different environments. Most studies mention that bicycle speed is related to longitudinal slope, road type, road width, age and gender of cyclists. The following will be discussed separately.

2.4.3 Influencing factors of bicycle riding speeds

2.4.3.1 Longitudinal slope

The main factors affecting bicycle climbing ability are the physical strength and endurance of cyclists and the performance of vehicles. Most bicycle traffic design guidelines, including the Traffic Engineering Manuals from China, AASHTO, ITE and FHWA, all agree that the maximum longitudinal slope of bicycle lanes should not exceed 5%.

Research results in Japan show that ordinary bicycles with gearless gears can climb up a slope of 7%–8%, and bicycles with gearshifts can climb up a slope of 12%. The empirical formula of up-and-down slope speed change proposed in China's TEM shows that the correction coefficient of bicycle speed should be 20% for every increase or decrease of 1% slope within the limit slope (<5%). For example, the original cycling speed is 15 km/h, the slope change is 1%, and the speed is 12 and 18 km/h. According to the observation in China, when the slope is 1%, the riding speed of young people up the slope is about 10–15 km/h; when the slope is 2%, the speed is about 7–12 km/h; and when the slope is 3%, the speed is about 5 km/h.

2.4.3.2 Facility types

In 1980, researchers in the Netherlands conducted a small sample survey of bicycle speeds on different roads (Table 2.3) in order to analyze whether the facilities types have an important impact on bicycle speeds [26]. It can be seen from the table that the bicycle has the highest speed in the bicycle lane and the lowest speed on the sidewalk. The possible reason is that the bicycle lane is separated from the motor vehicle traffic and the width of the lane is large, while the sidewalk should be more careful because it interacts with pedestrians. Given the limited sample size of the study, these statements are not conclusive.

Tanaboriboon *et al.* observed and analyzed the effects of different facility types (bicycle paths, bicycle lanes and mixed roads) on bicycle speeds in Shanghai, China and Sulakata, Indonesia [27]. It was found that the effects of the same road facilities on bicycle traffic in the two cities were quite different (see Tables 2.4 and 2.5).

In Shanghai, the speed on the bicycle paths is generally about 1 km/h higher than that of the bicycle lane, and the speed of the bicycle lanes is about 1 km/h higher than that of the mixed roads. This shows that the special bicycle facilities in Shanghai can improve the speed and service level of bicycles; but in Surakarta, the special bicycle facilities have little effect on the speeds of bicycles. The main

Table 2.3 Bicycle speed on different road types

Facilities types	**Sample size**	**Observed velocity (km/h)**		
		Average	**Maximum**	**Minimum**
Bicycle paths	14	20.26	39.18	4.38
Bicycle lanes	4	24.99	40.88	4.07
Pedestrian pavements	5	18.51	30.15	3.39
Mixed roads	5	19.07	36.91	8.06
Total	28	20.71		

Source: Reference [26].

Table 2.4 Comparisons of the influence of different types of roads on bicycle speed in Shanghai

Items	**1994**			**1992**		
	Bicycle path	**Bicycle lane**	**Mixed road**	**Bicycle path**	**Bicycle lane**	**Mixed road**
Average (km/h)	15.51	14.76	13.45	15.26	14.83	14.29
Standard deviation (km/h)	2.46	2.31	2.82	2.06	2.37	1.71
Sample size	1,007	993	1,116	430	424	408

Note: The difference of average speed between the two different facility types is significant under 95% confidence.

Table 2.5 Comparisons of the influence of different types of roads on bicycle speed in Surakarta

Items	Bicycle paths	Mixed road
Average speed (km/h)	11.87	11.80
Standard deviation (km/h)	1.84	2.24
Max. (km/h)	25.26	26.16
Min. (km/h)	7.03	6.78
Sample size	837	1,196

Note: The difference of average speed between the two types of roads is not significant under 95% confidence.
Source: Reference [27].

Table 2.6 Bicycle speed statistics

Location	Width (cm)	Average speed (km/h)	Standard deviation (km/h)	85% speed (km/h)
1 (narrow)	180	19.6	3.4	22.8
2 (wide)	240	19.0	3.1	22.3
3 (wide)	250	19.0	2.9	21.8
4 (wide)	270	18.9	2.5	21.4

Source: Reference [28].

reason is that the bicycle lanes in Surakarta have not been really "dedicated." There bicycle paths are occupied by many non-bicycle vehicles, such as street vendors, and illegal parking motor vehicles.

From this, they concluded that whether bicycle facilities can improve bicycle speed and service level largely depends on the way of facilities operation. If the corresponding supporting management measures are not coordinated, the benefits of bicycle facilities will be greatly reduced.

2.4.3.3 Bicycle path width

Dutch scholars Botma and Papendrecht observed bicycles on bicycle paths and obtained velocity distributions (Table 2.6) [28]. It can be seen from the table that the speed of bicycles in four different places is basically the same, which shows that the width of the road has no effect on the speed of bicycles. Generally speaking, the average speed of bicycles is 19 km/h, the standard deviation is 3 km/h and the distribution of speed basically conforms to the normal distribution.

Botma and Papendrecht also compared the data with the observation data of Vries (1979) in the Netherlands 10 years ago and found that the average speed of bicycles 10 years ago (19.2 km/h) was almost the same as the average speed of bicycles this time. It can be inferred that the speed of bicycles is a relatively stable data, and there has been no change in the past 10 years.

2.4.3.4 Age and gender

Tanaboriboon *et al.* also studied the cycling speed of cyclists of different ages and sexes in Shanghai and Surakarta, Indonesia, respectively. The specific values are shown in Tables 2.7 and 2.8 [27].

As can be seen from the previous tables, the average speed of male riders is about 1 km/h higher than that of female in any age group. The results of comparisons among different age groups show that the speed of young is the fastest, about 1 km/h higher than that of adults, and the speed of elder people is the slowest. It can be seen that age and gender still have a certain impact on the bicycle riding speed.

2.5 Braking characteristics

Braking is a kind of behavior that reduces the speed or stops the bicycle completely when the riders find obstacles or people in front of it. Emergency braking occurs when some obstacles suddenly appear in front of the vehicle. In theory, when the bicycle brakes completely, the wheel tires and the ground only do gliding motion, without considering the influence of air resistance. At this time, the theoretical value of bicycle deceleration is the maximum:

$$a = -(f + i) \times g \tag{2.2}$$

where a is the deceleration of bicycle m/s^2; f is adhesion coefficient between tire and road surface, generally 0.3–0.6; i is the longitudinal gradient of bicycle lane (%), positive on the uphill and negative on the downhill; g is gravity acceleration.

According to the formula (2.2), the range of bicycle deceleration is 2.94–5.88 m/s^2.

The value of parameter f has a great influence on the maximum braking deceleration of bicycles, so the selection of its value is the key to the problem.

Chinese scholars usually take f about 0.5; Japanese Police Science Research Institute uses 0.6 as safety value and 0.8 as maximum value; from the formula of braking distance, the former Soviet Union takes f about 0.428 (possibly because of the cold climate in the country, considering the ice on the road in winter, the value of f is relatively small). Table 2.9 lists the maximum braking deceleration of bicycles corresponding to different parameters f.

In addition to the friction formula, there are other methods for calculating the maximum braking deceleration of bicycles. As the bicycle rider bends forward to varying degrees when braking, and the center of gravity moves forward accordingly, the more severe the body bends, the greater the possibility of the rider turning over from the handlebar.

Based on this, Whitt and Wilson took preventing cyclists from turning over from handlebars as the limiting factor of the maximum braking deceleration of bicycles and calculated the theoretical extreme value of the braking deceleration of bicycles to be 0.56 g, about 5.5 m/s^2 [29].

In addition, in rainy weather, the friction coefficient between tire and ground may drop to about 0.4, and the maximum braking deceleration is about 4 m/s^2. In fact, in order to ensure safety, the deceleration of cyclists is lower than the theoretical value.

Table 2.7 Comparison of bicycles' speed of different ages and genders in Shanghai [27]

Item	Young			Adult			Elder		
	Male	**Female**	**Together**	**Male**	**Female**	**Together**	**Male**	**Female**	**Together**
Shanghai, 1994 (video)									
Average speed (km/h)	16.37	14.56	15.62	15.29	14.2	14.8	12.3	11.05	11.87
Standard deviation (km/h)	2.3	1.98	2.35	2.46	2.52	2.55	2.08	1.61	2.02
Max. (km/h)	24.97	22.68	24.97	23.8	22.47	23.8	18.5	16.47	18.5
Min. (km/h)	11.24	10.45	10.45	9.34	8.35	8.35	8.25	6.22	6.22
Sample size	509	361	870	1,002	750	1,752	315	183	498
Shanghai, 1992 (floating vehicle)									
Average speed (km/h)	16.17	15.03	15.66	14.76	14.32	14.57	13.72	13.02	13.66
Standard deviation (km/h)	2.62	2.01	2.44	2.04	2.05	2.06	1.85	0.82	1.79
Sample size	389	313	702	497	352	849	102	16	118

Note: Under 95% confidence, the difference of average speed between different genders in different age groups is significant.

Table 2.8 Comparison of bicycles' speed of different ages and genders in Surakarta

Item	Gender			Age		
	Male	**Female**	**Together**	**Young**	**Adult**	**Elder**
Average speed (km/h)	12.1	11.08	11.83	14.24	11.78	11.27
Standard deviation (km/h)	2.2	1.49	2.08	2.31	2.06	1.54
Max. (km/h)	26.16	18.29	26.16	24.59	26.16	17.35
Min. (km/h)	6.85	6.78	6.78	7.03	6.78	7.82
Sample size	1,486	547	2,033	504	1,274	255

Note: Under 95% confidence, the difference of average speed between different genders in different age groups is significant.

Table 2.9 The maximum braking deceleration corresponding to various values of f when i is 0

f	0.5	0.6	0.8	0.428
a (m/s^2)	4.9	5.88	7.84	4.195

Source: Reference [4].

2.6 Turning characteristics

In order to ensure the safety of bicycle driving on curved roads and the comfort and stability of cyclists, a minimum control value must be stipulated for the turning radius. The empirical and theoretical formulas can be found in the literature, and the values prescribed by various countries are quite different.

1. An empirical formula for the relationship between speed and radius is proposed in the Guiding Principles of Bicycle Lane Planning Standards in the United States.

$$R = 0.24v + 0.43(\mathrm{m}) \tag{2.3}$$

   The Maryland Transportation Administration stipulates that adults should use $R = 0.29v + 0.67(\mathrm{m})$ when turning at various speeds, which is basically consistent with the earlier formula.

2. In theory, the maximum centripetal force generated by bicycle turning is generated by static friction between bicycle and ground. The following formula can be obtained from physical kinematics:

$$R_{\min} = \frac{v^2}{f_{\max} g} \tag{2.4}$$

Table 2.10 Calculated values of different radius formulas (recommended and limit values)

Item	Speed (km/h)					
	5	**10**	**15**	**20**	**25**	**30**
$R_{min} = 0.24v + 0.43$	1.63	2.83	4.03	5.23	6.43	7.63
$R_{min} = 0.29v + 0.67$	2.12	3.57	5.02	6.47	7.92	9.37
$R_{min} = v^2/f_{max}g, f = 0.2$	0/98	3.94	8.86	15.75	24.61	35.43
$R_{min} = v^2/f_{max}g, f = 0.3$	0.65	2.62	5.91	10.49	16.40	23.62
Recommended value (m)	5	5	10	20	25	35
Limit value (m)	3	3	5	10	15	20

Source: Reference [4].

where f is the ground static friction force, thus the theoretical formula of minimum radius is obtained. Usually the maximum static friction f_{max} between bicycle and ground is 0.2 or 0.3. The turning radius and recommended values calculated by different formulas are listed in Table 2.10.

References

[1] W. Lai. Traffic Psychology, Wuhan, China, Central China University of Technology Press, 1988.

[2] Z. Wang, L. Nie, G. Xiao, and H. Yuan. Psychological Behaviors and Safety of Young Riders. Journal of North China Institute of Science and Technology, 2006, 3(2):61–63.

[3] S. Lin. Analysis and Research on Human-Machine Factors in Bicycle Design. Dissertation for Master, Kunming University of Science and Technology, 2008.

[4] J. Xu and X. Chen. Chapter 10 Pedestrian and Bicycle Traffic. In Traffic Engineering Manual, Beijing, China Communications Press, 1997.

[5] Northwestern University Traffic Institute. Bicycle Planning and Facility Workshop. Evanston, IL, May 1994.

[6] Centre for Research and Contract Standardization in Civil and Traffic Engineering. Sign Up for the Bike: Design Manual for a Cycle-Friendly Infrastructure. The Netherlands, Sep. 1994.

[7] American Association of State Highway and Transportation Officials. Guide for the Development of Bicycle Facilities, 1999.

[8] R.E. Miller and M.R. Ramey. Width Requirement for Bikeways: A Level of Service Approach. Davis, CA, Department of Civil Engineering, University of California, Dec. 1975, Report 75-4.

[9] Centre for Research and Contract Standardization in Civil and Traffic Engineering. Sign Up for the Bike: Design Manual for a Cycle-Friendly Infrastructure. The Netherlands, Sep. 1994.

[10] W.S. Homburger. Capacity of Bus Routes, and of Pedestrian and Bicycle Facilities. Berkeley, CA, Institute of Transportation Studies, University of California, Feb. 1976.
[11] F.P.D. Navin. Bicycle Traffic Flow Characteristics: Experimental Results and Comparisons. ITE Journal, 1994, 64(3):31–36.
[12] J.M. Yang. Bicycle Traffic in China. Transportation Quarterly, 1985, 39(1): 93–107.
[13] C. Liang. Study on Characteristics and Application of Bicycle Traffic Flow. Doctoral Dissertation, Changchun, Jilin University, 2007.
[14] S. Watanabe. Traffic Engineering. People's Communications Press, 1982 (*in Chinese*).
[15] J. Forester. Bicycle Transportation. (1st edition). Cambridge, MA, MIT Press, 1983.
[16] FHWA. U. S. Department of Transportation. Safety and Locational Criteria for Bicycle Facilities, User Manual Volume II: Design and Safety Criteria. Report FHWA-RD-75-114, 1976.
[17] S. Vagverk. Chapter 10: Bicycle Traffic Facilities (English Translation). In Swedish Capacity Manual. National Swedish Road Administration, 1977.
[18] H. Wei, F. Ren, and L. Liu. Research on the Relationship Between Bicycle Travelling State and Bicycle Road Capacity. China Journal of Highway and Transport. 1993, 6(4):60–64.
[19] D. Taylor. Analysis of traffic signal clearance interval requirements for bicycle-automobile mixed traffic. In Transport Research Record 1405, Washington, DC, TRB, National Research Council, 1993, pp. 13–20.
[20] W. Pein. Bicyclist performance on a multiuse trail. In Transport Research Record 1578, Washington, DC, TRB, National Research Council, 1997, pp. 127–131.
[21] X.M. Liu, L.D. Shen, and F.T. Ren. Operational Analysis of Bicycle Interchanges In Beijing, China. In Transportation Research Record 1396, Washington, DC, Transportation Research Board, National Research Council, 1993, pp. 18–21.
[22] Z. Tao. Study on Traffic Flow Characteristics of Machine-Non-Mixed Section. Changchun, Jilin University, 2007.
[23] L. Huang. Research on Micro Behavior of Bicycles at Signalized Intersections, Beijing, Beijing Jiaotong University, 2004.
[24] Z.-q. Xie, Q.-z. Chu, J. Chen, and F. Pan. Characteristics of Bicycle Traffic Flow on Bicycle-pedestrian Shared Paths. Journal of Transportation Engineering and Information, 2010, 8(2):93–99.
[25] D.I. Rubins and S.S. Handy. Times of Bicycle Crossings: Case Study of Davis, California. Transportation Research Record, 2005, 1939:22–27.
[26] K. Opiela, S. Khasnabis, and T. Datta. Determination of the Characteristics of Bicycle Traffic at Urban Intersections. In Transportation Research Record 743, Washington, DC, TRB, National Research Council, 1980, pp. 30–37.
[27] Y. Tanaboriboon, T. Agustin, and Y. Minfang. Experiences in Developing Countries with Impact of Exclusive Lanes for Non-motorized Transportation:

Case Studies of China and Indonesia. In Transportation Research Record 1487, Washington, DC, TRB, National Research Council, 1995, pp. 84–89.

[28] H. Botma and H. Papendrecht. Traffic Operation of Bicycle Traffic. In Transportation Research Record 1320, Washington, DC, TRB, National Research Council, 1991, pp. 65–72.

[29] F.R. Whitt and D.G. Wilson. Bicycling Science. Cambridge, MA, MIT Press, 1990.

Chapter 3

Bicycle microscopic behavioral characteristics at signalized intersections

Road intersections are the main manifestations of urban traffic congestions. Therefore, the key to making full use of road resources is the utilization of intersection resources. It is an effective and economic method to improve the capacity of road network and alleviate the urban traffic congestion, by improving the capacity of intersections.

Therefore, we analyzed of cyclists' microscopic behaviors at intersections (signalized and un-signalized ones).

The microcosmic behavior characteristics of bicycles at signalized intersections mainly include the speed of bicycles crossing intersections, the acceptance gap between bicycles and conflicting vehicle flows, the deceleration of encountering red lights, the acceleration of starting and the statistical characteristics of density.

There are two main purposes to study the microscopic behavioral characteristics of bicycles at signalized intersections:

1. Provide data basis for the simulation model of mixed traffic flow at signalized intersections.

   At present, the microscopic simulation model of motorcar traffic flow at urban intersections is relatively mature. However, with the development of bicycle traffic, mixed traffic flow is becoming more and more common in cities, especially at intersections, where non-motor vehicle flow and motor vehicle flow interact and influence each other.

   The microscopic mixed traffic flow simulation platform of urban intersection mainly simulates the driving (riding/walking) behaviors and delay of individual in mixed traffic flow, such as a motor vehicle (or bicycle/pedestrian). The establishment of this kind of micro-simulation behavior model needs a lot of basic data, and the study of bicycle microscopic behavior at signalized intersections is providing theoretical and data basis for it.
2. To assist in improving the management measures of urban signalized intersections, planning and design of bicycle-related transport facilities.

By observing and analyzing the bicycles' behavioral characteristics at signalized intersections, as well as their interactions between motor vehicles, bicycles and pedestrians, we can provide theoretical and behavioral basis for improving the management measures of urban signalized intersections, planning and design of bicycle-related traffic facilities.

In addition, the study of bicycle microscopic behavior at signalized intersections and the establishment of bicycle microscopic behavior database are also helpful to the design of bicycle transport facilities, road network planning and the formulation of bicycle-related traffic policies.

3.1 Field data collection

Accurate, reliable and extensive data are of great significance to scientific research. True and reliable data may not necessarily lead to correct conclusions and models, but without reliable data, it is impossible to achieve scientific and rigorous conclusions.

The basic behavioral models of traffic individuals (motor vehicle drivers, cyclists, pedestrians) are based on the analysis of a large number of databases collected on field and combined with some basic theories. Only by collecting a large amount of traffic individual behavior data can we ensure that the follow-up data analysis and modeling work is carried out on a correct and stable basis.

3.1.1 Data acquisition scheme based on video

1. Purpose

   The existing cameras are used to collect all relevant, useful, available and reliable video data related to the microscopic behaviors of bicycles at signalized intersections. The video data provide a solid data basis for bicycle behavior analysis, basic behavior modeling, parameter calibrations and model validations.
2. Expected data for microscopic behavioral research
   i. the cycling characteristics of bicycles at intersections, such as speed through intersections (straight, left and right turning), start acceleration and stop deceleration when encountering red lights;
   ii. the reaction time of cyclists at intersections;
   iii. acceptance lag and acceptance gap through conflicting motor vehicle flow, for both stopped and nonstopped bicycles;
   iv. bicycle queuing behaviors—density, queuing characteristics, etc.;
   v. the characteristics and distribution of bicycles arriving at intersections;
   vi. acceptance gaps and rejection gaps distribution of bicycles, and critical gap, etc.;
   vii. macroscopic characteristics of bicycle groups—the general size, width, length, overall speed of the group, etc.;
   viii. data on the interactions between bicycles and pedestrians.

3.1.1.1 Environmental requirements for video data acquisition

1. Intersections: Typical crossed signalized intersections should be selected without special geographical and geometric factors, such as slope and visual distance obstruction. In order to ensure adequate sample size and representativeness, the

intersection should have median to large mixed traffic flow volumes. There is a good shooting location at the intersections, such as a pedestrian overpass nearby.

2. Weather: Weather conditions should be sunny (or cloudy), windless weather. Such weather would not have a significant impact on the riding behavior and psychology of cyclists and pedestrians. The rain, snow, hail, gale and other abnormal weather should be avoided.
3. Time period: In order to ensure the representativeness of sample data, the working day time (Monday to Friday morning) should be selected for video data acquisition, and the data of morning and evening peak and nonpeak periods should be collected for future analysis and use. Generally, the morning peak time is 7.30–9.00, and the afternoon peak time is 17.30–18.30.
4. Equipment: Four digital cameras, four spare batteries, four tripods, four walkie-talkies and a tape ruler (30 m).
5. Personnel: There is at least one more investigator going out than the camera brought out. The camera should be taken care of while shooting. The rest are responsible for liaising and delivering goods.
6. Prepare: Require the camera local batteries, spare batteries and walkie-talkie batteries to be full the day before going out to collect video data; conduct simple training on the use of cameras for data acquisition personnel to ensure that everyone can use them; align the camera time before departure, requiring accurate to seconds.

3.1.2 Basic information of data acquisition process

We have carried out two video data acquisition of bicycle microscopic behavior at signalized intersections, respectively, at two different types of signalized intersections. The following is a summary of the basic situation of the two data acquisitions.

3.1.2.1 Video data acquisition at signalized intersection I

1. Time: December 5, 2002 Morning peak hours: 7.30–9.00
2. Location: Wantong Intersection, Haidian District, Beijing
3. Weather: sunny wind 2°–3°
4. Intersection overview

   Wantong Intersection is a typical cross signalized intersection, located in Fuchengmen, Beijing. There are many shopping malls, supermarkets, convenience stores, banks around the intersection, attracting large numbers of vehicles and people.

   The intersection has a large mixed traffic flow volume, including about 5,600 vehicles per hour in peak period, 4,100 bicycles per hour, and about 2,500 pedestrians per hour across the intersection (excluding the number of people using the overpass).

   The distance between the two stop lines in the north–south direction of the intersection is 58 m, and the distance between the two parking lines in the east–west direction is 25 m. There are various traffic management markings at the intersection, such as left-turn motor vehicle waiting area, left-turn bicycle secondary stop line. There are also overpass and crosswalks at intersections. During video data acquisition, two synchronous cameras are placed on the overpass, as shown in Figure 3.1.

5. North–south section

   The north and south sections of the intersection are South Lishi Road and North Lishi Road, respectively. They are arterial streets with eight lanes (two directions). Bicycle lanes and motorway lanes are separated by signs without physical separation. In addition, in Figure 3.1, the eastern bicycle lane of South Lishi Road is not a special bicycle lane, but a mixed lane shared by right-turning motor vehicles. Therefore, the conflict between mixed traffic flow in the north and south directions is obvious.

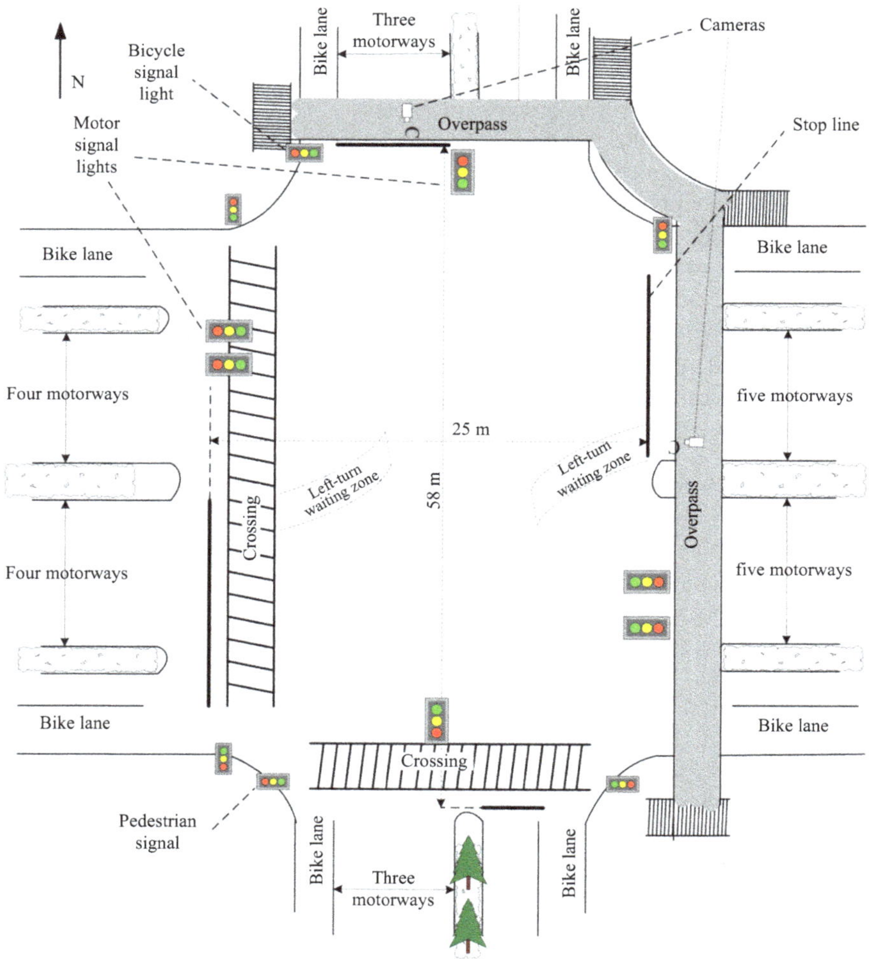

Figure 3.1 Video data acquisition diagram at Wantong Intersection

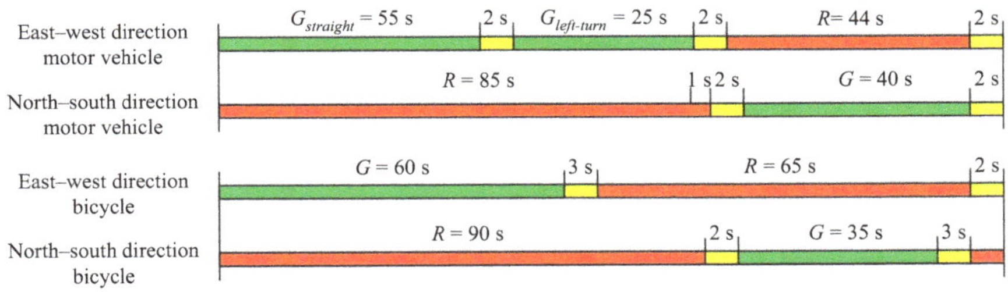

Figure 3.2 Wantong Intersection signal timing diagram

6. East–west road section

   The east–west sections of the intersection are Fuchengmen Inner Street and Fuchengmen Wai Street. They are arterial streets with ten lanes (two directions). The bicycle lanes and motorway lanes on both directions are physically separated. Therefore, the conflict between mixed traffic flow in the north and south directions is relatively less serious.
7. Signal configuration

   The intersection signal is set to three-phase, that is, the east–west direction signal has a special straight green light phase and left-turn green light phase, while the north–south direction signal has only one green light phase. In addition, there are special bicycle and pedestrian lights at intersections. The bicycle and pedestrian lights are set in exactly the same way. The signal timing at the intersection is shown in Figure 3.2.
8. Motor vehicle conflict points

   North–south direction
9. Motor vehicle–bicycle conflict points are shown in Figure 3.3.

3.1.2.2 Video data acquisition at signalized intersection II

1. Time: September 19, 2003 late peak hours: 5.30–7.00
2. Location: Zhanlanguan Intersection, Haidian District, Beijing
3. Weather: cloudy wind 2°–3°
4. Intersection overview

   The intersection is located 300 m south of Beijing Exhibition Hall. The main area around the intersection is residential district, not in the commercial center. There are a lot of bicycle traffic flow and medium pedestrian traffic flow at the intersection. The location of the intersection is ideal, and it can be used as a data acquisition site for bicycle-signalized intersection.

   In Figure 3.4, the distance between the two parking lines in the north and south directions of the exhibition hall intersection is 63 m, and the distance between the two parking lines in the east and west direction is 64 m.
5. North–south section

   The north and south sections of the intersection are Zhanlanguan Road (Figure 3.4). They are arterial streets with six motor vehicle lanes and two bicycle lanes (two directions). Bicycle lanes and motorway lanes are physically separated.

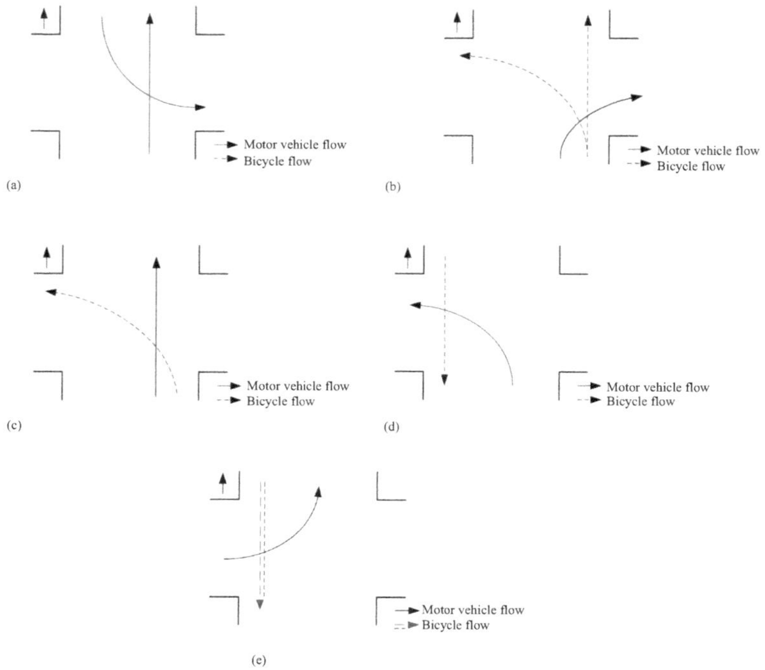

Figure 3.3 Conflicts between motor vehicles and bicycles in different signal phase for Wantong Intersection: (a) conflict between through vehicle and left-turn vehicle; (b) conflicts between right-turn motor vehicle and through and left-turn bicycle; (c) conflicts between through motor vehicle and left-turn bicycle; (d) conflicts between left-turn motor vehicle and through bicycle; (e) conflict between left-turn motor vehicle and north–south through bicycle running a red light

6. East–west section

   The east–west section of the intersection is Chegongzhuang Street. They are also arterial streets with six motor vehicle lanes and two bicycle lanes (two directions). Bicycle lanes and motorway lanes are physically separated.
7. Signal configuration

   The intersection signal is two phased, one for the east–west direction vehicles and the other for the north–south direction vehicles. In addition, there are special bicycle and pedestrian lights at intersections. The bicycle and pedestrian lights are set in exactly the same way. The signal timing at the intersection is shown in Figure 3.5.
8. Motor vehicle conflict points

   North–south direction
9. Motor vehicle–bicycle conflict points are shown in Figure 3.6.

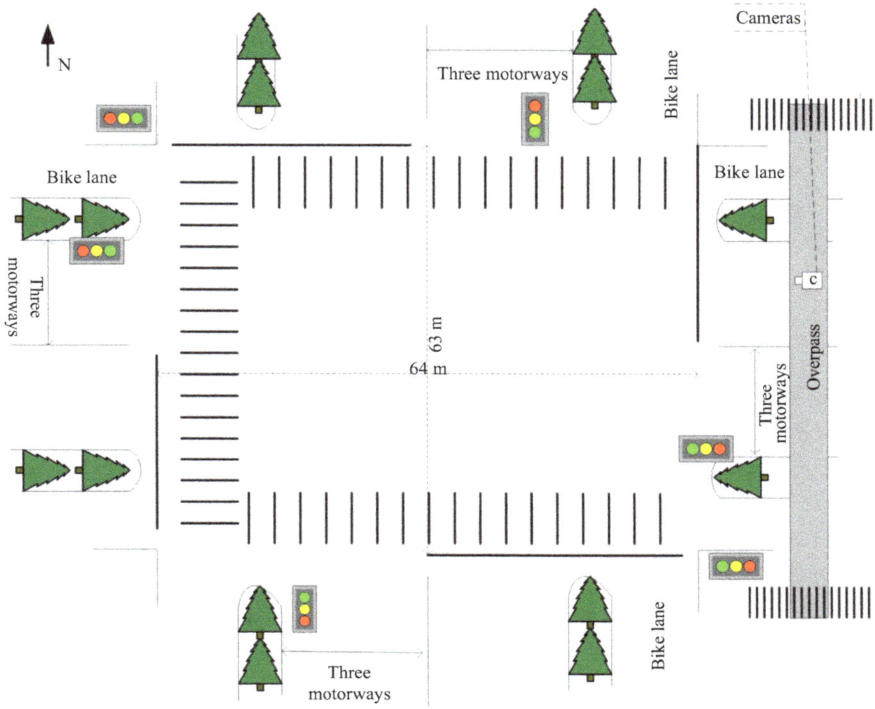

Figure 3.4 Video data acquisition diagram at Zhanlanguan Intersection

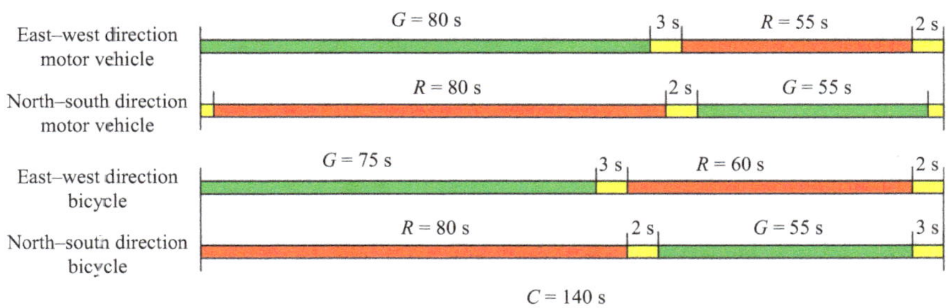

Figure 3.5 Zhanlanguan Intersection signal timing diagram

3.2 Micro behavior data extraction

3.2.1 Basic processing of video data

Video data collected by camera at intersection are a raw data containing a lot of redundant information. Therefore, it is necessary to sort and extract related information from the video data, to obtain the data needed for bicycle microscopic

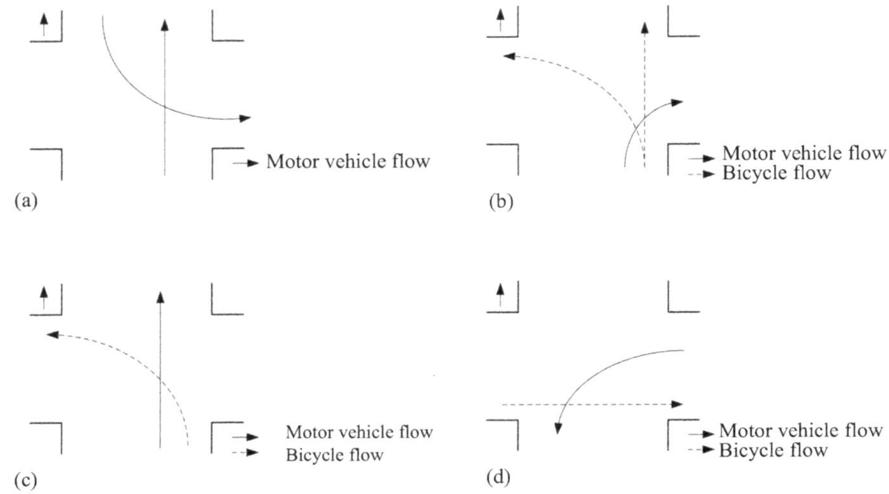

Figure 3.6 Conflicts between motor vehicles and bicycles in different signal phase for Zhanlanguan Intersection: (a) conflict between through vehicle and left-turn vehicle; (b) conflicts between right-turn motor vehicle and through/left-turn bicycle; (c) conflicts between through motor vehicle and left-turn bicycle; (d) conflicts between left-turn motor vehicle and through bicycle

behavior analysis and modeling at signalized intersection. The type and features of data determine the extracting methods.

In this chapter, the data mainly include two types: the data reflect the macroscopic traffic flow parameters and the microscopic riding behaviors. As the macroscopic traffic flow parameters data only provide some assistance for the analysis of microscopic behavior data, they are not the focus of the study. Most of the preliminary data processing work is to extract the microscopic riding behavior data.

3.2.1.1 Macroscopic traffic flow data extraction

Macroscopic data of traffic flow at intersections mainly include motorcar volume, bicycle and pedestrian volume during peak hours.

The traffic flow volume data extraction method is relatively simple: First, we count the number of vehicles passing through each intersection and the number of bicycles and pedestrians in each signal cycle, respectively. Take the signal period be T min, the traffic volumes at this intersection in peak period can be obtained by the following formula:

$$Vol_{peak} = \frac{60V_c}{T} \tag{3.1}$$

where Vol_{peak} stands for the traffic volumes of the intersection in peak hour, vehicle (bicycle or pedestrian)/h; V_c for the average volume of a signal cycle in peak hour at signalized intersection, vehicle (bicycle or pedestrians)/h; T is cycle time, min.

3.2.1.2 Bicycle microscopic behavior data extraction

The process flow of bicycle microscopic behavior data is shown in Figure 3.7.

It is worth noting that the video data acquisition time of bicycle microscopic behaviors at signalized intersection is year 2002 and 2003, which is recorded on overpass by tape recorder. The video quality is far from the current digital camera. Therefore, computer-aided video detection technology cannot be used to track the dynamic trajectory of bicycles directly, so in this section we used manual methods to extract data.

First, digitalize a large amount of video and image data stored on tape, which will be transformed into some kind of multimedia file, stored in the computer hard disk, and then analyzed and processed with appropriate software (if using a digital camera, this step can be skipped). At that time, we used Pinnacle's video capture card to convert tape images into audio video interleaved (AVI) files.

Adobe's Premiere software is used to frame the AVI file. The image in 1 s can be decomposed into 25 frames. In the actual processing process, we do not need to use Premiere software to decompose the AVI file into pictures. We just need to use Premiere software to open the AVI file. The time in seconds and the number of frames the picture is in the second in the status bar below the file. Using the data, we can extract the microscopic behavior data.

After the previous processing, the method of extracting time-related data from video image data is as follows: the lower part of the picture shows the date, hour, minute, second of the picture taken and the number of frames in 1 s. Therefore, as long as the starting and ending pictures of the observed event are determined, the starting and ending time interval of the event can be obtained by simply calculating the numbers under the two pictures.

The methods to extract distance-related data are as follows: there are always some "markers" in the middle and entrance of the intersection, such as the well cover on the ground, various markers (stop line, crosswalk marks, lane marks, etc.) and other facilities. These markers are used to help marking the starting and ending points of events, and the starting and ending distances of events are obtained by measuring the distance between these markers on field.

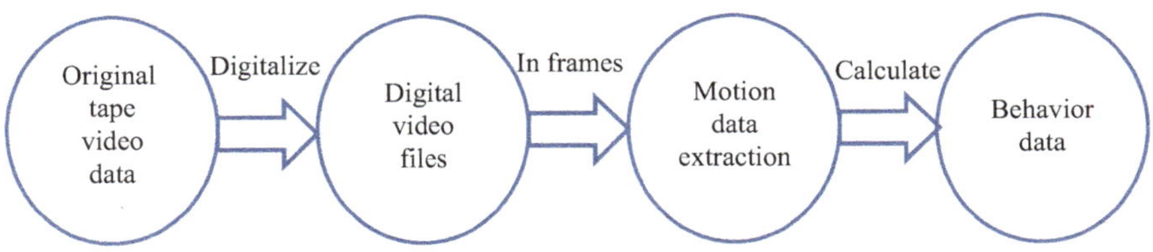

Figure 3.7 Procedure of bicycle microscopic behavior data extraction

Next, we will describe the methods of extracting the data of bicycle microscopic behavior at signalized intersections.

3.2.2 Specific extraction method of basic behavior data

The microscopic behavior data of bicycles at signalized intersections can be divided into four categories: speed, time, acceleration and density.

3.2.2.1 Speed data

Bicycle microscopic behavior speed data at signalized intersections mainly refer to the riding speed, including the speed of direct bicycle and left-turn bicycle. As right-turn bicycle does not actually enter the intersection, it is not included in our investigation.

The specific data extraction methods are as follows: first, the corresponding markers and parallel marking lines are determined, and the actual distance L is measured at the intersection, as long as the time interval ΔT between the bicycle front wheel contact point arriving at the starting marker (or parallel marking line) and arriving at the terminating marker (or parallel marking line) is determined. According to the formula $v = L/\Delta T$, the speed of the individual bicycle crossing the intersection can be obtained.

Figure 3.8 illustrates the process of extracting the speed data of bicycle samples (women in white) crossing intersections.

In the process of extraction, the corresponding cyclists' features (such as age, gender, whether stop at the stop line) of the speed sample are input into the database for further analysis and modeling.

3.2.2.2 Time-related data

In this book, a gap is the time between two successive vehicles in the conflicting traffic flow at intersections, beginning with the front of the leading vehicle and ending with the front of the following vehicle. A lag is the time from the moment

(a) (b)

Figure 3.8 The example of speed data extraction for bicycles crossing intersections: (a) the bicycle front wheel contacts the start marking line; (b) the bicycle front wheel contacts the termination marking line

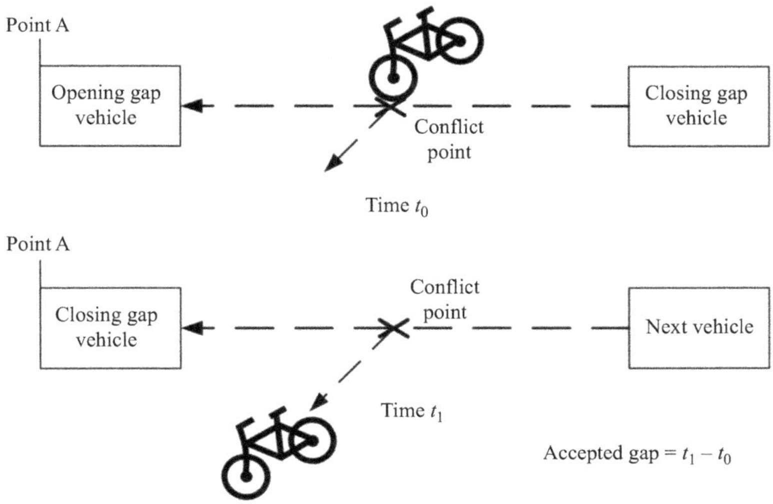

Figure 3.9 Schematic diagram of bicycle acceptance gap

the crossing bicycle reaches the point of conflict until the closing lag vehicle arrives.*

The duration of a gap is measured by the time internal between the beginning and end of the gap. The gap begins when the bicycle front wheel reaches the conflict point; at this moment t_0, the opening gap vehicle arrived at point A and the gap ends at t_1 when the closing gap vehicle reaches point A (see Figure 3.9). That is

$$gap = t_1 - t_0 \tag{3.2}$$

Likewise, we can measure an accepted lag in this way; the only difference is that point A locates at the conflict point in this case (see Figure 3.10).

As long as the acceptance gap and lag start and stop frame are determined, the specific data of acceptance gap and lag can be obtained, and the accuracy can reach 0.04 s. Examples are shown in Figures 3.11 and 3.12.

In the process of extraction, the corresponding cyclists' features (such as straight or left turn, gender, whether stop at the stop line, types of opening and closing gap vehicles) of the gap/lag sample are input into the database for further analysis and modeling.

3.2.2.3 Deceleration/acceleration data

The acceleration-related data mainly refers to the bicycle's deceleration in front of the stop line when it meets the red light at the intersection and the bicycle's starting

*Dean Taylor and Hani Mahmassani, Bicyclist and Motorist Gap Acceptance Behavior in Mixed-Traffic, presented at the 78th annual meeting of the Transportation Research Board and publication in Transportation Research Record November 1998.

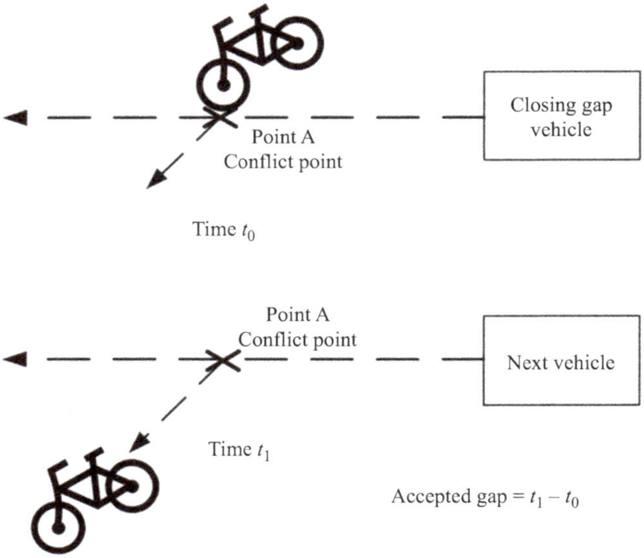

Figure 3.10 Schematic diagram of bicycle acceptance lag

(a) (b)

Figure 3.11 Example of bicycle acceptance gap data extraction: (a) start of the accepted gap; (b) end of the accepted gap

acceleration at the stop line when the signal lights turn green. The unit of acceleration-related data is m/s^2.

In video data extraction, we need to make further assumptions about the bicycles' deceleration and acceleration behavior. Here, we put forward two assumptions:

1. When bicycles meet a red light at signalized intersection, they decelerate evenly, that is, the deceleration is unchanged throughout the deceleration process.
2. The sign of bicycle deceleration is that the cyclist stops pedaling.

(a) (b)

Figure 3.12 Example of bicycle acceptance lag data extraction: (a) start of the accepted lag; (b) end of the accepted lag

In order to locate the bicycle's position, a parallel marking line is drawn every 2 m on the bicycle lane. Under the previous two basic assumptions, the method of measuring bicycle deceleration is as follows: first, the time t_1 of the nearest sign line L_1 arrived after the bicycle began to decelerate is determined, then the time t_2 of bicycle stops and the nearest sign line L_2 of front wheel are determined, and the calculation formula of bicycle deceleration to stop is as follows:

$$a = -2\frac{|L_2 - L_1|}{(t_2 - t_1)^2} \tag{3.3}$$

The method of extracting bicycle acceleration data is similar to that of deceleration. It is assumed that the acceleration of bicycle is uniform during the whole acceleration process, and the whole acceleration process is 20 m. In data extraction, attention should be paid to avoiding the impact of bicycle group and conflicting traffic flow. Therefore, bicycle individuals who are in the first row of waiting bicycle group and have not encountered conflicting traffic flow are selected as data samples, and the following acceleration formula is used to calculate:

$$a = 2\frac{20}{(t_2 - t_1)^2} \tag{3.4}$$

3.2.2.4 Intensity data

Intensity data of bicycle microscopic behavior at signalized intersections mainly refer to the static vehicle density (also called the waiting density—the number of bicycles waiting behind the stop line per meter square) when they encounter red lights at intersections. The dynamic bicycle density refers to the number of bicycles per meter square when bicycles cross intersections at green lights. Intensity-related data are usually obtained by dividing the number of bicycles by the area they occupy. The unit of intensity-related data is generally bicycle/m^2.

In the process of data extraction, the waiting density of bicycles at the stop line is defined as the waiting density in front of the intersection stop line at the

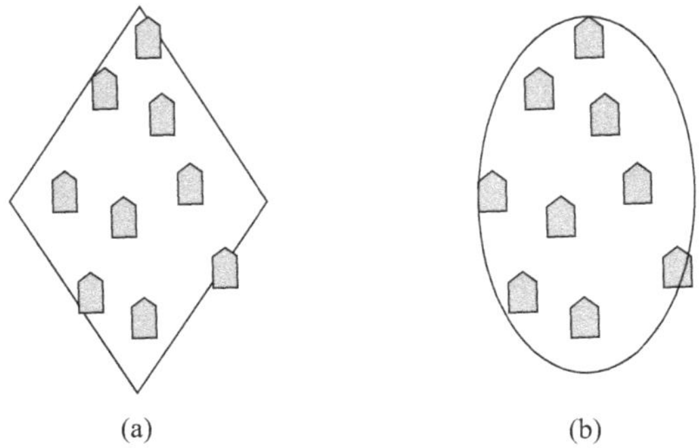

Figure 3.13 (a) Diamond-shaped bicycle cluster shape and (b) elliptical shape of bicycle groups

moment just before the signal lights turn green. The specific data extraction method is to divide the number of waiting bicycles by the area occupied by waiting bicycles. The ground in front of the stop line is divided into 2 m^2 rectangular squares in advance. The area is estimated by counting the number of rectangular squares. In this way, the total error of occupied area is no more than 2 m^2.

The dynamic vehicle density of bicycle groups crossing intersections is defined as the traffic density when the traffic lights turn green; the first bicycle of the bicycle group has not left the intersection and most of the bicycle group enter the intersection. The calculation method is also based on dividing the number of bicycles in group by the area occupied by the bicycles.

Because of the dynamic nature of the driving bicycle group, the method of extracting the static density data is not the same as the method of occupying the area of the bicycle group.

First, the width and length data of bicycle groups are extracted, and the geometric shape of bicycle groups at intersections is observed, and the driving area of bicycle groups is estimated. Details are shown in Figure 3.13.

According to observation, the geometry of bicycles passing through intersections is usually diamond or ellipse.

Strictly speaking, the density data corresponding to these two shapes are disjointed, and the two density data are compared and analyzed, and then the more reasonable density data results are selected.

For more detailed analysis, see Section 3.7.

3.3 Micro behavioral data analysis basis

The methods of mathematical statistics will be used in the analysis of traffic behavior data. Following is an overview of the basic concepts, principles and methods that will be used in the analysis of bicycle micro-behavior data.

3.3.1 Basic concepts

Random variables are variables used to describe random activities. Variables take a certain value according to a certain probability corresponding to random activities and get a certain result according to a certain probability.

Random variables are divided into random variables and continuous random variables. An important numerical characteristic of random variable is its mean, also known as **mathematical expectation**, which is a measure to describe the central intention of random variable.

For discrete random variables, if the distribution law is known to be $P(X = x_i) = p_i$, then the mean value is

$$E(X) = \sum_{i=1}^{\infty} x_i p_i \tag{3.5}$$

For discrete random variables, the probability density function is $f(x)$, then the mean value is

$$E(X) = \int_{-\infty}^{\infty} x f(x) dx \tag{3.6}$$

Another important numerical feature of random variables is their variance, which is a measure of the degree of deviation of random variables around their mean.

The variance of discrete random variables is

$$D(X) = \sum_{i=1}^{\infty} [x_i - E(X)]^2 p_i \tag{3.7}$$

The variance of a continuous random variable is

$$D(X) = \int_{-\infty}^{\infty} [x_i - E(X)]^2 f(x) dx \tag{3.8}$$

The median, 85% and 15% values are also used to characterize random variables. The median value refers to the x value at distribution function $F(x) = 0.5$, while the 85% and 15% values refer to the x value at $F(x) = 0.85$ and 0.15, respectively.

3.3.2 Distribution of random variables and parameter estimation

In data analysis, the basic method of mathematical statistics is usually used, that is, estimating the distribution form and parameters of a random variable by inferring the population from the sample. To infer population from samples mainly includes two aspects: inferring population distribution parameters from sample statistics, and inferring population distribution from sample distribution function.

Samples are representative and reflection of the population. After sampling, we can construct the sample function $\Phi(x_1, x_1, \ldots, x_n)$ to extract the useful information contained in the sample. Such a sample function is called a statistic, which involves inferring that the main distribution parameters of random variables are sample mean, sample variance, K-order origin moments of samples and K-order center moments of samples.

The function corresponding to the sample mean is

$$\bar{X}(n) = \frac{1}{n}\sum_{j=1}^{n} x_j \tag{3.9}$$

The functional formula corresponding to the sample variance is

$$S^2(n) = \frac{1}{n-1}\sum_{j=1}^{n} \left[x_j - \bar{X}(n)\right]^2 \tag{3.10}$$

To estimate the distribution form of a random variable (population) from the sample distribution function, it is usually desirable to fit a theoretical distribution, such as exponential distribution, normal distribution or Poisson distribution, with a set of sample observations. When a suitable theoretical distribution cannot be found, an empirical distribution can be defined directly by the observed values of samples.

The general steps to determine the probability distribution of any random variable (population) X are as follows:

1. according to the sampling theory to obtain a reasonable number of observation data, sample observation values $x_1, x_2, \ldots, x_n$;
2. the measured data are sorted out, and the hypothesis of the distribution of data variables is obtained according to the corresponding sorting results.
3. estimate the distribution parameters of data variables;
4. testing the distribution hypothesis according to the appropriate criterion;
5. if a suitable theoretical distribution function cannot be found, the empirical distribution function of random variables can be determined directly from the observed data collected.

The assumption of the distribution form of data variables is usually based on a set of sample observations. Here, we use the method of drawing frequency distribution histogram to obtain the approximate curve of probability density function $f(x)$ of data variables.

Then the distribution of random variable X can be assumed by comparing the obtained curves with the probability density curves obtained by common theory.

After assuming the theoretical distribution form of random variables, it is necessary to further estimate the parameters contained in the corresponding distribution in order to determine its distribution function comprehensively for sampling by computer simulation. In the analysis of bicycle micro-behavior data, because of the large sample size, the estimation of the parameters to be estimated is obtained by using the digital feature method.

3.3.3 *Hypothesis testing of parameters and distribution functions*

After assuming a theoretical distribution form for a random variable and estimating its parameters, it is necessary to test the parameters and check whether the observed values $x_1, x_2, \ldots, x_n$ are consistent with the fitted distribution.

3.3.3.1 Parameter testing

The parameter test of random variables can be divided into two main categories: the parameter test of normal distribution and the parameter test of non-normal distribution.

In the analysis of microscopic behavior data, *t*-test is used to test the mean of normal population, i.e., parameter test under unknown population variance.

3.3.3.2 Distribution test

To test the distribution of samples, we need to test hypothesis H_0:

$$F(x) = F_0(x) \tag{3.11}$$

Whether it is true or not, where $F(x)$ is the true distribution of random variables, $F_0(x)$ is the judgment distribution.

The sample size of bicycle micro-behavior data is relatively large, so the test method χ^2 can be used to test whether the overall X conforms to any given distribution function $F_0(x)$. The basis of the test method is that if the sample distribution $F_n(x)$ coincides with the judgment distribution $F_0(x)$, the difference of the theoretical frequencies of the samples in each subinterval should be relatively small. The specific steps of χ^2 inspection are as follows:

Assume H_0: $F(x) = F_0(x)$, where $F(x)$ is general theoretical distribution and $F_0(x)$ is distribution to be judged.

Step 1 According to the measured data, the real axis $(-\infty, +\infty)$ is divided into the k subintervals, and the boundary of each subinterval satisfies $-\infty = a_0 < a_1 < a_2 \cdots < a_n = +\infty$.

Step 2 The actual frequency f_i $(i = 1, 2, \cdots, k)$ of statistical sample value $x_1, x_2, \cdots, x_n$ falling into each subinterval i.

Step 3 When H_0: $F(x) = F_0(x)$ is true, the theoretical probability of total X falling into each subinterval i is

$$P = P\{a_{i-1} < X \leq a_i\} = F_0(a_i) - F_0(a_{i-1})\, i = 1, 2, \cdots, k$$

It can be concluded that the theoretical frequency of sample $x_1, x_2, \cdots, x_n$ in the subinterval is np_i.

Step 4 Build statistics $x^2 = \sum_{i=1}^{k} \left((f_i - np_i)^2/np_i\right)$, it can be proved that no matter what the distribution of $F_0(x)$ is, when H_0 is true, as long as n is large enough, the statistics χ^2 can obey the χ^2 distribution of degree of freedom $k-1$ (k is the number of groups). Therefore, given the test level α, the critical point g satisfying the relation $P\{\chi^2 > g\} = a$ can be found from the distribution table with degree of freedom $k-1$.

1. The f_i and np_i ($i = 1, 2, \cdots, k$) obtained from observation sample values $x_1, x_2, \cdots, x_n$ are put into $x^2 = \sum_{i=1}^{k} \left((f_i - np_i)^2 / np_i \right)$, and the experimental observations $\hat{\chi}^2$ are calculated.
2. To make a judgment: if $\hat{\chi}^2 > g$, reject H_0, that is to say, the overall distribution $F(x)$ and the judgment distribution $F_0(x)$ are not consistent; if $\hat{\chi}^2 < g$, accept H_0, that $F(x)$ and $F_0(x)$ are consistent.

3.4 Analysis of speed data

3.4.1 Basic steps of analysis

Speed microscopic behavior data mainly refer to the speed of bicycles (including tricycles and ordinary bicycles) passing through signalized intersections. The basic steps of the analysis are as follows.

3.4.1.1 Preliminary estimate of overall distribution

First, all the valid data are grouped into 10–15 groups, and the histogram is drawn. According to the common theoretical distribution density curve, the overall distribution of bicycle speed is preliminarily estimated.

3.4.1.2 Sample parameters estimate and test by groups

According to the corresponding characteristics of bicycle speed in the microscopic behavior database, the speed data are classified into two categories: tricycle and two-wheeled bicycle. Then the data are classified according to the riding direction of the vehicle (straight, left turn) and gender. The parameters of the classified sample data are estimated and tested, respectively.

3.4.1.3 Significance test between different groups of data

The purpose of classifying speed data by some feature, and estimating and testing parameters is to find out whether these features have a significant impact on the speed of bicycles. Therefore, it is necessary to test the difference significance of parameters of various sample data. If the test results are significant differences, it shows that the classification criterion has a significant impact on the speed of bicycles crossing intersections; if the test results are not significant, it shows that the classification criterion has no significant impact on the speed of bicycles crossing intersections.

3.4.2 Sample overall analysis

3.4.2.1 Judgment of overall distribution

The original sample size of bicycle speeds at signalized intersections is 548. In order to ensure the validity and representativeness of the data, data screening is

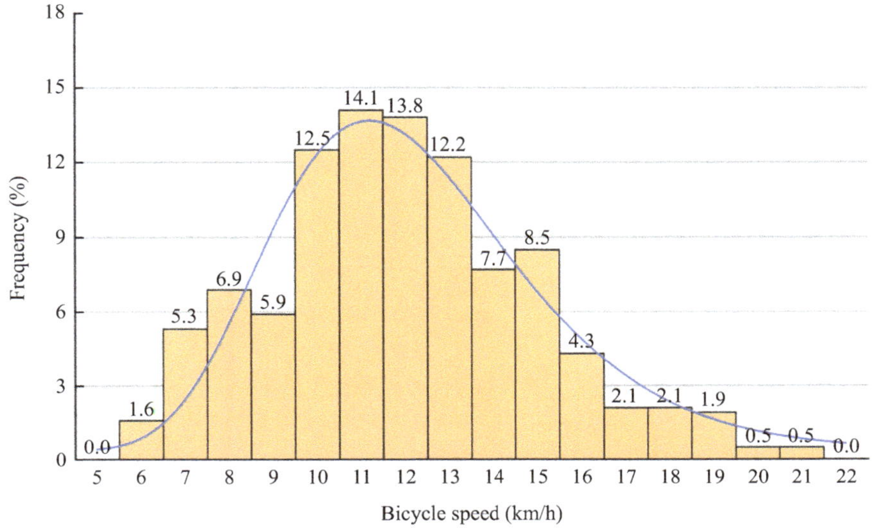

Figure 3.14 Bicycle speed histogram and normal distribution curve fitting

carried out before data analysis, and the three highest and lowest speeds of straight and left rotations are screened out, so there are 534 effective data.

Then, all the valid data of bicycles are grouped into frequency groups. Fifteen groups are selected and histograms are drawn. Compared with the probability density curves of the common theoretical distribution, it is found that the fitting of the sample histogram and the normal distribution curve is the closest (see Figure 3.14).

3.4.2.2 Samples groups

After estimating the overall distribution, the velocity data are classified according to the corresponding characteristics of the data.

In the previous section, when bicycle speed data were extracted, not only the data of the speed of bicycle samples, but also the gender of cyclists, the type of bicycles (tricycles or ordinary two-wheel bicycles), the direction of bicycles (left turn, straight) were obtained.

In order to investigate whether these factors have significant effects on the speed of bicycles passing through signalized intersections, the speed data are classified and estimated. Because the sample size of female tricycles is relatively small (there is only one female sample for left-turn tricycles and only seven female samples for direct tricycles), the gender impact is not examined. The groups of bicycle speed data are shown in Figure 3.15 (the number in parentheses is the corresponding sample size).

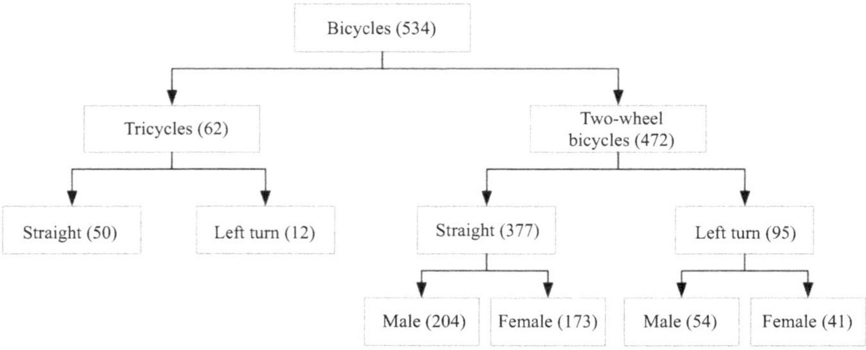

Figure 3.15 Grouping of speed sample data

3.4.3 Data analysis of two-wheeled bicycle

3.4.3.1 Basic analysis process

Because the bicycle speed data are grouped, it is necessary to analyze the data according to a certain sequence of processes in order to get effective results. According to the hierarchical classification of velocity data, the basic analysis process is as follows:

1. First, the speed data of straight and left-turn bicycles are divided into two groups and compared to see if there is significant difference between them, and then consider whether it is necessary to analyze the two groups of data separately.
2. If there are significant differences in the distribution or parameters between straight and left-turn bicycles, the speed distribution is analyzed separately.
3. Significant analysis was made on the influence of gender factors on direct bicycles and left-turn bicycles.
4. If there is no significant difference between the sample distribution or parameters of the speed of straight and left-turn bicycles, the data will be synthesized for distribution analysis, and the influence of gender factors will be analyzed.

Figure 3.16. shows the flowchart of bicycle speed data analysis.

3.4.3.2 Difference analysis between straight and left-turn bicycles

First, we use F-test to test whether the variance difference between straight and left-turn bicycle samples is significant, that is, whether the original hypothesis is H_0: $\sigma_1^2 = \sigma_2^2$ is valid.

In Table 3.1, statistical analysis shows that the statistic F is larger than the critical value of 0.7727 and is in the negative domain. Therefore, the original hypothesis H_0: $\sigma_1^2 = \sigma_2^2$ is rejected, i.e., the variance between the two types of samples is inconsistent.

After the analysis of variance, the difference between the two samples is compared. The original hypothesis H_0: $\mu_1 = \mu_2$ is tested. The t-test and the two-sample heteroscedasticity hypothesis are used. The test results are shown in Table 3.2.

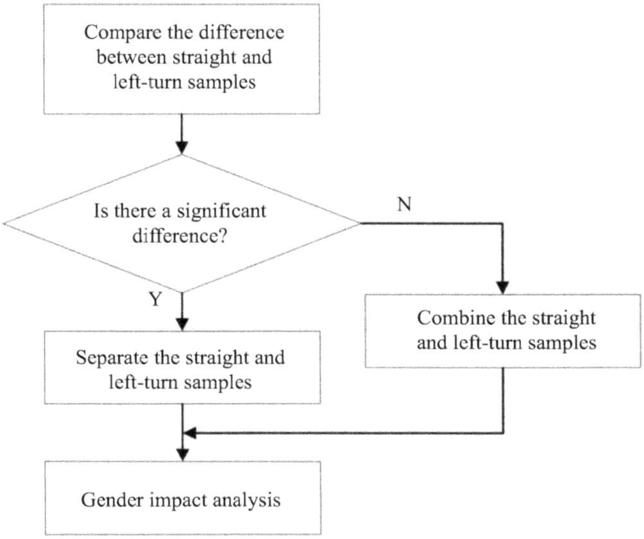

Figure 3.16 Basic flowchart of bicycle speed data analysis

Table 3.1 F-test two-sample variance analysis results

Item	Straight	Left turn
Minimum (km/h)	11.44	13.21
Variance σ^2	9.08	11.07
Sample size	377	95
Degree of freedom	376	94
F Statistics	0.8196	
F Single-tail critical value	0.7727	

Table 3.2 Significance test results of differences between direct and left-turn bicycles

Two-wheeled bicycle	Straight	Left turn
Average	11.44	13.21
Variance	9.077	11.075
Sample size	377	95
Assumed mean difference (μ_1–μ_2)	0	
t Statistic	–4.682	
t Double tail critical value g	1.977824	

Because the t statistic is -4.82, $|t|>g$, the original hypothesis H_0: $\mu_1 = \mu_2$ is overturned, that is, the variance and mean of the data of straight and left-turn bicycles are significantly different, and the speed data of straight and left-turn bicycles should be analyzed separately.

3.4.3.3 Two-wheeled bicycle

Overall speed distribution of straight bicycle

The parameters of the data distribution are estimated. The foregoing judgment shows that the general distribution is normal. The parameters to be estimated include mean value μ and standard deviation σ. The straight bicycle speed mean value μ is 11.44 km/h, and the standard deviation σ is 3.01 km/h.

Straight bicycles speed distribution is tested by χ^2-test. The original hypothesis H_0:

$$F(x) = \frac{1}{\sqrt{2\pi}} \int_{-\infty}^{x} e^{-(u-\mu)^2/2\sigma^2} du$$

where $\mu = 11.44$, $\sigma = 3.01$.

First, the real axis is divided into 20 subintervals. In order to facilitate calculation, we did not divide the real axis into equal intervals but divided the real axis into 20 subintervals by normal distribution and equal frequency. That is to say, the frequency of normal distribution in each subinterval is 5%.

Compare the statistical sample values $x_1, x_2, \cdots, x_n$ fall into the actual frequency f_i of each subinterval i ($i = 1, 2, \cdots$) and the theoretical frequency np_i of normal distribution falling into each subinterval i, the calculated results are shown in Table 3.3.

The experimental observation value $\hat{\chi}^2 = \sum_{i=1}^{20} Y_i = 14.776$. Under the condition of significant level $\alpha = 0.05$, the critical value $g = 30.14$, $\hat{\chi}^2 < g$; therefore, the original hypothesis is valid, that is, the speed of two-wheeled bicycle crossing intersection generally conforms to the normal distribution of $\mu = 11.44$, $\sigma = 3.01$.

Significance of gender factors in straight bicycle's speed

By comparing the speeds of male and female ***straight*** cyclists, the influence of gender factors on the speed of bicycle crossing intersection was analyzed. The data analysis results are shown in Table 3.4.

As can be seen from Table 3.4, for straight bicycles there is no significant difference between male and female cyclists. Gender factor is not a significant factor affecting straight bicycle crossing intersections.

3.4.3.4 Left-turn two-wheeled bicycles

Overall speed distribution of left-turn bicycle

The parameters of the data distribution are estimated. The foregoing judgment shows that the general distribution is normal. The left-turn bicycle speed mean value μ is 13.21 km/h, and the standard deviation σ is 3.33 km/h.

Table 3.3 Comparisons of actual and theoretical frequencies

No.	Actual frequency f_i	Theoretical frequency np_i	$Y_i = (f_i - np_i)/f_i^2$
1	19	18.85	0.001194
2	27	18.85	3.52374
3	13	18.85	1.815517
4	14	18.85	1.247878
5	18	18.85	0.038329
6	24	18.85	1.407029
7	23	18.85	0.91366
8	21	18.85	0.245225
9	19	18.85	0.001194
10	20	18.85	0.070159
11	18	18.85	0.038329
12	16	18.85	0.430902
13	22	18.85	0.526393
14	19	18.85	0.001194
15	16	18.85	0.430902
16	11	18.85	3.269098
17	20	18.85	0.070159
18	16	18.85	0.430902
19	20	18.85	0.070159
20	21	18.85	0.245225
		$\hat{\chi}^2 = \sum_{i=1}^{20} Y_i$	14.776

Table 3.4 t-Test results of differences in speed means between male and female cyclists

Straight bicycles	Female cyclist	Male cyclist
Mean (km/h)	11.27	11.58
Median (km/h)	11.05	11.45
Standard deviation (km/h)	2.72	3.24
Maximum (km/h)	19.8	20.47
Minimum (km/h)	5.28	5.56
Variance	7.398	10.492
Sample size	173	204
Is the variance consistent at a significant level of $\alpha = 0.05$?	Yes	
Combined variance	8.931	
Degree of freedom	375	
t Statistics	−1.01	
t Double-tail critical value g	1.966	
At a significant level of $\alpha = 0.05$, is the mean difference significant?	No	

Left-turn bicycles speed distribution is tested by χ^2-test. The original hypothesis H_0:

$$F(x) = \frac{1}{\sqrt{2\pi}} \int_{-\infty}^{x} e^{-(u-\mu)^2/2\sigma^2} du$$

where $\mu = 13.21$, $\sigma = 3.33$.

First, the real axis is divided into 14 subintervals. We divide the real axis into 14 subintervals by normal distribution and equal frequency. That is to say, the frequency of normal distribution in each subinterval is 7%. Compare the statistical sample values $x_1, x_2, \cdots, x_n$ falling into the actual frequency f_i of each subinterval i ($i = 1, 2, \cdots$), and the theoretical frequency np_i of normal distribution falling into each subinterval i.

The calculated results are shown observation value $\hat{\chi}^2 = 13.217$. Under the condition of significant level $\alpha = 0.05$, the critical value $g = 22.362$, $\hat{\chi}^2 < g$. Therefore, the original hypothesis holds, and the speed of left-turn bicycle crossing intersection generally conforms to the normal distribution.

Significance of gender factors in left-turn bicycle's speed

By comparing the speeds of male and female ***left-turn*** cyclists, the influence of gender factors on the speed of bicycle crossing intersection was analyzed. The data analysis results are shown in Table 3.5.

For left-turn bicycles, there is no significant difference between male and female cyclists. Gender factor is not a significant factor affecting straight bicycle crossing intersections.

Table 3.5 t-Test results of differences in speed means between left-turn male and female cyclists

Left-turn bicycles	**Female cyclist**	**Male cyclist**
Mean (km/h)	12.91	13.43
Median (km/h)	12.35	13.13
Standard deviation (km/h)	3.00	3.56
Maximum (km/h)	21.13	22.30
Minimum (km/h)	7.68	6.94
Variance	9.016	12.67
Sample size	41	54
Is the variance consistent at a significant level of $\alpha = 0.05$?	Yes	
Combined variance	10.60	
Degree of freedom	90	
t Statistics	−0.770	
t Double-tail critical value g	1.987	
At a significant level of $\alpha = 0.05$, is the mean difference significant?	No	

3.4.4 Tricycles speed data analysis

3.4.4.1 Basic analysis process

The basic analysis process of tricycle data is similar to that of two-wheeled bicycles, but the gender difference of tricycle speed is not considered. We first analyze the difference between the data of straight and left-turn tricycles, and then according to statistical test results, we analyzed the speed data separately or together.

3.4.4.2 Differences analysis between straight and left-turn tricycles

Comparing the straight and left-turn tricycles, the basic situation of the two types of data analysis is shown in Table 3.6.

There is a significant difference between the mean values of the speeds of straight and left-turn tricycles, and the speed data of straight and left-turn tricycles should be analyzed separately.

3.4.4.3 Speed distribution of straight tricycles

The parameters of the data distribution are estimated. The foregoing judgment shows that the general distribution is normal. The left-turn bicycle speed mean value μ is 8.72 km/h, and the standard deviation σ is 2.45 km/h.

Left-turn bicycles speed distribution is tested by χ^2-test. The original hypothesis H_0:

$$F(x) = \frac{1}{\sqrt{2\pi}} \int_{-\infty}^{x} e^{-(u-\mu)^2/2\sigma^2} du$$

where $\mu = 8.72$, $\sigma = 2.45$.

Table 3.6 t-Test results of differences in speed means between straight and left-turn tricycles

Tricycle	Straight	Left turn
Mean (km/h)	8.73	11.34
Median (km/h)	8.366834	11.6246
Maximum (km/h)	15.14	16.05
Minimum (km/h)	5.365	7.0341
Variance	9.75	6.01
Sample size	12	50
Is the variance consistent at a significant level of $\alpha = 0.05$?	Yes	
Combined variance	6.70	
Degree of freedom	60	
t Statistics	3.137	
t Double-tail critical value g	2.001	
At a significant level of $\alpha = 0.05$, is the mean difference significant?	Yes	

The calculated results are shown observation value $\hat{\chi}^2 < g$. Therefore, the original hypothesis holds, and the speed of straight tricycles crossing intersection generally conforms to the normal distribution.

Because the sample size is too small, the overall distribution analysis of left-turn tricycles is not carried out.

3.5 Time-related behavior data

3.5.1 *Basic steps of analysis*

Time-related microscopic behavioral data mainly refer to the accepted gap and accepted lag of bicycles (including tricycles and two-wheeled bicycles) crossing signalized intersections when crossing conflicting motor vehicle flows. The basic steps of analysis are similar to speed-based data.

3.5.1.1 Preliminary estimate of overall distribution

Acceptance gap and acceptance lag are two variables with different definitions. There must be significant differences between them. Therefore, the two types of data are completely separated in the process of data analysis.

Like speed data, all valid data of acceptance gap and acceptance lag are grouped into frequency groups, and histogram is drawn to preliminarily determine the overall distribution of acceptance gap and acceptance lag for bicycles, respectively.

3.5.1.2 Sample parameters estimate and test by groups

According to the characteristic quantities of bicycle acceptance gap and acceptance delay in the micro-database, the data are classified into two categories: tricycle and ordinary bicycle. Then, the data are classified according to the direction of the vehicle (straight, left turn), the type of opening and closing vehicles, whether to stop before crossing the motor vehicle flow and gender. The parameters of the classified sample data are estimated respectively.

3.5.1.3 Significance test between different groups of data

The purpose of estimating the classification parameters of data is to find out whether these features have a significant impact on the gap/lag behaviors of bicycles. Therefore, we need to test the difference significance of the parameters of various sample data.

If the test results are significantly different, it is considered that the classification criterion has a significant impact on the gap acceptance or lag acceptance of bicycles crossing intersections; if the test results are not significant, it is considered that the impact is not significant.

3.5.2 *Judgment of overall distribution*

There are 435 original bicycle time data at signalized intersections, including 110 receptive clearances and 325 receptive delays. First, in order to ensure the validity and representativeness of data, data screening is carried out before data analysis, and the highest and lowest accepted gaps and accepted delays are screened out. Therefore, 423 valid data are obtained, of which 104 are accepted gaps and 319 are accepted lags.

Then, all the valid data are grouped into frequency groups and histograms are drawn. Compared with the probability density curves of theoretical distribution, we think that the sample histograms of accepted gap and accepted delay are the closest to the curve fitting of lognormal distribution (see Figure 3.17).

3.5.3 Sample grouping

After judging the overall distribution, the time data are classified according to the corresponding characteristics of the data. In the previous stage, when extracting bicycle time-related data, not only the accepted gap and accepted lags of bicycle samples across intersections, but also other features, including the gender of cyclists, the type of bicycles (tricycles or ordinary two-wheel bicycles), the direction of bicycles (left turn and straight), type of opening and closing vehicles, whether to stop before crossing the flow of motor vehicles, are extracted.

In order to investigate whether these factors have significant effects on the gap/lag acceptance behaviors of bicycles crossing signalized intersections, accepted gaps/lags data are classified and estimated. There will be several different grouping criteria because there are many factors affecting the grouping:

1. according to the characteristics of acceptance gap/lag, such as the type of opening car and the type of closing car;
2. according to the riding characteristics of active bicycles, such as whether to stop before going straight or turning left or crossing the flow of motor vehicles;
3. according to the characteristics of bicycles, such as the gender of cyclists, the type of bicycles;
4. according to the characteristics of the active bicycle's vehicle flow, whether it is a single or a group of vehicles.

3.5.4 Bicycle-accepted gaps

3.5.4.1 Basic steps of analysis

As bicycle-accepted gaps data are grouped, data analysis needs to be carried out according to a certain sequence of processes, and the results are effective. According to the grouping of time data, the basic analysis process is as follows:

1. Because there are many grouping factors in accepted gaps/lags data and the sample size is relatively small, the distribution of accepted gap and lags is not grouped, but the overall distribution is estimated and tested first.
2. To classify and compare the accepted gaps according to the previous different grouping criteria, to investigate whether there are significant differences among the grouping data, and then to consider whether it is necessary to analyze the data separately.

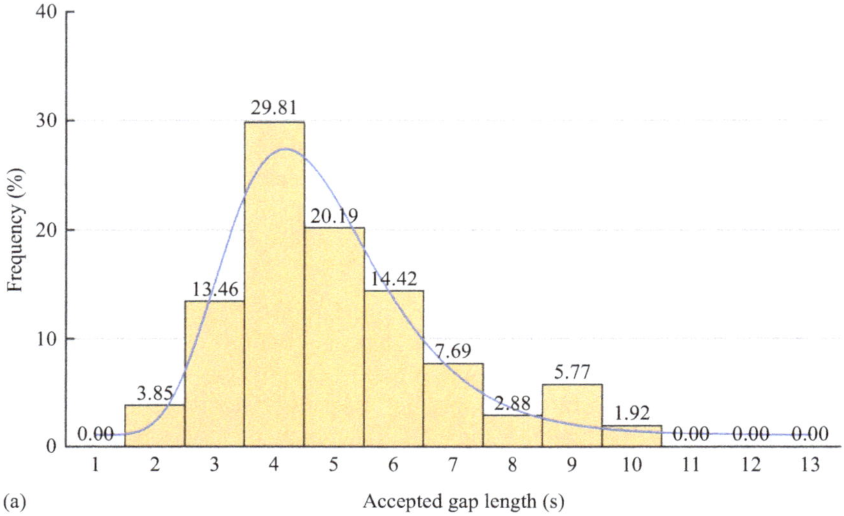

(a)

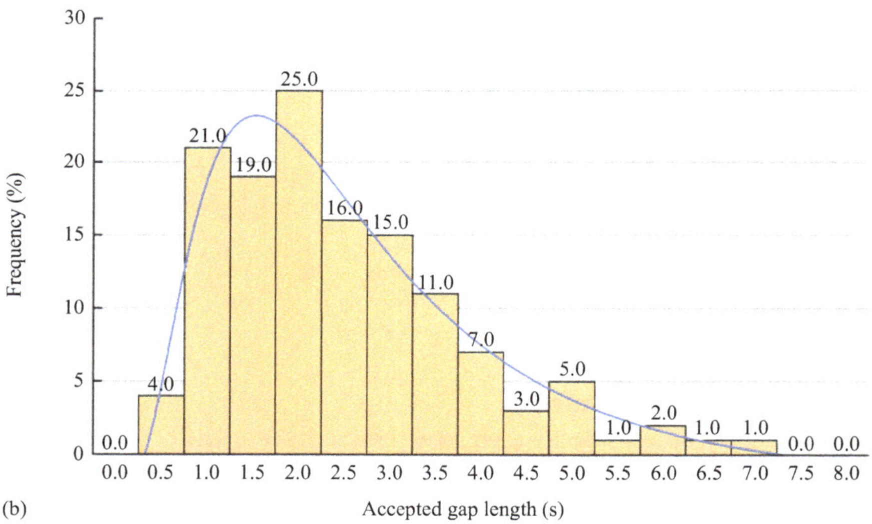

(b)

Figure 3.17 Accepted gaps/lags histogram and natural lognormal distribution curve fitting: (a) overall distribution of bicycle accepted gaps; (b) overall distribution of bicycle accepted lags

3. Finally, the comprehensive influence of the significant factors on the accepted gaps/lags is investigated.
4. If there is no significant difference in the distribution or parameters of the data according to some grouping criteria, the data will be analyzed together.

3.5.4.2 Overall distribution test

The parameters of the data distribution are estimated. The foregoing judgment shows that the general distribution is a natural lognormal distribution. The parameters to be estimated include the proportional parameter μ and the shape parameter σ. Statistical analysis shows that the average value of bicycle clearance sample is 4.5 s, the standard deviation is 1.81 s, the proportional parameter $\mu =$ 1.43 s and the shape parameter $\sigma = 0.39$.

The speed distribution of the straight bicycle is χ^2 tested. The result shows that the statistical value $\hat{\chi}^2 = \sum_{i=1}^{14} Y_i = 11.78$. Under the condition of significant level $\alpha = 0.05$, $g = 22.36$, $\hat{\chi}^2 < g$, so the original hypothesis is valid, that is, the acceptance gap length of bicycle crossing intersection generally conforms to the natural lognormal distribution of $\mu = 1.43$, $\sigma = 0.39$.

3.5.4.3 Comparisons among groups of data

Grouped by characteristics of accepted gap

Grouped by type of gap-opening vehicle: There are three categories of opening vehicles receiving clearance: cars, minibuses and buses. The other types of vehicles are classified into the previous three categories according to their size, such as jeeps into the van category. The acceptance clearance data of different types of opening vehicles are shown in Table 3.7.

First, the F-test is used to test whether the variance difference between car and bus is significant. Statistical analysis shows that the variance between the two types of samples is consistent, while the variance between the data of cars and buses is inconsistent.

The difference of gap data between cars and minibuses, cars and buses was tested by significance test. The original hypothesis H_0: $\mu_1 = \mu_2$ was tested by the double sample homovariance hypothesis t-test, while the latter was tested by the double sample heteroscedasticity hypothesis t-test. The conclusions are shown in Tables 3.8 and 3.9.

It is concluded that the opening vehicle is the van factor, which has no significant effect on the acceptance gap, but the test value $|t|$ between the car and the bus is far greater than the critical value, which indicates that if the opening vehicle is a bus, the accepted gap of the bicycle will be significantly influenced.

Table 3.7 Statistics of accepted gap for different opening vehicles

Accepted gaps	Car	Minibus	Bus
Mean (s)	4.34	4.03	6.81
Maximum (s)	9.08	5.48	8.44
85% value (s)	5.664	5.42	7.18
15% value (s)	2.928	2.58	6.38
Minimum (s)	1.56	2.40	5.48
Sample size	89	7	8

Table 3.8 Significance test of accepted gaps difference between car and minibus (grouped by opening cars)

Accepted gaps	Car		Minibus
Mean	4.34		4.027
Variance	3.13		1.88
Sample size	89		7
Combined variance	3.066		
Assumed mean difference		0	
Degree of freedom		94	
t Statistics		0.426	
t Double-tail critical value *g*		1.986	

Table 3.9 Significance test of accepted gaps difference between car and bus (grouped by opening cars)

Accepted gaps	Car		Bus
Mean	4.34		6.81
Variance	3.133		0.775
Sample size	89		8
Assumed mean difference		0	
Degree of freedom		11	
t Statistics		−6.466	
t Double-tail critical value *g*		2.228	

Table 3.10 Statistics of accepted gaps for different closing vehicle

Accepted gaps	Car	Minibus	Bus
Mean (s)	4.42	3.62	5.32
Maximum (s)	9.08	4.64	8.44
85% value (s)	6.66	4.416	6.72
15% value (s)	2.616	2.728	3.324
Minimum (s)	1.88	1.56	2.76
Sample size	77.00	9.00	18.00

Grouped by type of gap-closing vehicle: Similar to the gap-opening vehicle, the gap-closing vehicles of bicycle-accepted gaps can be divided into three categories: car, minibus and bus. The corresponding statistics of accepted gaps of each type of closing vehicle are shown in Table 3.10.

We tested the difference of accepted gap data between cars and minibus, cars and buses. The conclusions are shown in Tables 3.11 and 3.12.

Table 3.11 Significance test of accepted gaps difference between car and minibus (grouped by closing cars)

Accepted gaps	Car		Minibus
Mean	4.424		3.616
Variance	3.545		0.985
Sample size	76		9
Assumed mean difference		0	
Degree of freedom		16	
t Statistics		2.046	
t Double-tail critical value g		2.12	

Table 3.12 Significance test of accepted gaps difference between car and bus (grouped by closing cars)

Accepted gaps	Car		Bus
Mean	4.423684		5.315789
Variance	3.544797		2.457048
Sample size	76		19
Combined variance		3.334265	
Assumed mean difference		0	
Degree of freedom		93	
t Statistics		−1.905	
t Double-tail critical value g		1.99	

It can be seen that the statistical value $|t|$ of both cars and minibuses is slightly less than the critical value g, so the influence of the accepted gap between the closing minibus and the bus on the bicycle is *existent but not significant*.

Grouped by riding characteristics of active bicycles

Grouped by riding direction: According to this grouping criterion, the accepted gap data can be divided into two categories: straight and left turn. The corresponding accepted gaps are shown in Table 3.13.

Since $|t|<g$, it is concluded that the test cannot overturn the original hypothesis H_0: $\mu_1 = \mu_2$, so the direction of the bicycle has no significant effect on the acceptance gap.

Grouped by stopped or not before crossing the conflicting motor vehicles: According to this grouping standard, the acceptance gap data are divided into two groups: stopped and nonstopped. The corresponding statistics of accepted gaps are shown in Table 3.14.

The $|t|>g$ shows that stopped factors have a significant impact on the accepted gaps of bicycles.

Table 3.13 Significance test of accepted gap difference between straight and left-turn bicycles

Accepted gap	Straight	Left turn
Mean (s)	4.51	4.52
Maximum (s)	9.08	8.44
85% value (s)	7.01	6.354
15% value (s)	2.598	2.974
Minimum (s)	1.56	1.96
Variance	4.143	2.42
Sample size	54	50
Is the variance consistent at a significant level of $\alpha = 0.05$?	Yes	
Combined variance	3.31	
Degree of freedom	102	
t Statistic	0.033	
t Double-tailed critical value g	1.98	
Is the mean difference significant at a significant level of $\alpha = 0.05$?	No	

Table 3.14 Statistics of accepted gaps for stopped and nonstopped bicycles

Accepted gap	Nonstopped	Stopped
Mean (s)	4.33	5.27
Maximum (s)	8.84	9.08
85% value (s)	5.736	8.88
15% value (s)	2.972	2.6
Minimum (s)	1.56	2.16
Sample size	83	21
Is the variance consistent at a significant level of $\alpha = 0.05$?	Yes	
Combined variance	3.17	
Degree of freedom	102	
t Statistic	−2.17	
t Double-tailed critical value g	1.983	
Is the mean difference significant at a significant level of $\alpha = 0.05$?	Yes	

Grouped by characteristics of bicycles

Grouped by cyclists' gender: According to this grouping criterion, the data of accepted gap can be divided into two categories: male and female. The corresponding accepted gap is shown in Table 3.15.

The $|t|<g$ which shows that gender factors have no significant impact on the accepted gaps of bicycles.

Grouped by type of bicycles: According to this grouping standard, the accepted gap data can be divided into two categories: ordinary bicycle and tricycle. The corresponding accepted gap statistics and the test results of difference significance are shown in Table 3.16.

Table 3.15 Significance test of accepted gap difference between male and female cyclists

Accepted gaps	**Male**	**Female**
Mean (s)	4.52	4.50
Maximum (s)	9.08	9.08
85% value (s)	6.588	6.24
15% value (s)	2.942	2.97
Minimum (s)	1.56	2.08
Sample size	78	26
Is the variance consistent at a significant level of $\alpha = 0.05$?	No	
Degree of freedom	48	
t Statistic	−0.041	
t Double-tailed critical value g	2.011	
Is the mean difference significant at a significant level of $\alpha = 0.05$?	No	

Table 3.16 Significance test of gap acceptance differences between ordinary bicycles and tricycles

Accepted gaps	**Ordinary bicycles**	**Tricycles**
Mean (s)	4.56	3.66
Maximum (s)	9.08	5.32
85% value (s)	6.636	4.576
15% value (s)	2.948	2.776
Minimum (s)	1.56	1.96
Sample size	99	5
Is the variance consistent at a significant level of $\alpha = 0.05$?	No	
Degree of freedom	5	
t Statistic	−1.57	
t Double-tailed critical value g	2.571	
Is the mean difference significant at a significant level of $\alpha = 0.05$?	No	

The $|t|<g$ which shows that bicycle type factors have no significant impact on the accepted gaps of bicycles.

This result is different from the intuitive understanding. There is a big difference between the starting and braking performance of tricycles and bicycles. It seems that there should be a difference between the acceptance clearance of tricycles and bicycles. In fact, the sample size of tricycles is too small (only 5), so the result is not significant. Whether the fact is so requires larger sample data to be sure.

Grouped by the state characteristics of bicycles

To eliminate the possible influence of tricycles, all data are of ordinary bicycles with a sample size of 99. The test results of accepted gaps and significant difference

Table 3.17 Significance test of gap acceptance difference between single bicycle and groups

Accepted gaps	Single	In group
Mean (s)	6.53	4.29
Maximum (s)	9.08	8.48
85% value (s)	8.95	6.156
15% value (s)	4.55	2.748
Minimum (s)	3.08	1.56
Sample size	12	87
Is the variance consistent at a significant level of $\alpha = 0.05$?	No	
Degree of freedom	12	
t Statistic	3.255	
t Double-tailed critical value g	2.179	
Is the mean difference significant at a significant level of $\alpha = 0.05$?	Yes	

between single bicycles and those in a group when passing through motor vehicle flow are shown in Table 3.17.

The $|t|>g$ shows that group factors have a significant impact on the accepted gaps of bicycles.

3.5.4.4 Summary of bicycle gap acceptance influence factors

The bicycle gap acceptance influence factors are divided into four categories:

1. **Basic factors:** It refers to the default factors in each group, including opening and closing vehicle types of car, gender male, nonstop through conflicting traffic flow, single bicycle, straight bicycle.
2. **Significant influencing factors:** The mean difference tests of these factors are significant, including the bus as the gap-opening vehicle; the minibus as gap-closing vehicle; stop before crossing the motor vehicle flow and the group factor.
3. **Possible influencing factors:** The difference test values of the factors are slightly less than the critical values, including the bus as gap-closing vehicle.
4. **Un-significant influencing factors:** The mean difference tests of these factors are not significant, including the van as the gap-opening vehicle; tricycle; riding direction and gender factors.

However, we should be aware that the previous analyses of these factors are considered separately, that is, only considering their individual effects, without considering the comprehensive effects of these factors. For example, when the front car accepting clearance is a bus and the rear car is a bus, the impact may be different from the result of simply adding the effects of these two factors together. Therefore, it is necessary to analyze the comprehensive impact of factors.

Considering the complexity, we only consider the significant and possible influencing factors and only consider the comprehensive effects of two factors, not the comprehensive effects of more than three factors.

The comprehensive impact analysis of specific factors will be carried out in the next section of the modeling process.

3.5.5 Bicycle-accepted lags

The process and method of bicycle-accepted lags data analysis and accepted gap analysis are the similar. The main difference lies in the different definitions and influencing factors.

3.5.5.1 Overall distribution test

The parameters of the data distribution are estimated. The foregoing judgment is that the general distribution is natural lognormal distribution. The parameters to be estimated include the proportional parameter μ and the shape parameter σ. Statistical analysis shows that the average value of the sample for bicycle acceptance delay is 2.69 s, the standard deviation is 1.40, the proportional parameter $\mu = 0.86$ and the shape parameter $\sigma = 0.51$.

The experimental observation value $\hat{\chi}^2 = \sum_{i=1}^{20} Y_i = 5.2$. Under the condition of significant level $\alpha = 0.05$, the critical value $g = 30.14$, $\hat{\chi}^2 \ll g$, so the original hypothesis holds that the accepted lags of bicycle crossing intersection generally conforms to the natural lognormal distribution of $\mu = 0.86$, $\sigma = 0.51$.

3.5.5.2 Comparisons among groups of data

Grouped by types of lag-closing vehicle

As the gap-closing vehicles, the lag-closing vehicles can be divided into three categories: car, minibus and bus. The corresponding statistics of accepted lag of each type of closing vehicle are shown in Table 3.18.

Statistical analysis showed that the variance between car and minibus was inconsistent ($F = 0.97 > G = 0.64$), while the variance between the data of cars and buses was consistent ($F = 0.46 > G = 0.71$). Significant difference tests were conducted on the accepted lag data of cars and minibus, cars and buses. The conclusions are shown in Tables 3.19 and 3.20.

t-Test results show that the lag-closing vehicle as a minibus factor has no significant impact on the lag acceptance behavior, while the lag-closing vehicle as a bus has a significant impact on the lag acceptance behavior of bicycles.

Table 3.18 Statistics of accepted lags for different lag closing vehicle

Accepted lags	Car	Minibus	Bus
Mean (s)	2.61	2.34	3.27
Maximum (s)	7.52	7.16	7.88
85% value (s)	3.8	3.072	5.4
15% value (s)	1.39	1.32	1.84
Minimum (s)	0.52	0.56	1
Sample size	246	25	48

Table 3.19 Significance test of accepted lags difference between car and minibus

Accepted lags	Car		Minibus
Mean	2.61		2.34
Variance	1.62		1.67
Sample size	246		25
Assumed mean difference		0	
Degree of freedom		29	
t Statistics		1.003147	
t Double-tail critical value g		2.045	

Table 3.20 Significance test of accepted lags difference between car and bus

Accepted lags	Car		Bus
Mean	2.61		3.27
Variance	1.62		3.54
Sample size	246		48
Combined variance		1.931889	
Assumed mean difference		0	
Degree of freedom		292	
t Statistics		−2.98	
t Double-tail critical value g		1.97	

Table 3.21 Significance test of accepted lag difference between straight and left turn

Accepted lags	Straight	Left turn
Mean (s)	2.82	2.44
Maximum (s)	7.88	7.72
85% value (s)	4.08	3.45
15% value (s)	1.48	1.16
Minimum (s)	0.52	0.56
Sample size	209	110
Is the variance consistent at a significant level of $\alpha = 0.05$?	No	
Degree of freedom	229	
t Statistic	−2.39	
t Double-tailed critical value g	1.97	
Is the mean difference significant at a significant level of $\alpha = 0.05$?	Yes	

Grouped by riding characteristics of active bicycles

Grouped by riding direction: According to this grouping criterion, the accepted lag data can be divided into two categories: straight and left turn. The corresponding accepted gaps are shown in Table 3.21.

t-Test results show that the riding direction has a significant influence on the bicycles' accepted lags.

Grouped by stopped or not before crossing the conflicting motor vehicles: According to this grouping standard, the acceptance lag data are divided into two groups: stopped and nonstopped. The corresponding statistics of accepted gaps are shown in Table 3.22.

t-Test result shows that stopped factors have no significant impact on the accepted lags of bicycles.

Grouped by characteristics of bicycles

Grouped by cyclists' gender: According to this grouping criterion, the data of accepted lag can be divided into two categories: male and female. The corresponding accepted gap is shown in Table 3.23.

Table 3.22 Statistics of accepted lags for stopped and nonstopped bicycles

Accepted lags	Nonstopped	Stopped
Mean (s)	2.674384	2.872593
Maximum (s)	7.88	6.32
85% value (s)	3.894	4.112
15% value (s)	1.36	1.636
Minimum (s)	0.52	1.36
Sample size	292	27
Is the variance consistent at a significant level of $\alpha = 0.05$?	Yes	
Combined variance	1.97	
Degree of freedom	317	
t Statistic	−0.70	
t Double-tailed critical value *g*	1.97	
Is the mean difference significant at a significant level of $\alpha = 0.05$?	No	

Table 3.23 Significance test of accepted lag difference between male and female cyclists

Accepted gaps	Female	Male
Mean (s)	2.635854	2.710295
Maximum (s)	7.16	7.88
85% value (s)	3.776	3.968
15% value (s)	1.44	1.4
Minimum (s)	0.56	0.52
Variance	1.70	2.06
Sample size	82	237
Is the variance consistent at a significant level of $\alpha = 0.05$?	No	
Degree of freedom	154	
t Statistic	−0.43415	
t Double-tailed critical value *g*	1.975486	
Is the mean difference significant at a significant level of $\alpha = 0.05$?	No	

The $|t|<g$ which shows that gender factors have no significant impact on the accepted gaps of bicycles.

Grouped by type of bicycles: According to this grouping standard, the accepted lag data can be divided into two categories: ordinary bicycle and tricycle. The corresponding accepted gap statistics and the test results of difference significance are shown in Table 3.24.

Test result shows that bicycle type factors have no significant impact on the accepted lags of bicycles, which is consistent with result of accepted gaps.

Grouped by the state characteristics of bicycles

As the previous analysis shows that tricycles have no significant impact on acceptance delay, the following data include tricycle data. The test results of accepted gaps and significant difference between single bicycles and those in a group when passing through motor vehicle flow are shown in Table 3.25.

The results show that, as with the analysis of gap acceptance, the group factor also has a significant impact on the lags acceptance of bicycles.

3.5.5.3 Summary of bicycle lag acceptance influence factors

The bicycle lag acceptance influence factors are divided into three categories:

1. **Basic factors:** It refers to the default factors in each group, including the closing vehicle type of car, gender male, nonstop through conflicting traffic flow, single bicycle, straight bicycle.
2. **Significant influencing factors:** The mean difference tests of these factors are significant, including the bus as lag-closing vehicle; left-turn bicycle and the group factor.

Table 3.24 Significance test of lag acceptance differences between ordinary bicycles and tricycles

Accepted laps	Ordinary bicycles	Tricycles
Mean (s)	2.680282	2.779429
Maximum (s)	7.88	6.12
85% value (s)	3.902	4.376
15% value (s)	1.36	1.684
Minimum (s)	0.52	0.96
Variance	2.01	1.60
Sample size	284	35
Is the variance consistent at a significant level of $\alpha = 0.05$?	Yes	
Combined variance	1.97	
Degree of freedom	317	
t Statistic	−0.394	
t Double-tailed critical value g	1.967	
Is the mean difference significant at a significant level of $\alpha = 0.05$?	No	

Table 3.25 Significance test of lag acceptance difference between single bicycle and groups

Accepted lags	**Single**	**In group**
Mean (s)	2.52	3.06
Maximum (s)	7.64	7.88
85% value (s)	3.652	4.446
15% value (s)	1.32	1.794
Minimum (s)	0.52	0.8
Variance	1.78	2.18
Sample size	219	100
Is the variance consistent at a significant level of $\alpha = 0.05$?		No
Degree of freedom		175
t Statistic		−3.08
t Double-tailed critical value g		1.97
Is the mean difference significant at a significant level of $\alpha = 0.05$?		Yes

3. **Un-significant influencing factors:** The mean difference tests of these factors are not significant, including the van as the lag-closing vehicle; stop before crossing the motor vehicle flow; tricycle and gender factors.

However, we should be aware that the previous analysis of these factors are considered separately, that is, only considering their individual effects, without considering the comprehensive effects of these factors. For example, when the front car accepting clearance is a bus and the rear car is a bus, the impact may be different from the result of simply adding the effects of these two factors together. Therefore, it is necessary to analyze the comprehensive impact of factors.

Considering the complexity, we only consider the significant and possible influencing factors and only consider the comprehensive effects of two factors, not the comprehensive effects of more than three factors.

The comprehensive impact analysis of specific factors will be carried out in the next section of the modeling process.

3.6 Acceleration data analysis

Bicycle acceleration data mainly refer to the stop deceleration of bicycles (including tricycles and ordinary bicycles) when they meet red or yellow lights at signalized intersections, and the starting acceleration of bicycles waiting in front of the stop line when the signal lights turn green. In the analysis of deceleration data, for the sake of simplicity, the absolute values of data are analyzed.

3.6.1 Bicycle deceleration

3.6.1.1 Basic steps of analysis

In the analysis of bicycle deceleration data, the deceleration can be divided into comfortable deceleration and emergency deceleration according to the braking condition of the cyclist. *Comfortable deceleration* refers to that the cyclist knows in advance that he will stop before entering the intersection, so he begins to decelerate at a distance from the stop line, and the bicycle deceleration is relatively gentle.

Emergency deceleration is that the bicycle has no intention of stopping before entering the intersection. It only finds the change of signal lights when approaching the stop line or other incidents (e.g., to avoid a conflicting vehicle). Therefore, the deceleration in this kind of situation is rather hasty.

In data analysis, the average deceleration distance is used to classify the two kinds of deceleration situations. That is to say, we assumed that the emergency deceleration is shorter than the average braking distance, and the comfortable deceleration is longer than the average braking distance.

The basic steps of analysis are as follows:

1. First, the bicycle deceleration data are classified and compared according to gender, and test whether there are significant differences between the two groups of data, and whether it is necessary to analyze the data separately.
2. If there are significant differences in the overall distribution or parameters of deceleration between male and female cyclists, the distribution analysis will be carried out separately.
3. Then, the difference of comfortable deceleration and emergency deceleration between male and female cyclists is analyzed.
4. If there is no significant difference in deceleration between male and female cyclists, then we combine the data for distribution analysis and analyze the difference between comfort deceleration and emergency deceleration.

3.6.1.2 Overall distribution test

There are 414 bicycle deceleration data at signalized intersections, where 145 are female and 269 are male cyclists. In order to ensure the validity and representativeness of the data, data screening was carried out before data analysis, and the highest and lowest three were screened out. Therefore, there were 408 valid data, of which 144 were female and 264 were male cyclists. Then, all the valid data of bicycle are grouped into frequency groups, and histograms are drawn. By comparing the probability density curves of common theoretical distribution, it is found that the sample histogram of bicycle parking deceleration is closest to the curve of lognormal distribution (see Figure 3.18).

3.6.1.3 Gender difference analysis

The deceleration of male and female cyclists was compared, and the influence of gender factors was analyzed. The basic data of men and women are shown in Table 3.26.

The $|t| \gg g$ shows that there are significant differences in variance and mean value between male and female cyclists, which should be examined separately.

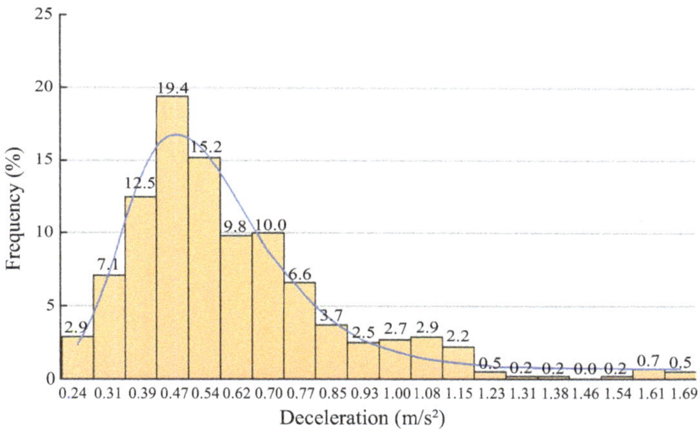

Figure 3.18 Bicycle deceleration histogram and natural lognormal distribution curve fitting

Table 3.26 Significance test of deceleration difference between male and female cyclists

Deceleration (m/s^2)	**Female**	**Male**
Mean	0.54	0.59
Maximum	0.25	0.27
85% value	1.69	1.92
15% value	0.71	0.88
Minimum	0.34	0.35
Variance	0.16	0.17
Sample size	144	264
Is the variance consistent at a significant level of $\alpha = 0.05$?	No	
Degree of freedom	312	
t Statistic	−2.05	
t Double-tailed critical value g	1.97	
Is the mean difference significant at a significant level of $\alpha = 0.05$?	Yes	

3.6.1.4 Comparisons of comfort and emergency deceleration

Male cyclists

According to distance classification, the basic data of comfort and emergency deceleration for male cyclists are shown in Table 3.27.

The $|t| \gg g$ shows that there is a significant difference between emergency and comfortable deceleration of male cyclists, as expected.

Female cyclists

According to distance classification, the basic data of comfort and emergency deceleration for female cyclists are shown in Table 3.28.

Table 3.27 Data on comfort and emergency deceleration of male cyclists

Deceleration (m/s^2)	Comfort	Emergency
Mean	0.40	0.77
Maximum	0.64	1.92
85% value	0.50	1.03
15% value	0.30	0.55
Minimum	0.17	0.40
Variance	0.07	0.01
Sample size	134	127
Is the variance consistent at a significant level of $\alpha = 0.05$?	No	
Degree of freedom	166	
t Statistic	16.12	
t Double-tailed critical value g	1.97	
Is the mean difference significant at a significant level of $\alpha = 0.05$?	Yes	

Table 3.28 Data on comfort and emergency deceleration of female cyclists

Deceleration (m/s^2)	Comfort	Emergency
Mean	0.36	0.64
Maximum	0.54	1.69
85% value	0.44	0.81
15% value	0.26	0.46
Minimum	0.16	0.28
Variance	0.01	0.06
Sample size	55	90
Is the variance consistent at a significant level of $\alpha = 0.05$?	No	
Degree of freedom	121	
t Statistic	9.81	
t Double-tailed critical value g	1.98	
Is the mean difference significant at a significant level of $\alpha = 0.05$?	Yes	

The $|t| \gg g$ showed that there were significant differences between comfort and emergency deceleration for female cyclists, as for male cyclists.

3.6.1.5 Overall distribution of various groups of decelerations

Previous data analysis shows that there are significant differences between comfort and emergency decelerations for cyclists of different genders, so it is necessary to analyze the overall distribution separately. After histogram judgment, all groups of data are close to the natural lognormal distribution, that is

$$F(x) = \int_{-\infty}^{x} \frac{1}{\sqrt{2\pi}\sigma x} e^{-\left((\ln x-\mu)^2/2\sigma^2\right)} du \tag{3.12}$$

Table 3.29 Summary of distribution test of classified data of bicycle decelerations

Group data	Parameter	Distribution test observation value $\hat{\chi}^2$	Critical value g	Whether conform to natural lognormal distribution?
Emergency—male	$\mu = -0.30$ $\sigma = 0.30$	12.57	16.92	Yes
Emergency—female	$\mu = -0.50$ $\sigma = 0.33$	12.67	16.92	Yes
Comfort—male	$\mu = -0.95$ $\sigma = 0.2$	9.85	16.92	Yes
Comfort—female	$\mu = -1.06$ $\sigma = 0.30$	15.36	16.92	Yes

Among them, μ and σ are proportional parameters and shape parameters, respectively.

In Table 3.29, statistical analysis results show that all groups of deceleration data are in accordance with the natural lognormal distribution.

3.6.2 Bicycle starting acceleration

3.6.2.1 Overall distribution

There are 201 bicycle starting acceleration data at signalized intersections, of which 86 are female and 115 are male cyclists. In order to ensure the validity and representativeness of the data, data screening was carried out before data analysis, and the highest and lowest three were screened out. Therefore, 195 valid data were obtained, of which 84 were female and 111 were male cyclists. Then, all the valid data are grouped into frequency groups and histograms are drawn. By comparing the probability density curves of theoretical distribution, it is found that the sample histogram of bicycle starting acceleration is closest to the lognormal distribution curve (see Figure 3.19).

3.6.2.2 Bicycle acceleration influencing factors analysis

Gender factor analysis

The starting acceleration of male and female cyclists was compared, and the influence of gender factors was analyzed. The basic data of male and female cyclists are shown in Table 3.30.

The $|t|>g$ shows that there is a significant difference between the average acceleration of male and female cyclists.

Comparison of average acceleration and first half-range acceleration

In order to verify the hypothesis that the starting acceleration of cyclists is uniform at signalized intersections, we recorded the time of the first half-range acceleration distance (about 8 m) and the time of the whole acceleration distance (about 16 m),

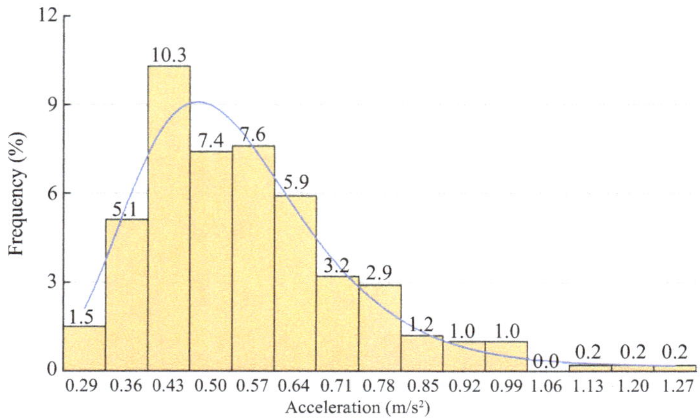

Figure 3.19 Bicycle start acceleration histogram and natural lognormal distribution curve fitting

Table 3.30 Significance test of start acceleration difference between male and female cyclists

Start acceleration (m/s^2)	**Female**	**Male**
Mean	0.46	0.58
Maximum	0.88	1.23
85% value	0.58	0.76
15% value	0.33	0.39
Minimum	0.23	0.27
Variance	0.017	0.036
Sample size	84	111
Is the variance consistent at a significant level of $\alpha = 0.05$?	Yes	
Combined variance	0.02787	
Degree of freedom	193	
t Statistic	−4.71145	
t Double-tailed critical value g	1.97233	
Is the mean difference significant at a significant level of $\alpha = 0.05$?	Yes	

and then the first half-range starting acceleration and the total average acceleration are calculated according to the data. And we test whether there are significant differences between the two accelerations.

If there is a significant difference which shows that the average acceleration of the rider in the first 8 m is different from that of the whole range. That means our premise of calculating acceleration is wrong; if there is no significant difference, it shows that the premise is basically correct. Since there is a significant difference between male and female cyclists in starting acceleration, the correctness of the premises should be analyzed separately.

Table 3.31 Average and first half-range acceleration of male cyclists

Male cyclists	**Average acceleration**	**First half-range acceleration**
Mean (m/s^2)	0.58	0.62
Maximum (m/s^2)	1.23	1.07
85% value (m/s^2)	0.76	0.67
15% value (m/s^2)	0.40	0.46
Minimum (m/s^2)	0.30	0.39
Variance	0.03	0.03
Sample size	94	111
Is the variance consistent at a significant level of $\alpha = 0.05$?	No	
Degree of freedom	95	
t Statistic	−0.72	
t Double-tailed critical value g	2.20	
Is the mean difference significant at a significant level of $\alpha = 0.05$?	No	

Male cyclists: The data of the average acceleration and the first half-range acceleration for men cyclists are shown in Table 3.31.

The $|t|<g$ shows that we cannot overturn the original hypothesis H_0: $\mu_1 = \mu_2$. As expected by the premise of calculating acceleration, the acceleration of cyclists is relatively uniform and does not change significantly with time in the acceleration distance.

Female cyclists: The data of the average acceleration and the first half-range acceleration for men cyclists are shown in Table 3.32.

Because $|t|<g$, the statistical test results are the same as those of male cyclists, that is, we cannot overturn the original hypothesis H_0: $\mu_1 = \mu_2$, and the starting acceleration of female cyclists at intersections is relatively uniform and does not change significantly with time in the acceleration distance.

Other influencing factors analysis

When collecting the data of bicycle starting acceleration, besides the factors of gender and acceleration distance, other factors such as bicycle type (tricycle, ordinary bicycle), whether the cyclist gets off when stop, whether the bicycle loading cargo or not are also investigated. The results of these factors significance test are summarized in Table 3.33.

Overall distribution of acceleration of male and female cyclists

Previous data analysis shows that there are significant differences in acceleration between male and female cyclists, so it is necessary to analyze the overall distribution separately. By histogram judgment, all kinds of data are close to natural lognormal distribution. Statistical distribution test results are shown in Table 3.34.

Table 3.32 Average and first half-range acceleration of female cyclists

Male cyclists	Average acceleration	First half-range acceleration
Mean (m/s^2)	0.45	0.59
Maximum (m/s^2)	0.88	0.81
85% value (m/s^2)	0.55	0.75
15% value (m/s^2)	0.32	0.40
Minimum (m/s^2)	0.23	0.39
Variance	0.014	0.027
Sample size	76	84
Is the variance consistent at a significant level of $\alpha = 0.05$?	No	
Degree of freedom	76	
t Statistic	−2.03398	
t Double-tailed critical value g	2.306006	
Is the mean difference significant at a significant level of $\alpha = 0.05$?	No	

Table 3.33 Significance test results of various influencing factors

Influencing factors		Parameters (m/s^2)				Sample size	Is the effect significant?
		Mean	Variance	Maximum	Minimum		
Bicycle type	Tricycle	0.45	0.24	0.97	0.27	17	No
	Ordinary bicycle	0.53	0.17	1.23	0.23	178	
Get off	Get off	0.51	0.15	0.81	0.26	25	No
	Not get off	0.53	0.18	1.24	0.23	170	
Loading	Loading cargo	0.52	0.31	0.97	0.31	10	No
	No loading	0.53	0.17	1.23	0.23	185	

Table 3.34 Acceleration distribution tests for men and women cyclists

Distribution	Parameters	Distribution test observation value $\hat{\chi}^2$	Critical value g	Does it conform to natural lognormal distribution?	Sample size
Male cyclists' acceleration	$\mu = -0.59$ $\sigma = 0.32$	9.21	23.7	Yes	111
Female cyclists' acceleration	$\mu = -0.81$ $\sigma = 0.28$	13.6	23.7	Yes	84

3.7 Density data analysis

Density data mainly refers to the dynamic density of bicycles when bicycles crossing signalized intersections and the static density of bicycles waiting in front of the stop line at red lights. Compared with the earlier speed and time data, the process of density data analysis is simpler.

3.7.1 Dynamic group density

3.7.1.1 Determine the bicycle group shape

As mentioned in Section 3.2.2.4, the value of dynamic density data of bicycle groups is actually related to the selection of approximate shape of bicycle groups. According to the observation, the bicycle groups crossing the intersection generally present elliptical and diamond shapes. Therefore, in Table 3.35 obtained from these two shapes are analyzed and compared.

The maximum size of the bicycle is 1.9 m in length, 0.6 m in width, 2.25 m in height and 1.2–1.8 m^2 in static parking area. Because the bicycle is a snake-like movement, its width is 100 cm (handlebar width is 60 cm plus snake-like track width is 40 cm); its height can be 250 cm, of which 25 cm is safe space. The formula for calculating road area occupied by vehicles in operation is generally as follows:

$$S_O = W \times h_v \tag{3.13}$$

where S_O represents area occupied by the running bicycles in m^2; W represents the width of the vehicle and h_v represents headway time under speed v.

According to formula (3.13), the road area occupied by bicycles at different speeds is shown in Table 3.36.

According to the data analysis in Section 3.4.3, the average speed of ordinary bicycles cross through signalized intersections is 11.44 km/h. Here, the speed is approximated to 10 km/h and the road area occupied by each bicycle is 5.2 m^2.

Table 3.35 Density data of diamond and ellipse shape group

Group shape	Average density (bicycle/m^2)	Average area occupied by bicycles (m^2/bicycle)
Diamond	0.356	2.959
Ellipse	0.227	4.645

Table 3.36 Road area occupied by bicycles at different speeds

Speeds (km/h)	5	10	15	20	25	30
Road area occupied (m^2)	4.1	5.2	8.1	10	12	16

By comparison, it seems that an ellipse should be chosen as its result in Table 3.35 is closer to bicycle road occupied area.

However, it is worth noting that the riding characteristics and mechanism of bicycles on road sections are different from those in bicycle groups. The bicycle's headway on the road section is obviously larger than that in the group of bicycles. From this point of view, the shape of the ellipse is not appropriate for bicycle groups.

Therefore, after analysis and comparison, we decide to choose the diamond shape with small average vehicle area as the approximate shape of the bicycle group at the intersection and calculate the density data of the bicycle group at the intersection based on this shape.

3.7.1.2 Significance test of grouped data difference

In the process of collecting density data of bicycle groups, bicycle groups passing through intersections are divided into two categories: one is that bicycle groups do not encounter conflicting motor traffic when crossing signalized intersections, which is called non-disturbed bicycle groups; the other is that bicycle groups encounter conflicting motor traffic when crossing signalized intersections (such as left-turn motor traffic), which is called disturbed bicycle group.

There are 56 bicycle group density data at signalized intersections, of which 26 are non-disturbed and 30 are disturbed bicycle group samples. The results of the two types of bicycle groups and the comprehensive density data analysis are shown in Table 3.37.

The $|t|>g$ shows that there is a significant difference between the density of disturbed and non-disturbed bicycle groups.

Table 3.37 Data on bicycle group density

Bicycle group density	**Non-disturbed bicycle groups**	**Disturbed bicycle group**	**Overall**
Mean (m^2/bicycle)	3.34	2.52	2.96
Maximum (m^2/bicycle)	4.64	3.82	4.64
85% value (m^2/bicycle)	4.108	2.938	3.852
15% value (m^2/bicycle)	2.636	2.188	2.25
Minimum (m^2/bicycle)	2.25	1.67	1.67
Variance	0.46	0.18	–
Sample size	30	26	56
Is the variance consistent at a significant level of $\alpha = 0.05$?		Yes	–
Combined variance		0.326	
Degree of freedom		54	
t Statistic		−5.36961	
t Double-tailed critical value g		2.00	
Is the mean difference significant at a significant level of $\alpha = 0.05$?		Yes	

3.7.2 Static group density

Static waiting density data of bicycle groups in front of the stop line were collected. The density data of bicycle groups in front of the stop line at the moment before the signal light turned green were used. One sample was collected in a cycle, and 36 samples were collected in peak hours. The basic situation of static wait density data is shown in Table 3.38.

Generally, bicycles occupy about 4–6 m^2 of road area on sections, but when there is cluster in front of the stop line at intersections, their density is rather high, usually 2–4 m^2/bicycle, sometimes even larger. In the 1980s, the maximum density of stop line was observed on Zhujiang Road in Nanjing, and the average density of bicycles was as high as 0.63 bicycle/m^2 (i.e., 1.59 m^2/bicycle). In Table 3.39, the data of eight intersections in Beijing showed that the density of bicycles was very high. The traffic density is as high as 0.54 bicycle/m^2 (i.e., 1.85 m^2/bicycle) [1].

From the previous data, we can see that the maximum static density (0.67 bicycle/m^2) of bicycle group in front of stop line has not changed much compared with that of more than 20 years ago (0.63 bicycle/m^2), and the average (0.46 and 0.54 bicycle/m^2, respectively) has not changed much. This proves that the static waiting density of bicycle groups in front of the stop line at signalized intersections is stable and does not change with time.

Table 3.38 Bicycle static group density data

Item	Bicycle waiting density (bicycle/m^2)	Static waiting area per bicycle (m^2/bicycle)
Mean	0.46	2.30
Maximum	0.67	1.49
85% value	0.56	2.78
15% value	0.36	1.79
Minimum	0.29	1.50
Sample size	36	

Table 3.39 Bicycle static density at stop lines in Beijing

Intersection	Lane width (m)	Average density (bicycle/m^2)	Maximum density (bicycle/m^2)
Xidan	8.00	0.54	0.63
Dongdan	3.75	0.50	0.64
Chongwenmen	6.50	0.47	0.60
Dongsi	5.00	0.52	0.59
Shuangjing	6.00	0.66	0.70
Ganjiakou	4.50	0.46	0.53
Di'anmen	3.20	0.53	0.69
Zhushikou	3.80	0.55	0.63
Average	–	0.54	0.63

3.8 Summary

This chapter applies field survey data of bicycle microscopic behavior at signalized intersections in Beijing.

Four categories of bicycles microscopic behavior data at intersections are analyzed, including speed, time related, acceleration and density.

These provide the data basis for the next chapter of the modeling of bicycle crossing conflicting motor vehicle flow at signalized intersections.

Reference

[1] J. Xu and X. Chen. Chapter 10 Pedestrian and Bicycle Traffic. In Traffic Engineering Manual, Beijing, China Communications Press, 1997.

Chapter 4
Cyclists' crossing behavior model at signalized intersection

4.1 Introduction

This chapter mainly focuses on modeling the acceptance gap behavior of bicycles when they encounter the conflict motor traffic flow at signalized intersections.

There are often some conflicts between bicycles and motor vehicles at intersections, so bicycles have to cross the conflict motor traffic flow or vice versa. The time headway between two successive vehicles in the motor traffic flow (priority traffic flow) is called a gap.

It is generally believed that critical gap is the minimum gap duration that a traffic individual (including motor vehicles, bicycles and pedestrians) will accept in a specific situation. The value of this critical gap is subjective, so the numerical value varies from person to person, but some of its statistical percentile values can be used for analysis and design. Usually, the distribution center of the observed data of the accepted gap is regarded as the critical gap in the design [1].

4.1.1 Distribution of accepted gaps

Smith in 1976 collected the data of vehicle convergence and crossing bicycle flow at un-signalized intersection of two motorway lanes and obtained the following average accepted gap (i.e., design critical clearance). For vehicles, side-to-side through bicycle flow is 3.5 s, and left-turn through bicycle flow is 2.6 s. As expected, due to the low speed and low impact threat of bicycles, the design critical gap of motor vehicle through bicycle flow is much smaller than that through motor vehicle flow [1].

Smith also estimated that the critical design gap for bicycles crossing through the motor traffic flow was 4.4 s. This value is still shorter than the minimum design value for motor vehicles (5 s in AASHTO). Our intuition shows that bicycles are more likely to be injured in collisions, and cyclists should have a higher gap than motorists, unless the speed of the motor vehicle is very low. However, Smith's article does not mention the speed of the motor vehicle.

Ferrara and Lam inspected the accepted gaps for nonstopped and stopped bicycles of bicycles passing through two traffic flows (at two 4.3-m lanes) at Davis, Florida un-signalized intersections. They found that there was little difference in the accepted gaps between the two conditions (nonstopped and stopped) [2].

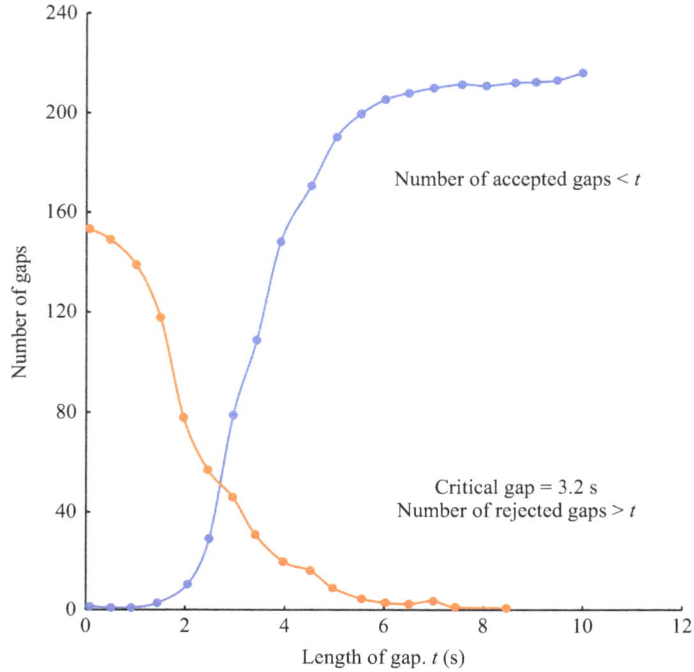

Figure 4.1 Distribution of accepted gaps and rejected gaps [3]

Opiela *et al.* recorded and analyzed the gap acceptance behavior of 260 bicycles crossing one-way motor vehicle flow (two lanes). They found that the distribution of "accepted gaps" of bicycles fitted well with the exponential normal distribution, and the average accepted gap was 3.9 s, the minimum value was 1.1 s, and the average value of rejection gap was 2.6 s, the maximum value was 8.1 s [3].

The distributions of accepted and rejected gaps are shown in Figure 4.1, in which the "critical gap" is 3.2 s (the critical gap definition here is different from the common definition we mentioned earlier, which is the intersection of accepted gap distribution curve and rejected gap distribution curve). Opiela *et al.* also observed that if a cyclist did not stop and pass through, his accepted gap was much smaller than that in the case of stopping and restarting [3].

4.1.2 Cyclists' gap acceptance behavior model

Taylor and Mahmassani studied the crossing and merge behavior between bicycles and motor vehicles at three low-speed stopped control intersections near campus. They estimated the drivers' and cyclists' individual gap acceptance probability model when they crossed or merged into the mixed traffic flow (motor vehicles and bicycles) at intersections, respectively [4].

By these two models, Taylor and Mahmassani compare the critical gaps of motor vehicles and bicycles with different gap acceptance situations. Though the estimates derived from such a small sample size are not accurate enough, yet they can also explain the influence of model parameters, the range of average critical gap in a certain situation and the limitation of relying only on a total average critical gap [4].

They not only estimate the average value of the accepted gaps distribution but also estimate the magnitude and direction of the influence of various important factors on the average critical gap according to the model. Important factors include the following.

4.1.2.1 The difference between gap and lag

As expected, the parameters of the model show that the critical accepted lag of both motor vehicle drivers and cyclists is larger than the critical accepted gap. This may be due to the fact that it takes a little time for the drivers and cyclists to get familiar with the intersection environment when they arrive at the intersection.

4.1.2.2 Type and size of gap-closing vehicle

An important inference can be drawn from the model: when bicycles and motor vehicles turn left laterally, compared to the accepted gap of gap-closing vehicle of cars and assembly vehicles (i.e., general size), if the approaching vehicle is a bicycle, the accepted gap is relatively small; if the approaching vehicle is a bus or heavy truck (i.e., large vehicles), the accepted gap is relatively large.

4.1.2.3 Active vehicle's maneuver

The model shows that for motor vehicles and bicycles, the gap required for lateral left turn and opposite left and right turn is generally larger, and the gap required for lateral right turn maneuver is smaller than that required for left-turn maneuver.

4.1.2.4 Location of active bicycles

If the bicycle on the right side of the lane wants to turn left, it has to go through a longer distance and more than one motor vehicle lane (in the same direction) than the bicycle on the left or in the middle of the lane, so turning is more difficult. As expected, the influence of this parameter in the model will increase the average critical gap (1.03 s). The results show that if the behavior of bicycle is changed, the passing capacity of bicycle can be improved and the delay of bicycle can be reduced.

Next, on the basis of Taylor's and Mahmassani's research, combined with field data collection, we will model the gap acceptance behavior of bicycles crossing conflict motor vehicle flows at signalized intersections in Beijing.

4.2 Gap acceptance choice behavior model

4.2.1 Model construction and basic formula

There are often some conflicts between bicycles and motor vehicles at intersections, so bicycles have to cross the conflict motor traffic flow. The time headway between two successive vehicles in the motor traffic flow (priority traffic flow) is called a gap.

It is generally believed that critical gap is the minimum gap duration that a road user will accept in a specific situation. The value of this critical gap is subjective, varying from person to person. The critical gap can be regarded as a random variable, so we decided to use statistical methods to model it.

We will build a cyclist's gap acceptance behavior probabilistic model at signalized intersections when he/she cross the conflict motor vehicle flow. The following are the main ideas of model construction.

First, the factors that may influence the cyclist's gap acceptance behavior at signalized intersections are investigated and analyzed, and these factors are divided into four categories:

- **Inherent characteristics of gap:** Such as the gap-opening and gap-closing vehicle type; there are three types of vehicles: cars, minibus and buses. Other types of vehicles are classified into the three categories according to their size, such as minivans into minibus and jeeps and cars.
- **Behavioral factors of active bicycles:** Factors related to the active cyclist's movement, including the direction of the bicycle (such as turning left or going straight) and whether the initiative vehicle stops before crossing the flow of motor vehicles.
- **Individual characteristics of bicycle:** Such as the gender of the cyclist, the type of bicycle.
- **The state of the bicycle crossing the conflict flow:** It refers to when crossing the conflict motor traffic flow, whether it is a single bicycle or in a group. The main purpose of this factor is to investigate whether bicycle groups have an impact on the individual cyclist's gap acceptance behavior.

These four categories of factors may have their own effects on cyclists' gap acceptance behavior, but they may have additional comprehensive effects when combined. Therefore, comprehensive factors should be considered in the modeling.

Prior to this, some researchers applied discrete choice modeling methods to individual gap acceptance situations, such as Mahmassani and Sheffi [5].

As researchers pointed out, two problems can be avoided when the discrete choice modeling method is used to estimate the parameters of the gap acceptance equation of individual road user:

1. Other methods can only estimate the average critical accepted gap of samples and can no longer be used for further analysis on gap acceptance behavior.
2. Because most of the methods use only one accepted gap value to ensure the representativeness and unbiasedness of samples, leading to the problem of data loss.

It should be noted that Mahmassani and Sheffi [5] have also studied the rejected gap of individual cyclist and the effect of the rejected gap on the gap acceptance behaviors. In this book, as the data of rejection gap are not extracted due to the time limitation, we did not analyze the influence of rejected gaps.

The construction principle and formula deduction process of discrete choice model describing the gap acceptance behavior are described next.

The precondition of this model is *assuming that all the gap acceptance behaviors of an individual are independent and have no influence on each other, even if the same individual at the same intersection carries out the same riding behavior, the gap acceptance decisions he/she makes are independent.*

The model that uses this precondition to estimate parameters is therefore called independent decisions model.

Considering the critical acceptance gap T_{gvma} of a bicycle individual v at intersection, it can be considered that a bicycle individual v is faced with a decision-making situation of crossing the gap of motor traffic flow, which is determined by the **behavioral factors of active bicycles** m, the **inherent characteristics of gap** g and **the state of the bicycle crossing the conflict flow** a.

The critical accepted gap varies with different **individual of bicycle** and decision-making situations and can be modeled as follows:

$$T_{gvma} = \mu_0 + \xi_{gvma} \tag{4.1}$$

where μ_0 is the average critical accepted gap for all cyclists and all situations, and ξ_{gvma} is a random interference term.

From the bicycle microscopic behavior data analysis results in Chapter 3, we can see that the natural lognormal distribution of ξ_{gvma} is observed in different cyclists and gap acceptance situations.

Therefore, for simplicity of analysis, we perform natural logarithmic operations on sample data $x_1, x_2, \ldots, x_n$, then obtained data $\ln x_1, \ln x_2, \ldots, \ln x_n$ will show normal distribution. Formula (4.1) becomes

$$\ln T_{gvma} = \mu_1 + \xi'_{gvma} \tag{4.2}$$

where ξ'_{gvma} is normal distribution, and μ_1 is the average value of natural logarithm of acceptance gaps of all individual bicycles and all cases, that is, the average value of $\ln x_1, \ln x_2, \ldots, \ln x_n$, which is also the proportional parameter of lognormal distribution.

The relationship between μ_1 and the average critical gap μ_0 of all cyclists and in all cases is as follows:

$$\mu_0 = e^{\mu_1 + (\sigma_1^2/2)} \tag{4.3}$$

where $\sigma_1{}^2$ is the total variance of data after natural logarithmic operation, it is also the shape parameter of lognormal distribution.

Here, the acceptance criterion of individual bicycle crossing the motor vehicle flow is defined as if the actual gap duration L_{gvma} is shorter than the critical acceptance gap, the gap will be rejected, otherwise, the gap will be accepted and the bicycle will cross the motor vehicle flow.

This criterion can be described by the following formula:

$$Y_{gvma} = T_{gvma} - L_{gvma} = (\mu_0 + \xi_{gvma}) - L_{gvma} \tag{4.4}$$

or

$$Y_{gvma} = \ln T_{gvma} - \ln L_{gvma} = \left(\mu_1 + \xi'_{gvma}\right) - \ln L_{gvma} \tag{4.5}$$

If $Y_{gvma} \leq 0$, it means that the gap is accepted and the active bicycle crosses the motor vehicle flow; on the contrary, it means that the gap is rejected.

Therefore, a probability equation can be used to model this gap acceptance behavior. The formula is as follows:

$$\Pr(accepting) = \Pr(Y_{gvma} \leq 0) = \Pr(\mu_1 + \xi'_{gvma} - \ln L_{gvma} \leq 0) =$$

$$\Pr(\xi'_{gvma} \leq \ln L_{gvma} - \mu_1) = \Phi\left[\frac{\ln(L_{gvma}) - \mu_1)}{\sigma_1}\right] = \Phi\left[\frac{\ln(L_{gvma}) - \left(\ln \mu_0 - \frac{{\sigma_1}^2}{2}\right)}{\sigma_1}\right] \tag{4.6}$$

where $\Phi()$ is the normal distribution function, μ_1 and σ_1 are the proportional parameter and shape parameter of the natural logarithmic distribution, respectively. And μ_0 and σ_0 are the average value and standard deviation of the total original data, respectively.

According to the previous investigation and grouping of influencing factors for accepting gap decision-making and the preliminary analysis of comprehensive impact, all possible influencing factors are included in formula (4.6). By the significance test of the coefficient of formula (4.6), the factors that have systematic impact on gap decision-making are obtained.

Formula (4.7) is to model the influencing factors of the original acceptance gap data

$$Y_{gvma} = \beta_0 + \beta_{gap}X_{gap} + \beta_{man}X_{man} + \beta_{vo}X_{vo} + \beta_a X_a + \beta_{\text{int}}X_{\text{int}} + \xi_{gvma} \tag{4.7}$$

where Y_{gvma} is the case of gap g, the average accepted gap of individual v with behavior m; β_0 is the average critical gap when all factors are 0; X_{gap} is the vector attributes for the inherent characteristics of gap; X_{man} is the vector attributes for behavioral factors of active bicycles; X_{vo} is the vector attribute for the individual characteristics of bicycle; X_a is whether bicycles are in the group when crossing; X_{int} is the comprehensive influencing factor of the previous four categories of factors; β_{gap}, β_{man}, β_{vo}, β_a and β_{int} are vector coefficients to be estimated.

When formula (4.7) is introduced into (4.6), the cyclist's gap acceptance decision probability model is changed to

$$\Pr(accepting) = \Phi\left[\frac{\ln L_{gvma} - \left[\ln(\beta_0 + \beta_{gap}X_{gap} + \beta_{man}X_{man} + \beta_{vo}X_{vo} + \beta_a X_a + \beta_{\text{int}}X_{\text{int}} + \xi_{gvma}) - \frac{\sigma_1{}^2}{2}\right]}{\sigma_1}\right]$$

$$= \Phi\left[\frac{\ln L_{gvma} - \ln\left(\beta_0 + \beta_{gap}X_{gap} + \beta_{man}X_{man} + \beta_{vo}X_{vo} + \beta_a X_a + \beta_{\text{int}}X_{\text{int}} + \xi_{gvma}\right)}{\sigma_1} + \frac{\sigma_1}{2}\right]$$

$$= \Phi\left[\frac{\ln\left\{\frac{L_{gvma}}{\left(\beta_0 + \beta_{gap}X_{gap} + \beta_{man}X_{man} + \beta_{vo}X_{vo} + \beta_a X_a + \beta_{\text{int}}X_{\text{int}} + \xi_{gvma}\right)}\right\}}{\sigma_1} + \frac{\sigma_1}{2}\right] \tag{4.8}$$

The definitions of symbols in the formulas are the same as those in the earlier formulas.

4.2.2 Model calibration and optimization

4.2.2.1 Model calibration

The results of data analysis in Chapter 3 show that the accepted gaps of bicycle crossing through motor vehicle flow conforms to the natural lognormal distribution of $\mu = 1.43$ and $\sigma = 0.39$. Therefore, the parameters to be calibrated for the cyclist's gap acceptance decision probability model at signalized intersections are mainly the vector coefficients β_{gap}, β_{man}, β_{vo}, β_a and β_{int} in formula (4.7).

The results of data analysis in Chapter 3 show that the significant influencing factors of bicycle acceptance gap include (1) bus as gap-opening vehicle; (2) minibus as gap-closing vehicle; (3) stopped before crossing the motor vehicle flow and (4) group factor. On the other hand, the possible influencing factors include (5) bus as gap-closing vehicle. Finally, the nonsignificant factors include (6) minibus as gap-opening vehicle; (7) the tricycle type; (8) the riding direction and (9) the gender factors.

For the sake of simplicity, all the single factors are considered in linear regression; only the comprehensive factors between the significant and the possible factors are considered, excluding the nonsignificant factors. The model also considers the comprehensive effects of only two factors, not more than three factors.

In theory, the previous five significant influencing factors (1)–(5) can be combined to obtain nine comprehensive factors, respectively:

①bus as gap-opening vehicle; + bus as gap-closing vehicle;
②bus as gap-opening vehicle; + minibus as gap-closing vehicle;

③bus as gap-opening vehicle; + stopped;
④bus as gap-opening vehicle; + in group;
⑤minibus as gap-closing vehicle; + stopped;
⑥minibus as gap-closing vehicle; + in group;
⑦bus as gap-closing vehicle; + stopped;
⑧bus as gap-closing vehicle; + in group;
⑨stopped; + in group.

However, due to the incomplete representation of sample data, only the following four situations of comprehensive factors exist in the sample data:

①minibus as gap-closing vehicle; + stopped;
②bus as gap-closing vehicle; + stopped;
③bus as gap-closing vehicle; + in group;
④stopped; + in group.

Therefore, the basic process of multiple regression process for the original data of bicycle gap acceptance behavior at signalized intersections is as follows:

1. Preliminary process: Since the Y values of the linear regression model are the average critical accepted gaps in various situations, the average value of accepted gap in different situations is calculated according to the classification of situations, and then the regression analysis of these results' data is carried out.
2. Multiple linear regressions: The previous results were analyzed by multiple linear regressions by mathematical statistics software, and the regression effect and the significance of each independent variable were tested.
3. Model optimization: The independent variables and coefficients of the regression equation are optimized and obtain the optimal model.

For the sample data, the average accepted gaps are calculated according to the situations. There are 12 cases in which 8 are meant for single-factor situations and 4 for two-factor comprehensive situations. The results of the preliminary regression of average accepted gaps are shown in Table 4.1.

After solving the multiple regression equation, it is necessary to test the significance of regression effect, that is, whether there is a linear correlation between Y and various influencing factors X. The main problem is to test whether all regression coefficients beta are zero. If all regression coefficients beta are zero, the linear relationship is not significant; if not all are zero, the linear relationship is significant. Therefore, the null hypothesis is $H_0 : \beta_1 = \beta_2 = \beta_3 = \cdots = \beta_{12} = 0$.

Calculate the test statistic:

$$F = \frac{S_R/m}{S_e/(n - m - 1)} \tag{4.9}$$

$$S_R = \Sigma(\hat{y}_i - \overline{y}_i)^2 \tag{4.10}$$

$$S_e = \Sigma(y_i - \hat{y}_i)^2 \tag{4.11}$$

Table 4.1 Preliminary regression results of raw data

Interception and independent variables	**Coefficients (s)**	**Standard error**	***t*-Statistics**
β_0	3.693	0.489	7.556
Minibus as gap-opening vehicle	0.312	0.748	0.418
Bus as gap-opening vehicle	2.947	0.542	5.438
Bus as gap-closing vehicle	−0.345	0.489	−0.705
Stopped	0.194	0.710	0.273
Left turn	0.175	0.401	0.437
Gender	−0.043	0.345	−0.124
Minibus as gap-closing vehicle	−0.255	0.537	−0.475
In group	0.784	0.704	1.113
Minibus as gap-closing vehicle and stopped	0.809	1.215	0.665
Bus as gap-closing vehicle and stopped	2.501	0.806	3.104
Bus as gap-closing vehicle and in group	0.992	0.827	1.199
Stopped and in group	−2.198	0.817	−2.691

Table 4.2 Fitting results of multiple regression equations

R^2	0.768
Standard error	0.877
Sample size	35

Table 4.3 Analysis of variance table (accepted gaps)

Source	**Degree of freedom (df)**	**Sum of squares (SS)**	**F' Statistics**	**$F_{0.05}(12, 22)$**
Regression	12	55.85	6.058	0.00014
Error	22	16.90		
Total	34	72.75		

where m is the number of independent variables; N is regression sample size; y_i is the ith original data and $\hat{y}_i$ is the ith regression value.

When H_0 is true, $F \sim F(m, n-m-1)$.

If $F \geq F_{1-\alpha}(m, n-m-1)$ at the significant level α, the null hypothesis H_0 is rejected and the linear regression effect of Y on X is significant.

At the significant level $\alpha = 0.05$, the analysis of variance of multiple regression equation is shown in Tables 4.2 and 4.3.

Because of the statistical value $F' \gg F$, the null hypothesis is rejected, that is, the linear relationship between dependent variable Y and independent variable X is significant. However, in multivariate linear regression, it is not enough to test the overall significance of linear regression effect. It is also necessary to clarify that each variable $x_1, x_2, x_3, \ldots, x_{12}$ has a significant linear effect on Y. Secondary and insignificant variables should be removed from the regression equation and obtain a refined regression equation containing only significant variables.

If the linear effect of factor x_i on Y is not significant, then there should be $\beta_i = 0$ in the linear regression model. Therefore, to test whether the variable x_i has a significant effect on Y is to test null hypothesis H_{0i}:

$$H_{0i}: \beta_i = 0; i = 1, 2, 3, \ldots, 12 \tag{4.12}$$

by t-test methods with t-statistic:

$$t_i = \frac{\hat{\beta}_i \sqrt{l_{xx}}}{\sqrt{\frac{S_e}{n-2}}} = \frac{\hat{\beta}_i}{\sqrt{\frac{S_e}{l_{xx}(n-2)}}} \tag{4.13}$$

where $\hat{\beta}_i$ is the estimation coefficient of independent variable x_i, $\sqrt{S_e/l_{xx}(n-2)}$ is standard error for x_i, and

$$t_i \sim t(n - m - 1)$$

If $|t_i|$ is greater than the critical value g, the null hypothesis is rejected, indicating that the linear effect of factor x_i is significant; if $|t_i|$ is less than the critical value g, the null hypothesis cannot be rejected, that is to say, the linear effect of this factor is not significant and should be removed from the linear equation.

At significance level $\alpha = 0.10$, the critical value $g = t_{1-\alpha/2}(22) = 1.71$.

The significance analysis results for each factor x of the multiple regression equation are shown in Table 4.4.

Table 4.4 Summary of significant analysis results of various factors (accepted gaps)

Independent variables	**Coefficients (s)**	**Standard error**	***t*-Statistics**	**Significant or not?**
Minibus as gap-opening vehicle	0.312	0.748	0.418	N
Bus as gap-opening vehicle	2.947	0.542	5.438	Y
Bus as gap-closing vehicle	−0.345	0.489	−0.705	N
Stopped	0.194	0.710	0.273	N
Left turn	0.175	0.401	0.437	N
Gender	−0.043	0.345	−0.124	N
Minibus as gap-closing vehicle	−0.255	0.537	−0.475	N
In group	0.784	0.704	1.113	N
Minibus as gap-closing vehicle and stopped	0.809	1.215	0.665	N
Bus as gap-closing vehicle and stopped	2.501	0.806	3.104	Y
Bus as gap-closing vehicle and in group	0.992	0.827	1.199	N
Stopped and in group	−2.198	0.817	−2.691	Y

Results in Table 4.4 shows that there are only three significant independent variables in the linear regression equation: single factor "bus as gap-opening vehicle"; comprehensive factor: "bus as gap-closing vehicle+stopped" and "stopped+in group."

4.2.2.2 Regression model optimization

In practical application, we always hope that the regression equation used to predict is the optimal linear regression equation. The so-called optimal linear regression equation includes all the independent variables that have a significant impact on Y on the one hand, and the number of independent variables in the equation should be as few as possible. When more than one equations satisfy the previous two requirements, we think that the equation with the minimum estimate of variance σ^2 is the best.

Generally there are three main methods to select the optimal equation:

Method 1. All comparison method

This method first proposes all the independent variables that may affect Y according to the expertise of specific problems, then obtains the linear regression equation for all the possible combinations of independent variables, and then chooses the best one.

This method can always find an optimal regression equation, but when the number of independent variables m is a bit larger, the amount of calculation is very large. So this method is not used here.

Method 2. Only out and not in method

As in method 1, assume that all the selected independent variables are $X_1, X_2, \ldots, X_m$. First, a linear regression equation with all independent variables is established, and the significance test is made for the equation. If the regression effect is significant, then the significance test is made for each independent variable, excluding variable X_k that has the least influence on Y in the nonsignificant variables, i.e.,

$$F_k = \min_{1 \le j \le m} F_j \tag{4.14}$$

Then a new regression equation is established after removing the variable X_k. The significance test of each independent variable of the new regression equation is done again. The least variable affecting Y is removed from the nonsignificant variable. The process is repeated continuously until the variables contained in the regression equation are significant, and the optimal regression equation is obtained.

This method can be used when there are few independent variables, especially when there are few insignificant variables.

But when there are many independent variables, the workload is still considerable. The fatal disadvantage of this method is that once the independent variables are eliminated, they can no longer enter the regression equation.

In practical problems, it often happens that when one variable X_k is excluded as well as other variables are excluded, X_k may become a variable that has a significant impact on dependent variables.

Therefore, the regression equation obtained by the "only out and no in" method is not necessarily the real optimal regression equation.

Method 3. Stepwise regression

As in method 1, assuming that all independent variables have been selected, the correlation coefficient r between each independent variable X and dependent variable Y is calculated. According to the absolute value of the correlation coefficient $|r|$ from large to small, the order of influence of all independent variables X on dependent variable Y is obtained. The independent variable X_k with the largest $|r|$ and Y were analyzed by simple linear regression and taken the significant test.

If X_k is not significant, it can be considered that all the independent variables selected are not the main factors affecting the dependent variable Y. If the regression effect is significant, the independent X_i is introduced into the regression equation one by one by the order of influence.

When every new independent variable is introduced, significance test of each independent variable contained in the new regression equation must be taken.

If there are unsignificant factors in the new regression equation, by eliminating an independent variable that is not significant and has the smallest influence on Y, then a new regression equation will be obtained.

Every independent variable of the new regression equation should be tested and eliminated until every independent variable contained in the obtained regression equation is significant and then introduce a variable that has the greatest influence on Y.

Repeat the previous steps until neither the selected variables can be eliminated nor the new ones can be selected.

Stepwise regression cannot guarantee that the final equation is the optimal regression equation, but in practical application, the regression equation obtained by stepwise regression method has better prediction effect, and the amount of calculation is not too large, so it is currently widely used.

Next, we choose the stepwise regression method to optimize the regression model.

In practical applications, we should properly choose the significance level α when using the stepwise regression method. The larger α is selected (i.e., the critical value $F_{1-\alpha}(n-m-1)$ is smaller), more independent variables will be selected, whereas α is smaller (i.e., the critical value $F_{1-\alpha}(n-m-1)$ is larger), will lead to some important independent variables are eliminated.

After analysis and comparison, we use *method 3* to optimize the regression equation, and the significance level α is 0.1.

According to the statistical definition, the correlation coefficient r_i of each independent and dependent variables is calculated as follows:

$$r_i = \frac{l_{x_i,y}}{\sqrt{l_{x_ix_i}l_{yy}}} = \hat{b}_i\sqrt{\frac{l_{x_ix_i}}{l_{yy}}} \tag{4.15}$$

The optimization process is as follows:

1. Perform multiple linear regression on all selected independent variables and sort according to the absolute value of the correlation coefficient $|r|$. The results are shown in Table 4.5.

Table 4.5 Independent variables ordered by |r| (accepted gaps)

Independent variables	Coefficients (s)	l_{xx}	l_{yy}	$\|r\|$
Bus as gap-opening vehicle	2.947	4.2857	72.75335	0.511716
Bus as gap-closing vehicle and stopped	2.501	4.9714	72.75335	0.427448
Stopped and in group	−2.198	2.7428	72.75335	0.182187
Bus as gap-closing vehicle and in group	0.992	4.2857	72.75335	0.057966
In group	0.784	6.171	72.753	0.052
Bus as gap-closing vehicle	−0.345	8.400	72.753	0.014
Minibus as gap-closing vehicle and stopped	0.809	0.971	72.753	0.009
Minibus as gap-closing vehicle	−0.255	5.600	72.753	0.005
Stopped	0.194	7.143	72.753	0.004
Left turn	0.175	8.400	72.753	0.004
Minibus as gap-opening vehicle	0.312	1.886	72.753	0.003
Gender	−0.043	8.571	72.753	0.0002

Table 4.6 Results of a regression analysis of "Bus as gap-opening vehicle"

R^2	**0.317**				
Standard error	**1.227**				
Variance analysis	**Degree of freedom (df)**	**SS**	**MS**	***F***	**Threshold *g***
Regression	1	23.073	23.073	15.326	0.000428
Error	33	49.681	1.505		
Total	34	72.753			
	Coefficient	Standard error	*t*-Statistics		
Intercept	4.332	0.224	19.337		
Bus as gap-opening vehicle	2.320	0.593	3.915		

2. Regression analysis is performed on $|r|$, the largest independent variable *Bus as gap-opening vehicle*, and the results are shown in Table 4.6.
   From the previous, we get $F > g$, the regression effect is significant.
3. Add the independent variable *Bus as gap-closing vehicle and stopped* and perform regression analysis. The results are shown in Table 4.7.
   The *t*-statistics of both independent variables are greater than the critical value $t'_{0.95}(32) = 1.69$, so they are not eliminated.
4. Add the independent variable *Stopped and in group* and perform regression analysis. The results are shown in Table 4.8.
   The *t*-statistics of the factor *Stopped and in group* is smaller than the critical value $t'_{0.95}(31) = 1.69$, and the absolute value of the correlation index is also close to zero, so this independent variable is eliminated.

Table 4.7 Results of two independent variable regression analysis

R^2	**0.613**				
Standard error	**0.938**				
Variance analysis	**Degree of freedom (df)**	**SS**	**MS**	***F***	**Threshold *g***
Regression	2	44.619	22.309	25.375	2.5E−07
Error	32	28.135	0.879		
Total	34	72.753			
	Coefficient	Standard error	*t*-Statistics		
Intercept	3.908	0.191	20.418		
Bus as gap-opening vehicle	2.744	0.461	5.953		
Bus as gap-closing vehicle and stopped	2.119	0.428	4.950		

Table 4.8 Regression analysis results of three independent variables (1)

R^2	**0.477**				
Standard error	**1.107**				
Variance analysis	**Degree of freedom (df)**	**SS**	**MS**	***F***	**Threshold *g***
Regression	3	34.732	11.577	9.439	0.000139
Error	31	38.021	1.226		
Total	34	72.753			
	Coefficient	Standard error	*t*-Statistics	l_{xx}	$/r/$
Intercept	3.859	0.257	15.015	–	–
Bus as gap-opening vehicle	2.323	0.539	4.308	4.286	0.564
Bus as gap-closing vehicle and stopped	1.175	0.388	3.031	4.971	0.307
Stopped and in group	0.031	0.684	0.046	2.743	0.006

5. Using the two independent variables in step (3), add the independent variable *Bus as gap-closing vehicle and in group* for regression analysis. The results are shown in Table 4.9.

   The *t*-statistics of each independent variable is greater than the critical value $t'_{0.95}(32) = 1.69$, so the three independent variables are not eliminated.
6. Based on the three independent variables in step (5), add the independent variable *In group* and perform regression analysis. The results are shown in Table 4.10.

The *t*-statistic of the independent variable "In group" is smaller than the critical value $t'_{0.95}(30) = 1.70$, so this independent variable is eliminated. Since the

Table 4.9 Regression analysis results of three independent variables (2)

R^2	**0.658**				
Standard error	**0.896**				
Variance analysis	**Df**	**SS**	**MS**	***F***	**Threshold *g***
Regression	3	47.861	15.954	19.869	2.26E−07
Error	31	24.892	0.803		
Total	34	72.753			
	Coefficients	Standard error	*t*-Statistics	l_{xx}	$/r/$
Intercept	3.795	0.191	19.829	–	–
Bus as gap-opening vehicle	2.857	0.444	6.434	4.286	0.693
Bus as gap-closing vehicle and stopped	1.930	0.420	4.600	4.971	0.505
Bus as gap-closing vehicle and in group	0.905	0.450	2.010	4.286	0.220

Table 4.10 Results of regression analysis of four independent variables

R^2	**0.658**				
Standard error	**0.911**				
Variance analysis	**Df**	**SS**	**MS**	***F***	**Threshold *g***
Regression	4	47.879	11.970	14.436	1.11E−06
Error	30	24.874	0.829		
Total	34	72.753			
	Coefficients	Standard error	*t*-Statistics	l_{xx}	$/r/$
Intercept	3.783	0.209	18.075	–	–
Bus as gap-opening vehicle	2.869	0.458	6.265	4.286	0.696
Bus as gap-closing vehicle and stopped	1.939	0.431	4.503	4.971	0.507
Bus as gap-closing vehicle and in group	0.830	0.687	1.208	4.286	0.201
In group	0.083	0.566	0.147	6.171	0.024

regression of the earlier three independent variables has been analyzed, it is not necessary to do so, and other independent variable can be directly added.

In the remaining optimization process, the added independent variables are not significantly linear with the dependent variable. The following is a summary of the remaining independent variables of the linear regression equation in Table 4.11.

The final optimized regression equation obtained is

$$Y_{gvma} = 3.795 + 2.857X_{g,fb} + 1.930X_{\text{int},\ rb,\ s} + 0.905X_{\text{int},\ rb,\ a} \tag{4.16}$$

Table 4.11 Summaries of the remaining independent variable test process and results (gaps)

Optimization steps	Independent variables	*t*-Statistics	Threshold *g*	Significant or not?
(7)	Bus as gap-closing vehicle	0.121	1.70	No
(8)	Minibus as gap-closing vehicle and stopped	0.731	1.70	No
(9)	Minibus as gap-closing vehicle	−0.667	1.70	No
(10)	Stopped	−0.353	1.70	No
(11)	Left turn	0.784	1.70	No
(12)	Minibus as gap-opening vehicle	0.417	1.70	No
(13)	Gender	−0.218	1.70	No

where Y_{gvma} in the case of the gap situation g, the average value of the accepted gap of the bicycle individual v crossing through the motor flow in state a; $X_{g,fb}$ stands for the "Bus as gap-opening vehicle" factor; $X_{\text{int},rb,s}$ stands for "Bus as gap-closing vehicle and stopped" factor; $X_{\text{int},rb,a}$ stands for "Bus as gap-closing vehicle and in group" factor.

4.2.3 Model result analysis

After the calibration of the main parameters of the model is completed, that is, after the coefficients of the linear regression equation model are calibrated, the model is not completely completed and can be put into practical use.

We should be aware that the previous linear model is established by statistical regression calculation based on the sample data at the signal intersection. The statistical regression itself can only explain some statistical characteristics and relationships of the sample data. Therefore, the obtained model may be statistically reasonable, but it is unreasonable in the real traffic situations.

Therefore, after establishing and optimizing the linear regression model, it is necessary to analyze the validity and rationality of each parameter (independent coefficient) of the model from the perspective of traffic behavior.

After the plausibility of the model parameter values is confirmed, the model needs to be verified using actual data.

The more rigorous approach is to apply the model to a computer simulation program to simulate a signalized intersection. This requires collecting all relevant data of the subject intersection, including intersection geometry data, signal timing data, traffic flow macroscopic properties, including the flow and steering ratio of motor vehicles, bicycles and pedestrians; and other basic data needed; then it may be necessary to collect the total delay time of the bicycle during peak hours and off-peak hours.

The signalized intersection is simulated, and the actual collected data of the simulated bicycle macroscopic traffic flow results are compared to verify whether the bicycle behavior model is effective.

In this book, the linear simulation model of the average critical acceptance gap is analyzed only from the perspective of traffic behavior as the mixed flow microscopic simulation program at the signalized intersection has not been fully developed.

4.2.3.1 Model intercept analysis

The intercept of the linear regression model refers to the average (design) critical accepted gap length of the bicycle rider under all default factors.

The default factor in the model is that when the bicycle crosses the motor vehicle flow at signalized intersection, the gap-opening and gap-closing vehicles are all cars, the bicycle is straight through the intersection and not stopped before crossing the motor vehicle flow, the rider is male and riding the ordinary two-wheeled bicycle, the bicycle alone crossing the motor vehicle flow, not in a bicycle group.

After optimization, the average intercept gap linear model has an intercept $\beta_0 = 3.995$ (s), and the *t*-statistic is 19.828, which is much larger than the critical value g.

In a two-lane un-signalized intersection study of bicycles crossing a motor traffic flow, Smith found that the bicycle through the vehicle flow in design critical accepted gap (i.e., the distribution center of the cyclist's accepted gap) is 4.4 s; the motor vehicle driver's design critical accepted gap (i.e., the distribution center of the driver's accepted gap) is 3.5 s. Both parameters are lower than the minimum design value (5 s) of the pedestrian crossing through the car flow in the US Highway Capacity Manual [1,6].

Opiela *et al.* [3] studied the behavior of 260 bicycles crossing through a one-way motor vehicle flow (two lanes) and found that the distribution of the "accepted gaps" of the bicycle fits the lognormal distribution well. The average value of the "accepted gap" is 3.9 s and the minimum value is 1.1 s. They also examined the "rejected gaps" of the bicycle crossing the motor vehicle flow, resulting in the "rejected gap" average of 2.6 s and a maximum of 8.1 s.

By observing, they also found that if the rider does not stop before crossing the motor vehicle flow, his accepted gap is much smaller than the accepted gap when he stopped and restarted. However, they define the critical gaps differently from the standard definition. The standard definition is to use the distribution center of the accepted gaps as the design critical gap, and they used the intersection of the accepted gap distribution curve and the rejected gap distribution curve as the design critical gap. In this way, the critical gap is 3.2 s [2].

Comparing with the previous similar research results, it was found that the intercept of the model = 3.795 is very close to the average of the accepted gap of 3.9 s by Opiela *et al.* [3]; the difference between the results of Smith [1] is 0.5 s. Considering other influencing factors, it can prove that the intercept of the linear model is still quite effective and representative.

4.2.3.2 Significant factor analysis

After accepted gap average linear model optimization, there are only three significant factors: the single factor *Bus as gap-opening vehicle*, the comprehensive factors *Bus as gap-closing vehicle and stopped* and *Bus as gap-closing vehicle and in group*. The coefficients of these three factors are analyzed separately next.

- Bus as gap-opening vehicle

  In the linear model, the factor *Bus as gap-opening vehicle* is the most significant and most relevant independent variable of the dependent variable Y, which increases the average accepted gap Y by nearly 3 s.

  From the perspective of traffic analysis, it is reasonable to have a positive parameter coefficient, because most of the buses at the signalized intersection of the data acquisition are articulated bus, and the length of the body is relatively large, about 20 m. Therefore, the gap would be larger since it is the time headway between the gap-opening and gap-closing vehicles.

  According to the previous macroscopic data analysis in Chapter 3, it is known that the average speed of the motor vehicle crossing the signalized intersection is about 21.6 km/h, that is, 6 m/s, then it can be calculated that the extra time of the gap-closing vehicle arrival point A due to the length of the vehicle is

$$\Delta t = \frac{L_{bus} - L_{car}}{\bar{v}} = \frac{(20 - 4)}{6} \approx 2.67(s)$$

  where Δt indicates extra time due to the larger size of bus, L_{bus} is bus length, L_{car} is car length, and $\bar{v}$ the average speed of the gap-closing car crossing the intersection.

  The result is 2.67 s which is quite close to the coefficient of this independent variable. This is consistent with our common sense of traffic behavior, because the rider is concerned about the gap between the end of the gap-opening vehicle and the head of the gap-closing vehicle. The effect of the behavior may only be that the length of the gap-opening vehicle will affect the time when the gap-closing car arrives at point A, thus affecting the average length of the accepted gap. Generally, the gap-opening vehicle does not have any special influence on the cyclist's gap acceptance behavior of the rider. Therefore, we believe that this independent coefficient is reasonable.

- Bus as gap-closing vehicle and stopped

  The second significant independent variable is the comprehensive factor *Bus as gap-closing vehicle and stopped*, which has the effect on the dependent variable Y second only to the *Bus as gap-opening vehicle* factor. This independent variable increases the average accepted gap Y by about 2 s.

  From the perspective of traffic analysis, this is reasonable, because it is intuitive to think that if the gap-closing vehicle is large, the psychological threat to the rider may be relatively large, so he/she will choose a larger accepted gap. Yet there may be another explanation that the speed of the buses that crosses the intersection is relatively low, thus extending the length of the accepted gaps.

Similarly, it takes a certain amount of time (about 0.5 s) to restart the bicycle when it is crossing the motor vehicle flow, which may also be a reason for extending the average accepted gap.

As for why the significant independent variables are comprehensive factors rather than two separate factors, the author believes that it may be because the trends of the two independent variables are consistent, but the effect on the dependent variable *Y* is not particularly significant, and when the two are integrated the effect becomes significant.

- Bus as gap-closing vehicle and in group

The last significant independent variable is the comprehensive factor "Bus as gap-closing vehicle and in group," which increases the average accepted gap *Y* by about 1 s.

The influence of the single factor *In group* on the dependent variable *Y* is increased by 0.78 s, while the single factor *Bus as gap-closing vehicle* has a shortened effect on *Y* of 0.3 s, yet it cannot be simply assumed that the coefficient of this independent variable is wrong.

As mentioned earlier, considering that the impact of the composite factor occurs as the combination of the effects of the two single factors may not be a simple linear superposition, which is a good proof of this.

This situation may indicate that the *In group* factor has a relatively large impact when the gap-closing vehicle is a bus.

The author thinks that a reasonable explanation is that because the speed of the bus crossing the intersection is relatively low, the bicycle group is easier to cross in front of it, and because of the group protection or herd mentality, the rider's accepted gap will be reduced. At the same time, the presence of the bicycle group has caused the bus driver to slow down, resulting in the accepted gap extension.

This is due to the impact of the bicycle group on the driver of the gap-closing bus, resulting in an extension in the accepted gap. From this perspective, the coefficient of this comprehensive independent variable is reasonable.

4.2.3.3 Insignificant factor analysis

There are nine factors that are eliminated in the linear model optimization results. It cannot be considered that the factors that are eliminated by the optimization process are all factors that are insignificant for the dependent variable. Some of them have a certain influence on the dependent variable *Y*, so they will be eliminated next. The influencing factors were analyzed separately.

- "In group" (+0.8 s)

The biggest characteristic of urban bicycle traffic in China is the large traffic volume. Therefore, the bicycle group in the traffic flow is very common, and the bicycle group has some impact on the rider's behaviors. The linear model shows that if the bicycle is crossing a motor vehicle in a group, the average accepted gap will be larger (+0.8 s). A reasonable explanation is that when the bicycle group crosses the motor vehicle flow, the gap-closing vehicle will have to slow down, thereby increasing the accepted gap.

- "Bus as gap-closing vehicle" (−0.3 s)
  The effect of this influencing factor on the dependent variable Y is shortening 0.3 s. This seems to have some discrepancies with our intuitive judgment. It feels that the bus size is relatively large, and the psychological pressure on the rider should be relatively large. Therefore, it should be reasonable to choose a larger gap. However, from another point of view, it may be that the speed of the bus is relatively low; it seems safe and easy to brake, so the rider may accept a smaller gap.
- "Minibus as gap-opening vehicle" (+0.3 s)
  The effect of this factor on the dependent variable Y is an increase of 0.3 s. It is the same principle as the bus as gap-opening vehicle: it takes a longer time for the car to arrive at point A because of the longer vehicle length. It can be calculated that the extra time when the gap-closing car arrives at point A due to the length of the minibus is

$$\Delta t = \frac{L_{minibus} - L_{car}}{\bar{v}} = \frac{(6-4)}{6} \approx 0.33\ (s)$$

  where Δt indicates the extra time due to the larger size of minibus, $L_{minibus}$ is the minibus's length, L_{car} is the car's length, and $\bar{v}$ the average speed of the gap-closing car crossing the intersection. The result of the calculation (0.33 s) is quite close to the coefficient of this independent variable, indicating that our explanation is reasonable.
- "Minibus as gap-closing vehicle" (+0.2 s)
  The effect of this factor on the dependent variable Y is an increase of 0.2 s, which may be similar to the factor of *Bus as gap-closing vehicle*. The difference may be that the length of the minibus is between the car and the bus, so its impact is not as significant as a bus. And it further shows that the size of the gap-closing vehicle has a certain impact on the rider's gap acceptance behavior, and the degree of influence is also proportional to the size of the gap-closing vehicle.
- "Stopped, and in group" (−2 s)
  The effect of this factor on the dependent variable Y is quite significant because in the un-optimized linear model it is a significant independent variable, and the t-statistic exceeds the critical value g. A more reasonable explanation is that when the phenomenon of *Stopped and in group* occurs, it is likely that quite a lot bicycles are stopped at the stop line to wait for the appropriate gap. Therefore, due to the mentality of the crowd and waiting too long, once the motor vehicle flow appears a gap that may be crossed, although relatively small, which is accepted by bicycle riders.
- "Minibus as gap-closing vehicle and stopped" (+0.8 s)
  This combined effect factor increases the average acceptance gap Y by approximately 0.8 s. The reason for the influence of this independent variable on the dependent one may be the same as the comprehensive factor *Bus as gap-closing vehicle and stopped*, but the effect is not so significant because the body of the van is not as long as the bus.

For the small effect of the dependent variable *Y*, the almost negligible independent variables include *Stopped*, *Left turn* and *Gender*.

- "Stopped" (+0.19 s)

  This factor may have two effects that offset each other, so it appears to have no effect on the dependent variable *Y*: one effect is the time it takes for the stopped bicycle to be restarted when it is ready to cross the motor vehicle flow. The other effect is that when the bicycle is stopped waiting for the gap of the motor vehicle, it may be psychologically prepared or impatience, so that the gap is relatively small. The two effects are opposite, so they are likely to offset each other and the effect is not obvious.
- "Left turn" and "Gender" (+0.18 and −0.004 s)

These two factors have no particular effect on the mean value of the acceptance gap. Especially for the *Gender* factor, the effect is almost equal to zero.

In summary, the three independent variable coefficients in the optimization model are reasonable, and there may be related factors among some factors that are eliminated. Therefore, these factors cannot be completely excluded in future research.

4.3 Lag acceptance choice behavior model

4.3.1 Model construction and basic formula

The lag acceptance choice behavior model of bicycle individuals crossing the motor vehicle flow at the signalized intersection is almost the same as the gap acceptance choice behavior model principle and basic formula. The difference is that the influencing factors of the acceptance delay do not include the gap-opening vehicle type. The model formula is as follows:

$$
\begin{aligned}
&\Pr(accepting)= \\
&\Phi\left[\frac{\ln L_{gvma}-\left[\ln\left(\beta_0+\beta_{gap}X_{gap}+\beta_{man}X_{man}+\beta_{vo}X_{vo}+\beta_a X_a+\beta_{\text{int}}X_{\text{int}}+\xi_{gvma}\right)-\frac{\sigma_1^2}{2}\right]}{\sigma_1}\right] \\
&=\Phi\left[\frac{\ln L_{gvma}-\ln\left(\beta_0+\beta_{gap}X_{gap}+\beta_{man}X_{man}+\beta_{vo}X_{vo}+\beta_a X_a+\beta_{\text{int}}X_{\text{int}}+\xi_{gvma}\right)}{\sigma_1}+\frac{\sigma_1}{2}\right] \\
&=\Phi\left[\frac{\ln\left\{\frac{L_{gvma}}{\left(\beta_0+\beta_{gap}X_{gap}+\beta_{man}X_{man}+\beta_{vo}X_{vo}+\beta_a X_a+\beta_{\text{int}}X_{\text{int}}+\xi_{gvma}\right)}\right\}}{\sigma_1}+\frac{\sigma_1}{2}\right]
\end{aligned}
\tag{4.17}
$$

4.3.2 Model calibration and optimization

The results of data analysis in Chapter 3 show that the accepted lags of bicycle crossing through motor vehicle flow conforms to the natural lognormal distribution of $\mu = 0.906$ and $\sigma = 0.51$. Therefore, the parameters to be calibrated for the cyclist's gap acceptance decision probability model at signalized intersections are mainly the vector coefficients β_{gap}, β_{map}, β_{vo}, β_a and β_{int} in formula (4.7).

The influencing factors of bicycle acceptance lag include (1) minibus as gap-closing vehicle; (2) in group; (3) the riding direction; (4) bus as gap-closing vehicle; (5) stopped before crossing the motor vehicle flow; (6) the gender and (7) the tricycle type factor.

The results of data analysis in Chapter 3 show that the significant influencing factors of bicycle acceptance lag include (1) bus as gap-closing vehicle; (2) stopped before crossing the motor vehicle flow; (3) in group factor. On the other hand, nonsignificant factors include (4) minibus as gap-closing vehicle; (5) the riding direction; (6) the tricycle type and (7) the gender factor.

In theory, the previous three significant factors can be combined to obtain three comprehensive factors: "Bus as gap-closing vehicle and stopped," "Bus as gap-closing vehicle and in group" and "Stopped and in group."

The basic flow of model optimization for accepted lag data is consistent with the basic flow of accepted gaps in Section 4.3.2. The regression results are shown in Tables 4.12–4.14.

Table 4.12 Preliminary regression independent coefficients

Interception and independent variables	Coefficients (s)	Standard error	t-Statistics
β_0	2.115	0.557	3.797
In group	1.152	0.679	1.696
Stopped	0.708	0.611	1.158
Bus as gap-closing vehicle	0.559	0.678	0.824
Gender	0.308	0.417	0.74
Stopped and in group	−0.519	0.863	−0.601
Left turn	−0.238	0.406	−0.586
Bus as gap-closing vehicle and in group	−0.495	0.871	−0.568
Minibus as gap-closing vehicle	−0.229	0.606	−0.379
Tricycle	0.218	0.579	0.376
Bus as gap-closing vehicle and stopped	0.114	0.866	0.132

Table 4.13 Fitting results of multiple regression equations

R^2	0.305
Standard error	1.134
Sample size	37

Table 4.14 Analysis of variance table (accepted lags)

Source	Degree of freedom (df)	Sum of squares (SS)	F' Statistics	$F_{0.05}(12, 22)$
Regression	10	14.672	1.140	0.372065
Error	26	33.460		
Total	36	48.133		

Table 4.15 Summary of significant analysis results of various factors (accepted lags)

Independent variables	Coefficients (s)	Standard error	*t*-Statistics	Significant or not?
In group	1.152	0.679	1.696	Y
Stopped	0.708	0.611	1.158	N
Bus as gap-closing vehicle	0.559	0.678	0.824	N
Gender	0.308	0.417	0.74	N
Stopped and in group	−0.519	0.863	−0.601	N
Left turn	−0.238	0.406	−0.586	N
Bus as gap-closing vehicle and in group	−0.495	0.871	−0.568	N
Minibus as gap-closing vehicle	−0.229	0.606	−0.379	N
Tricycle	0.218	0.579	0.376	N
Bus as gap-closing vehicle and stopped	0.114	0.866	0.132	N

The results of the significance analysis of each factor of the multiple regression equation are shown in Table 4.15.

The regression model optimization process is as follows:

1. Perform multiple linear regression on all selected independent variables and sort according to the absolute value $|r|$ of the correlation coefficient. The significant level $\alpha = 0.20$. The results are shown in Table 4.16.
2. The remaining factors of the linear regression equation are summarized in Table 4.17.
3. The regression results of significant independent variables are shown in Table 4.18.

The final optimized regression equation for lags obtained is

$$Y_{gvma} = 2.371 + 0.997X_a + 0.518X_{man,\ s} \tag{4.18}$$

where Y_{gvma} is, in the case of the gap situation g, the average value of the accepted lag of the bicycle individual v crossing through the motor flow in state a; X_a stands for the "In group" factor; $X_{man,\ s}$ stands for the "Stopped" factor.

Table 4.16 Independent variables ordered by |r| (accepted lags)

Independent variables	Coefficients (s)	l_{xx}	l_{yy}	$\|r\|$
In group	1.152	8.703	48.133	0.490
Stopped	0.708	8.108	48.133	0.290
Bus as gap-closing vehicle	0.559	8.919	48.133	0.241
Gender	−0.495	6.270	48.133	0.179
Stopped and in group	−0.519	3.568	48.133	0.141
Left turn	0.308	9.081	48.133	0.134
Bus as gap-closing vehicle and in group	−0.238	9.243	48.133	0.104
Minibus as gap-closing vehicle	−0.229	5.030	48.133	0.074
Tricycle	0.218	5.030	48.133	0.071
Bus as gap-closing vehicle and stopped	0.114	4.324	48.133	0.034

Table 4.17 Summaries of the remaining independent variable test process and results (lags)

Optimization steps	Independent variables to be tested	*t*-Statistics	Threshold *g*	Significant or not?	Existing significant independent variable
(2)	"In group"	2.662	1.307	Y	In group
(3)	"Stopped"	1.396	1.307	Y	In group; Stopped
(4)	Bus as gap-closing vehicle	1.008	1.307	N	In group; Stopped
(5)	Stopped and in group	0.853	1.307	N	In group; Stopped
(6)	Gender	0.845	1.307	N	In group; Stopped
(7)	Left turn	−0.464	1.307	N	In group; Stopped
(8)	Minibus as gap-closing vehicle	−1.105	1.307	N	In group; Stopped
(9)	Tricycle	0.568	1.307	N	In group; Stopped
(10)	Bus as gap-closing vehicle and stopped	0.538	1.307	N	In group; Stopped

4.3.3 Model result analysis

4.3.3.1 Model intercept analysis

As with the bicycle gap acceptance model, the intercept of the bicycle lag linear regression model refers to the cyclist's average accepted lag under all default situations. After optimization, the average accepted lag linear model has an intercept $\beta_0 = 2.115$ (s).

Table 4.18 Results of regression analysis of two independent variables (lags)

R^2	**0.213**				
Standard error	**1.055**				
Variance analysis	**Df**	**SS**	**MS**	***F***	**Threshold *g***
Regression	2	10.274	5.137	4.613	0.016881
Error	34	37.859	1.114		
Total	36	48.133			
	Coefficient	Standard error	*t*-Statistics	l_{xx}	$\|r\|$
Intersect	2.371	0.255	9.294	–	–
In group	0.997	0.358	2.782	8.703	0.424
Stopped	0.518	0.371	1.396	8.108	0.213

There are few studies on the lag acceptance behavior of bicycles crossing the motorway at the intersection. Only Taylor and Mahmassani took lag as an influencing factor when they investigated the gap acceptance behaviors of bicycle and motor vehicle crossing at the un-signalized low-speed intersection. Research shows that if the bicycle arrives at the intersection and accepts a lag instead of a gap, the length of lag will be significantly larger than the gap length (about 2–4 s depending on the situation). Taylor and Mahmassani explain that it takes time for the cyclist to get familiar with the environment when he/she just reaches the intersection [4].

However, in our research, we found that the accepted lags are significantly smaller than the accepted gaps. The reason may be that Taylor and Mahmassani studied the un-signalized intersection. And it takes a certain time for the cyclists to adapt to the intersection environment when entering the intersection. While we are studying cyclists' gap acceptance behaviors at signalized intersections, cyclists that conflict with motor traffic flow generally wait for the red light at the stop line. They can use this time to familiarize themselves with the intersection environment without taking up the time in the accepted lags to do this. In addition, due to the definition in our study, the accepted lag is smaller than the accepted gap.

Here, the default average of the accepted lag can be roughly estimated on the basis of the default average of the accepted gap $\beta_{0\ gap}$, assuming that the distance between gap-opening vehicle and bicycle is the same as the distance from that between the bicycle and gap-closing vehicle; and the motor traffic flow advances at a constant speed, then the default average of the accepted lag $\beta_{0l\ ag}$ would be

$$\beta_{0l\ ag} = \frac{\beta_{0gap}}{2} \approx 1.9(\mathrm{s})$$

As expected, the results are very close to the intercept of the linear model. The reliability and effectiveness of the model are demonstrated from this perspective.

4.3.3.2 Significant factor analysis

After the optimization of the linear model of the average accepted lag, there are only two significant factors in the model: the single factor of *In group* and *Stopped* respectively. The coefficients of these two factors are analyzed separately next.

- "In group"
  In the linear model, the factor *In group* increases the average accepted lag Y by more than 1 s. From the perspective of traffic analysis, this is reasonable, because when the bicycle group appears in front of the motor vehicle flow, the gap-closing vehicle often has to slow down, sometimes even stop, and wait until the bicycle group has passed.
- "Stopped"

The other significant independent variable of the linear model of the lag average is the *In group* factor, which increases the average accepted lag Y by about 0.7 s.

Why *Stopped* can be a significant factor in the lag model, and it is an insignificant factor in the gap model above?

In the bicycle gap acceptance behavior, the *Stopped* factor has two effects that offset each other. One is the time it takes for the stopped bicycle to be restarted when it is ready to cross the motor vehicle flow. The other effect is that when the bicycle is stopped waiting for the gap of the motor vehicle, it may be psychologically prepared or impatience, so that the gap is relatively small. The two effects are opposite, so they are likely to offset each other and the effect is not obvious.

However, in the lag acceptance model, the *Stopped* factor seems to have only the first effect, and there is no second effect, so that the *Stopped* factor is highlighted in the bicycle's lag acceptance behavior. The coefficient of the independent variable is quite close to the restarting time of bicycle (about 0.5 s), so it can be considered reasonable.

4.3.3.3 Insignificant factor analysis

There are eight factors that are eliminated in the linear model optimization results. The influencing factors were analyzed separately.

- "Bus as gap-closing vehicle" (+0.6 s)
  This factor prolongs the length of the bicycle accepted lag, and among the factors removed is the factor that has the largest correlation coefficient $|r|$ with the dependent variable Y. The size of the bus is larger, and the critical accepted lag of the rider is correspondingly increased, which is in line with the general logic.
- "Gender" (+0.3 s)
  Different from the gap model, the *Gender* factor has a certain influence on the dependent variable Y. It may be because the risk of bicycle passing through the motor vehicle is relatively large, so the gender difference is highlighted. Female cyclists' psychology is considered more conservative, and they are more willing to choose a safer way to cross the motor vehicle flow. The male cyclists psychologically are more risky, so the accepted lags are relatively small.

- "Left turn" (−0.2 s)
  Same as the gap regression model, the *Left turn* factor has no particular effect on the mean value of the accepted lag.
- "Minibus as gap-closing vehicle" (+0.2 s)
  The effect of this factor on the dependent variable *Y* is an increase of 0.2 s, which is the same to that of the gap acceptance model (+0.2 s).
- "Tricycle" (+0.2 s)
  In the lag acceptance model, the effect of *Tricycle* is an increase of 0.2 s. Since the tricycle is slightly larger than the ordinary bicycle, the riding speeds and the required lag will be larger.
  The comprehensive factors that have a certain impact in the lag linear model include *Stopped and in group*, *Bus as gap-closing vehicle and in group* and *Bus as gap-closing vehicle and stopped*.
- "Stopped and in group" and "Bus as gap-closing vehicle and in group"
  The effect of these two factors on the dependent variable *Y* is shortened by about 0.5 s. The analysis of these two comprehensive factors is the same as the analysis of the previously gap acceptance behavioral model.
- "Bus as gap-closing vehicle and stopped"

It has little effect on the acceptance delay (increased by about 0.1 s), but the single factor, respectively, of both *Bus as gap-closing vehicle* and *stopped* increases the accepted lag, which may indicate that the factor has no combined effect on the dependent variable *Y*.

4.4 Summary

In this chapter, the acceptance gap theory is used to model the gap/lag acceptance choice behaviors of the bicycle crossing the conflicting motor traffic flow at the signalized intersection. And the parameter calibration and model optimization are carried out by using the field data collected in the previous chapter. The models can be used as the basic behavior models of the mixed traffic flow microscopic simulation platform at signalized intersections in China.

References

[1] D. Smith. Safety and Locational Criteria for Bicycle Facilities. In Report FHWA-RD-75-112. FHWA, U.S. Department of Transportation, 1976.

[2] T.C. Ferrara and T.N. Lam. Analysis of Bicycle Delays at Intersections and Crossings by Computer Simulation. In Transportation Research Record 706, Washington, DC, TRB, National Research Council, 1979, pp. 36–44.

[3] K. Opiela, S. Khasnabis, and T. Datta. Determination of the Characteristics of Bicycle Traffic at Urban Intersections. In Transportation Research Record 743, Washington, DC, TRB, National Research Council, 1980, pp. 30–37.

[4] D. Taylor and H. Mahmassani. Bicyclist and Motorist Gap Acceptance Behavior in Mixed-Traffic. Presented at the 78th Annual Meeting of the Transportation Research Board and publication in Transportation Research Record, Nov. 1998.
[5] H. Mahmassani and Y. Sheffi. Using Gap Sequences to Estimate Gap Acceptance Functions. Transportation Research Part B Methodological, 1981, 15(3):143–148.
[6] TRB. Highway Capacity Manual (HCM 2000), Washington, DC, Transportation Research Board, National Research Council, 2000.

Chapter 5
Bicycle microscopic behavior analysis patterns

As people have free will and individual differences, which leads to a complex diversity of human behavior, we have not been able to fully understand the regularity of this field.

However, in the long process of exploration, human beings cannot fully reveal their mysteries to their own behaviors. But after all, they have accumulated a considerable amount of understanding. At least one thing which can be confirmed is that human behavior has internal and objective rules.

As Nobel pointed out that all kinds of sciences are intrinsically linked to each other, to solve problems in a certain subject area, we should rely on the knowledge of other related disciplines.

It is not enough to study the microscopic traffic behavior of bicycles at un-signalized intersections under mixed traffic flow situations, relying solely on knowledge in the transportation field.

Therefore, this chapter aims to use the overall concept of the system, introduce multiple research methods, and draw on the achievements of various disciplines, especially basic theories such as psychology, behavioral science, microeconomics, operations research, to explore and develop suitable for mixed traffic flow. To explore a suitable behavior analysis pattern for bicycles at un-signalized intersections under mixed traffic flow situations.

This chapter first describes the general pattern of behavior analysis, and then summarizes the characteristics of human behavior and analyzes the behavior analysis patterns and research from the fields of psychology, sociology, economics, traffic engineering and other fields. Then, in the following chapter, based on the research results of various disciplines, an analytical framework suitable for the microscopic behavior of bicycles at un-signalized intersections under mixed traffic flow situation is proposed.

5.1 General pattern of behavior analysis

Behavior is the way in which an organism lives. It is an activity based on reality and affecting reality. Human behavior is one of the most complex and difficult characteristics to recognize phenomena in the world.

Human behavior is quite different from animal behavior.

Animal behavior is a passive response based on physiological needs or external stimuli. It is an instinctive activity and a negative adaptation. The change in its behavioral pattern is mainly the result of natural selection and survival of the fittest.

Human behavior is not the case. This is a conscious activity based on the transformation and utilization of the objective world. Human labor, the ability to make and use tools, increase the initiative of human behavior. So human behavior has largely broken through the limitations of their own action body.

With the help of language and thinking, human behavior has risen to a conscious level of rationality, which can coordinate, correct and improve each other. The effect of behavior is perceptible and controllable. Such human behavior has already entailed the possibility of infinite enrichment and high development.

What is more, the essence of human beings is also social. Real people always live in a certain social environment. Human behavior is inevitably influenced, restricted and guided by various social factors.

Social life is rich and varied, and the diversification of social politics, economy and culture is vividly reflected by the differences in human behavior.

In this way, human behavior forms social behavior.

Although people's conscious initiative has never been buried, it can only be confined to certain social and historical conditions, in-line with the corresponding social standards and norms.

In this case, uniform human behavior is almost impossible to find, and the behaviors between people are always different.

So is there any law underneath human behavior? Can these laws be recognized and grasped by observation and thinking? Humans have been questioning themselves for thousands of years.

Although it is still far from completely reveal the mystery of human behavior, after all, it has accumulated a considerable amount of understanding. At least it can be confirmed that human behavior has rules to follow and has its inherent objective regularity. Revealing the secrets of human behavior is a common task of many disciplines.

It is clear that social sciences or humanities are directly studying human and human behavior.

For example, philosophy is an attempt to illustrate the close relationship between human behavior and worldview and outlook on life; psychology focuses on revealing the correlation between the generation, development, change and behavior of human psychological activities; anthropology illustrates the origin of human beings, differences in constitution, ethnic differentiation and changes in language and culture. Economics demonstrates how the pursuit of material interests translates into different behavioral orientations; law describes the guiding and restricting behavior of social morality and legal norms; history is an effort to trace the past human behavior process and explore its inherent behavioral laws; literature and art use various image means to summarize the social life and behavioral personality of people at all levels.

There are also some disciplines in the natural sciences that directly study human behavior.

For example, biology analyzes the influence of physiological structure on behavior; medicine concentrates on revealing the objective basis of various pathological behaviors and provides various ways of diagnosis and treatment to return behavior to normal.

As far as the content of research is concerned with human behavior, the disciplines are even more numerous.

After all, any discipline is the crystallization of human thinking creation, which is the product of rational analysis, observation and experimentation. It must satisfy the specific needs of human beings, and ultimately it is necessary to test whether it is right or wrong by human behavior practice.

It can be asserted that almost all of human's existing knowledge systems are related to specific human behaviors, and the investigation of human behavior is a huge part.

From different angles, the various disciplines that study human behavior in different ways have formed a large group of disciplines, which has enabled us to have a rich and multifaceted understanding of complex human behavior.

It should be pointed out that it is not entirely consistent to look at the trees and understand the forest.

It is necessary and reasonable to conduct detailed analysis in various limited subject areas according to the characteristics, objects and conditions of human behavior. However, this is not enough and difficult to reveal the original purpose of human behavior in this way.

As any discipline has a clear scope of research, it can only reveal a certain field or a certain part of human behavior and cannot extend the human behavior that extends to the whole. Otherwise, there will be a situation of paradoxical and partial, like a blind man feels an elephant, only touching some part of it, and concluding what the elephant is like.

Many scholars have reached a consensus that the accurate understanding of human behavior must be based on the basic reality that human behavior is infinitely complex and diverse. Only by using the concept of system as a whole, introducing multiple research methods, and drawing on the achievements of various disciplines, the interdisciplinary scientific research can be competent [1,2].

5.2 Human behavior characteristics

What is the cause of human behavior? This is an old and new topic.

It is said to be old because many sages at home and abroad have done various explorations and discussions on the relationship between "knowing and doing" and "thinking and doing."

It is said to be new because it is a systematic study of behavior as a science and truly forms a "behavioral science" in the 1950s.

Behavioral scientists have shown that human behavior is different because of differences in human motives, goals and perceptions. But there are still some common points in the vast differences, including the following:

- Behaviors are what people say and do, are the individual's dynamic actions, not static features; behaviors can be open actions or hidden thoughts.
- Human behavior is targeted, such as pursuing a certain life goal.
- Human behavior is directed, such as the stimulation of the external environment or the drive of intrinsic motivation.
- Human behavior is carried out by the perception, judgment and instruction of the brain.
- People behave a bit like steam locomotives, and steam locomotives are driven by the pressure of steam, and human behavior is driven by "pressure."
- Human behavior is governed by natural laws and social laws. Human behavior has a certain pattern to follow. If we can use scientific procedures to guide our behavior, then our chances of success will be greater, and the setbacks will be smaller.
- Behavior has more than one measurement scale that measures the frequency, duration and intensity of behavior.
- Behaviors can be observed and recorded by others or by the subject himself.
- The occurrence of behavior is systematically affected by environmental events, and behavior also affects the external environment.
- Behavior is an action involving temporal and spatial movements. Because of the continuity of time and space, behaviors are interactive.

5.3 Behavior analysis patterns in psychology and sociology

The humanities such as psychology and sociology are clearly subjects that directly study the behavior of people and people.

Psychology focuses on revealing the relationship between the generation, development, change and psychology and behavior of human psychological activities.

Sociology focuses on revealing the correlation between social environmental impacts and human psychology and behavior, the interaction of people's behaviors, and the behavior of social organizations.

The research on human behavior in the field of psychology and sociology mainly has the following related aspects: (1) individual differences and common characteristics of human behavior; (2) general classification and corresponding models of behavior; (3) behavior pattern (model) research. These three aspects will be introduced separately later.

5.3.1 Individual behavior differences and common characteristics

5.3.1.1 Individual behavior differences

The same behavior can come from different reasons, and the same stimulus or situation can also produce different behaviors, depending on the individual differences. The main reasons for the individual differences in behavior are as follows [3]:

1. **Genetic factors.** Human physical characteristics such as height, body size, physical and physical strength, and temperament and intelligence are largely influenced by race and parental inheritance. Behaviors determined by genetic factors are often difficult to change.
2. **Environmental factors.** Environment is the most influential factor on human behavior. The impact of environmental factors is reflected in the following aspects.
   i. **Family.** The family has a significant and profound impact on human behavior. Especially in childhood, family education and parents' words and deeds have a great influence on the development, growth and formation of human behavior.
   ii. **Education.** Education has an important influence on the development and formation of people's personalities and attitudes.
   iii. **Working environment and social experience.** The working environment has a great influence on people's habitual behaviors, such as people's habitual behaviors with their professional characteristics.
   iv. **Cultural background.** Different cultural backgrounds affect people's concepts and values to some extent.
3. **Psychological factors.** Psychological factors mainly refer to psychological processes and personality psychology. Although the psychological process is a common psychological phenomenon of human beings, it is often individualized, but it often shows various characteristics, which leads to differences in individual behavior. Moreover, because each person's ability, personality, temperament is different, needs, motivation, interests, ideals, beliefs and world views are different, which constitutes different characteristics of individuals. Therefore, it is decided that each person has his or her own behavioral patterns, thus bringing about a wide variety of individual differences in behavior.
4. **Physiological factors.** People's physical conditions are different, and their behaviors are also very different, such as height, physique, physical strength, health status, etc. Because each of these factors is different, human behavior (including safe behavior) must also have different, thus showing the characteristics of individual differences.

5.3.1.2 Individual behavior common characteristics

Although human behavior varies widely among individuals, there are some common behavioral characteristics. According to the results of psychologists, at least the common characteristics of human behavior include *spontaneous action* (automatically, rather than passively, externally induced behavior); behavior for a reason (behaves for a certain reason); *purposeful behavior* (behavior to achieve a certain goal); *persistent behavior* (before the target is not reached, the behavior may sometimes change the way, but will not terminate); *changeable behavior* (behavior that changes through learning or training).

The following mainly discusses the common characteristics of people's behaviors related to traffic behavior and the like.

1. **Human territory.** Psychologists have discovered that human beings have a behavioral characteristic of *personal territory*, which is centered on one's own body and maintains a certain distance from others. When infringing, there will be reactions such as avoidance, and embarrassment, sometimes causing unhappiness, squabbles and battles. There are four types of distances related to *personal territory*:
   i. **Intimate distance.** It refers to the distance close to the body of another person. There are two kinds of distances; one is the close state, which refers to the distance between caress, comfort, protection, contact and communication between intimate people. At this time the body can be touched or approached. The second is the normal state (15–45 cm), the head and foot do not touch each other, but the hand can hold or touch each other.
   ii. **Personal distance.** It refers to the elastic distance between the individual and others. There are also two kinds of distances. One is the close state (45–75 cm), which is the distance that the intimate person allows the other party to enter without causing difficulty and avoiding. However, non-intimate people have a strong reaction when entering this range. The second is the normal distance (75–l00 cm), which is the distance between the two people and the fingertips can just touch.
   iii. **Social distance.** It refers to the distance expressed when participating in social activities. The approaching state is 120–210 cm, which is usually the distance to work together. The normal state is 210–360 cm. Formal talks and etiquette are carried out according to this distance.
   iv. **Public distance.** It refers to the distance of public events such as speeches and performances. The approximate state is 360–750 cm, and the normal state is 7.5 m or more.

2. **Side bias.** The structure of the human body is perfect. The shape of the human body is symmetrical through the sagittal plane of the sternum and the midline of the spine. However, in terms of internal organs, the heart is slightly to the left, the liver is to the right, and the right lung is composed of the upper, middle and lower limbs. The left lung has only the upper and lower leaves. Therefore, the right half is heavier than the left half.

   Due to increased breathing during exercise, the right lung consisting of three leaves is more dilated than the left lung. The rib on the right side is also higher than the left side, making the liver that is already right-left more right-handed, so the center of gravity is more and more inclined to the right.

   Therefore, some scholars believe that in order to balance, people often use the right foot as the pillar, causing the body to bend slightly right, so they are used to the right hand.

   But the most important thing is the human brain. The brain consists of identical hemispheres constructed from the left and right sides, because most people's dominant hemispheres are the left hemisphere and the left hemisphere dominates the right side, so the habitual side of most people is the right side.

The side bias of behavior is also reflected in the choice of stairs.

Japanese applied psychologist Fujisawa Shinsuke observed the T-shaped stairs of a building (the distance between the left and right stairs can reach the same place). He found that people who climbed the stairs accounted for 66% of the left-turners and 34% of the right-turners. And factors such as gender, carrying items or not, which side the items are held, and which feet are taken first are not the decisive factors in choosing the right and left direction.

He believes that as the heart is on the left side, in order to protect the heart, while the right-handed person is used to maintaining a balance with a strong right hand, the left hand is often used to hold the stairs (or the left side leaning against the building, so that psychological support).

In addition, with the right hand, the right foot is powerful, and the performance in the walking form is the left side [4].

Therefore, both physically and psychologically, the left side is stable and ideal for people.

In this way, the stairs on the left side are good for people's evasive behavior in the event of a disaster such as a fire.

3. **Shortcut response.** In daily life and work, people often exhibit a shortcut response, that is, the shortest distance behavior is adopted in order to achieve maximum effect by consuming less energy.

   For example, when you reach for a hand, it is often a straight line that extends to the object and often crosses the open space and walks diagonally. But the shortcut response sometimes does not reduce energy consumption, but only a psychological factor.

   For example, if you can cross the road from the overpass or the underpass at the same time, even if the energy consumed by the two passes is similar, most people prefer to go by underpass.

   Another example is that when people take the bus, people would rather squeeze in the doorway, as the crowds consume more energy, yet they do not want to enter the middle of the carriage where much less crowded.

   These are psychological shortcuts that do not actually save energy.

4. **Evasive behavior.** When disasters and accidents occur, people have some common actions, and these action characteristics constitute evasive behavior. In order to seek their own safety, those who panic will rush to seek a few exits to escape. Psychologists have shown through experimental research that returning along the path of coming in and going out to the population is a prominent feature of disaster and accident evasive behavior. For example, the characteristics of fire evasive behavior are as follows:

   i. rushing to the exits with the shortest route;
   ii. escape toward the direction of the fire smoke extension;
   iii. choose the route with the least obstacles to go;
   iv. along the wall to the bright place;
   v. follow the direction of the left turn;

vi. return along the path of the incoming;
vii. moving with people;
viii. go along the roads and exits that are customary;
ix. go toward the ground (high buildings down, basement up).

Regarding the avoidance of the flying objects, psychologists have tried to attack the flying objects in front. About 80% of them avoid the object, and 20% cannot respond or fail to avoid the object. The study found that no matter in which direction the object flies (left front, right front, or front), people's leftward avoidance behavior is higher than the rightward avoidance behavior. This is because most people are used to the right side, the right hand and the right foot are stronger, so the tendency to avoid to the left is more obvious.

5. **Herd behavior.** When people encounter sudden events, many people are often difficult to judge the situation and take action immediately. Therefore, the attitudes and behaviors of the people are consistent with those people around them. This kind of behavior is called herd behavior or synchronous behavior. Some researchers have experimented: when a dangerous object suddenly flies in front, if the two people avoid in one side at the same time, the follower will consciously avoid in the same side. When the current two sides evade from different sides, the third person often avoids in the same side as the second person.

5.3.2 Classification of behaviors

5.3.2.1 Complex behavior and simple reaction behavior

A widely accepted view by psychologists argues that behavior is the result of interactions between the external environment and stimuli and subject characteristics (such as goals, personality, interests) [5]. Figure 5.1 shows the process of people's behavior change [6]. From the figure we can see that the sensation of stimuli makes the behavior subject choose the subjective utility maximum behavior from the behavioral alternative according to its own goals and interests and generate corresponding behavior changes [6].

According to the external environment and the complexity of the stimulus, psychologists roughly divide human behavior into two categories: complex and simple (Table 5.1): simple situations or standard situations that are easy to predict correspond to simple behaviors. And the existence of randomness, ambiguity, and complex situations or new situations that have never occurred before are complex behaviors. The complexity levels of these two types of behavior are different, and the behaviors of different levels of complexity have different behavioral modeling methods.

Helbing and Molnár argue that in the case of pedestrians, because pedestrians are often accustomed to the vast majority of situations encountered while walking, their dynamic behavior is usually quite "automatic" and he/she will be subconscious. Choose the best option known in the experience as a response [6]. From this, it can be analogized that bicycle riding behavior, like pedestrian behavior, is also a simple behavior of this kind.

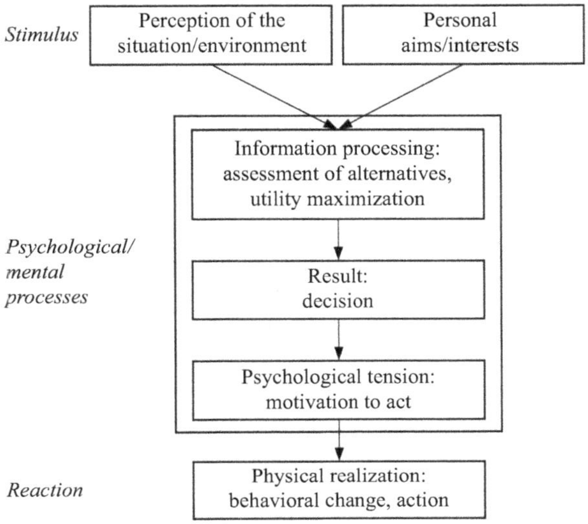

Figure 5.1 Schematic representation of processes leading to behavioral changes [6]

Table 5.1 Classification of behaviors according to their complexity

Stimulus	Simple/standard situations	Complex/new situations
Reaction	Automatic reaction, "reflex"	Result of evaluation, decision process
Characterization	Well predictable	Probabilistic
Modeling concept	Social force	Decision theoretical model, etc.
Example	Pedestrian motion, cyclists' conflict avoidance	Destination choice by pedestrian, pedestrian route choice

5.3.2.2 Reactive behavior and deliberative behavior

In addition to the simple division of behavior according to the complexity of the environment, psychologists and behaviorists also divide the behavior of human beings in more detail and analyze various behaviors and behavior patterns. The following is a description of Bernd Schmidt's classification and modeling of human behavior in his book *Modeling of Human Behavior – The PECS Reference Model* [7].

Different forms (types) of behavior have different preconditions, so the behavior is divided and then modeled. For example, learned behavior requires individuals to have the ability to absorb and process information, but instinct behavior does not. To establish a relatively complete reference model for behavior, we must first characterize the functioning modes of different types of behaviors, the preconditions of behavior, and how these behaviors are described in the model.

Behaviors are divided into two basic types based on behavioral control: *reactive behavior and deliberative behavior*.

Reactive behavior is behavior that follows a fixed, predefined rule of conduct. This means that no obvious thinking processes are required in these behavioral control processes.

The focus of modeling this type of behavior is the regulation set that behavior follows.

Conversely, conscious behavior usually has a goal, through the subject's conscious thinking, modeling the external environment, and trying and testing the model in consciousness.

The component of conscious behavior is the action plan established to achieve the goal.

Since the preconditions of behavior directly affect the behavioral model structure, the two basic types of behaviors (reactive behavior and deliberative behavior) can be further classified according to the preconditions of behavior. Then analyze and model the different types of behavior separately.

Reactive behavior

Instinctive behavior: Instinctive behavior follows a simple stimulus–state–reaction (i.e., S-S-R) pattern. The organism receives stimulation and responds to the stimulus according to its internal state. Instinctive behavior is not only related to external stimuli, but also related to the internal state of the individual. If the individual lacks internal functional mechanisms or necessary internal resources, it will not produce instinctive behavior. This law applies not only to instinctive behavior, but also to reactive behavior. The focus of modeling instinct behavior is to clarify what is an important internal behavioral mechanism of an individual and an externally observable behavioral manifestation of an instinctive behavior. For example, the behavior of animals generally belongs to the category of instinctive behavior, and humans also have instinctive behaviors, as in the following:

- Hit a person's knee with a small hammer, and the calf will bounce.
- The crying of the child will automatically trigger the mother's protective response.
- A strong stimulus (such as a loud sound) can cause people to have an automatic oriental reaction.

Instinctive behavior is a behavior that is fixed, responds to stimuli and does not have subjective initiative.

Learned behavior: From external performance, the learned behavior is quite similar to the instinctive behavior. Both behaviors seem to follow a fixed rule and react "automatically" to the stimulus. But the difference between learned behavior and instinctive behavior is that the acquired behavior is not born but acquired through learning. As far as human beings are concerned, learned behavior plays an important role in social life. If you carefully observe people's daily activities, you will soon find that most people's behavior does not require deliberation and

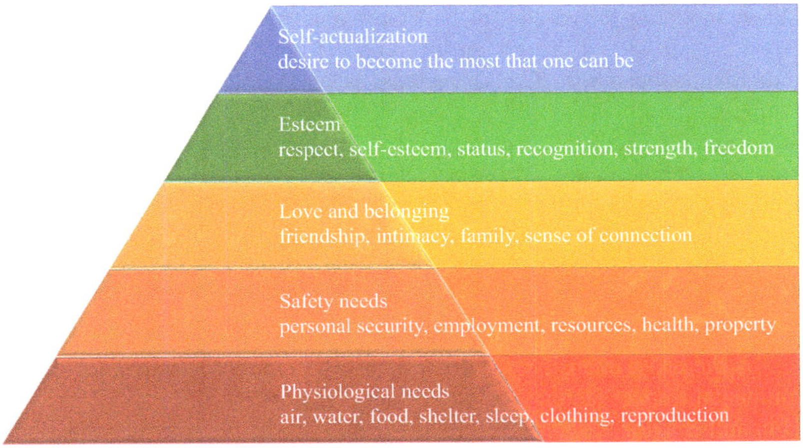

Figure 5.2 Maslow's hierarchy of needs

reflection. Only in a few critical moments, people use their brains to think and conduct deliberative behavior. There are many examples of learned behavior in daily life, as follows:

- When the car driver sees a child running across the street, he/she will automatically brake.
- People will naturally use chopsticks (or forks and knives) when dining without having to think about how to use them.

Drive-controlled behavior: Drive-controlled behavior is a major advance in the evolution of organisms. Unlike instinctive behavior, the organism no longer passively waits for external stimuli but is driven by internal drives to actively find situations (environments) that meet their needs. The emergence of drives makes appetence behavior possible.

The emergence of drives is to meet the needs. Regarding human needs, Abrahan H. Maslow (1943) conducted a systematic study and proposed a hierarchy of needs. He divides the various needs of human beings into five different needs and prioritizes the hierarchy of needs: self-actualization, esteem, love and belonging, safety needs and physiological needs (see Figure 5.2).

For the control of behavior, the needs of the bottom layer are usually more important and important than the needs of the upper layer, so that there is greater drive strength. At present, the research on the behavior-oriented function of driving and motivation is still not perfect, and there are still many uncertainties and academic disputes.

Therefore, in the process of modeling drive control behavior, Schmidt [7] considers a drive as an inner force associated with a need, and this force drives the individual to make actions. A simple example of drive is hunger.

When modeling drive control behavior, it is believed that each need has a corresponding drive and drive intensity, and the most powerful drive determines the individual's behavior. The drive tension is directly related to the strength of need.

It is generally believed that the greater the difference between the individual (internal or external) state and the predetermined or desired level, the greater the corresponding drive intensity.

In addition, environmental impacts and other possible influencing factors will also work. Therefore, the drive control behavior model is as follows:

$$T = f(N, E, X) \tag{5.1}$$

where T is the drive intensity; f is drive intensity function; N stands for need; E is environmental impact factor; X is other impact factors.

Human behavior is determined by the form of the drive intensity function. The steeper the curve of the function, the greater the response of a person to a need.

If there are multiple needs coexisting at the same time, we can determine the drive intensity corresponding to each need, and select the drive with the strongest intensity as the motivation of action, and make the corresponding behavior. Here is the simple thumb principle [7]. In addition, the drive intensity changes over time, which means that at different times, different drives guide the person's actions.

Emotionally controlled behavior: Unlike the internal drivers of drive control behavior mentioned earlier, emotions are caused by external stimuli and are accompanied by a cognitive process. For example, a sudden emergence of a lion (external stimuli) causes people's fear (emotion) because people recognize that it is a dangerous carnivore (cognitive process).

When modeling emotional control behavior, some scholars believe that emotion is an autonomous entity that determines the characteristics of an individual's internal state. Other scholars do not agree that he believes that emotion is only a mechanism for behavioral control [2].

Similar to drive emotion also causes an individual to make an action, but unlike drive, emotion requires an external stimulus and a cognitive process to identify (evaluate) the stimulus. Psychology roughly divides the following six basic emotions: anger, disgust, fear, happiness, sadness and surprise.

Deliberative behavior

Deliberative behavior refers to the behavior that is consciously pursued by the goal. The individual decides the most desirable action plan to achieve the goal.

The action plan contains the sub-targets necessary to achieve the ultimate goal. Each subgoal contains a series of actions, each of which (usually reactive behaviors) is reactively executed according to its own basic rules.

In order to develop an action plan and determine the sequence of actions necessary to achieve each subgoal, an individual must have an *optimization procedure*.

Conscious behavior can be divided into two subclasses: *constructive behavior* and *reflective behavior*.

Constructive behavior refers to that the target is known and unchangeable, and the individual consciously formulates an action plan to achieve the stated goal.

Reflective behavior refers to an individual's self-regulation ability, which can set his own goals and adjust or modify existing goals.

Constructive behavior: Constructive behavior consists of two aspects: First, the individual can create a model to describe the environment around it. The model shows how the individual changes the surrounding environment according to his or her own will through certain behaviors.

Second, the individual can use the environmental model in the brain to "preview" different action alternatives.

Individual's internal models of the external environment may have different levels of complexity. For simple situations, it may simply absorb external static data (information), while a more powerful model may contain dynamic information, so that individual not only knows static data, but also predicts how the surrounding environment changes, and thus develops and adjusts his/her action plan.

However, because the individual's model of the external environment cannot fully conform to the objective reality, there are more or less errors or deviations.

This explains that wrong behaviors and inaccurate knowledge arise when an individual's internal environmental model does not match the objective reality around him/her.

Reflective behavior: Reflective behavior is a unique behavioral pattern of human beings. Three characteristics of human behavior can be described as reflective behavior:

- Human beings can not only think but also observe, supervise, and control their own thinking processes.
- Human behavior is not bound by fixed targets. They can set new goals for themselves and think about and justify them.
- Human have self-consciousness. They not only model the external environment but also establish corresponding models for themselves. The self-image reflected by this model has a decisive influence on people's behavior patterns.

Human behavior depends not only on his real situation but also on what he considers to be "real situations."

This "self-model" includes not only cognitive variables, but also physical, emotional and social state variables. The behavior of this person is determined in part by the physical, emotional and social factors that he believes he has. There is currently no convincing and universally valid conceptual model for reflective behavior.

5.3.3 Major behavior patterns

Since the twentieth century, many psychologists, social psychologists and biologists have a keen interest in exploring the mysteries of human behaviors. They have devoted themselves to human behavior research, trying to reveal the objective rules behind the complex human behavioral phenomena. Among the many research results, we introduce the research results of American behavioral psychologist B.P. Skinner and psychologist Kurt Lewin.

5.3.3.1 Stimulus-reaction pattern

Early behavioral psychology argues that the nature of behavior is an externally observable response caused by stimuli. For example, muscle contraction, gland secretion can be simply attributed to the stimulation (S)–reaction (R) pattern.

Later, "radical behaviorism" regards all psychological activities as behaviors. For example, the American psychologist Skinner also distinguishes behaviors into S-type (stimulation responsive behavior) and R-type (reaction operant behavior), and the former refers to a reaction provoked by a special observable stimulus or situation. The latter refers to a reaction that occurs without any external stimuli or conditions that can be observed. On this basis, the psychologist R.P. Maier proposed the following behavior patterns:

$$\text{S(Stimulus/situation)} \rightarrow \text{O(Organism)} \rightarrow \text{R(Reaction)} \rightarrow \text{A(Action)}$$

Maier believes that the *stimulus* or *situation* is inseparable. In the production environment, such as light, sound, temperature, and even the behavior of the people around them who can form a stimulus that is perceived by people, it becomes a situation.

Organisms include the individual's uniqueness due to genetic and acquired conditions, maturity of personality development, learned technology and knowledge needs, motivation, attitudes, values and so on.

The *reaction–action* includes physical exercise, language, expression, emotions, thinking and so on. Action completion includes changing situations, living activities, escaping dangers, disasters and others.

Meyer thinks that the same behavior can come from different stimuli. On the other hand, the same stimuli can produce different behaviors in different people.

5.3.3.2 Social field pattern

Psychologist K. Lewin disagreed the stimulus–response model of behavioral psychology and proposed the field theory of psychology [5]. He believes that man is a *Field* – "the living space (LSP), including this person and his psychological environment," the behavior is determined by this field, the basic pattern is

$$B = f(P, E) \tag{5.2}$$

where B (behavior) indicates the individual's behavior; P (personal) indicates the individual's intrinsic condition and intrinsic characteristics, including feeling and perception, learning and memory, motivation, attitude and emotion, and the age, gender and personality of the individual; E (environment) indicates that the external environment in which an individual is located is also called a social factor that affects individual behavior; it can represent various factors that constitute the environment, such as cultural status, economic development level, scientific and technological strength, and institutional model.

The model shows that human behavior is an outcome of the interaction between the individual and the environment. At the same time, the model further shows that human behavior, orientation and intensity are mainly influenced and restricted by two major factors, namely, individual internal factors and external environmental factors.

Among them, the internal factors of the individual include the basic factors of physical and psychological, while the external environmental factors include the natural environment and the social environment.

According to Lewin's theory, a person has a need (including physical and spiritual), which creates a state of psychological tension, and the person will take some kind of behavior to achieve his/her goal. When the goal is achieved, and the need is satisfied, the state of psychological tension is released, and then there will be new needs to motivate people to achieve the new goals, as shown in Figure 5.3.

However, the previous model does not clearly express the model concept in (3.1). According to Levin's theory, Tolman improved the Levin's model by using various factors as independent variables and behavior as dependent variables. The model is shown in Figure 5.4 [8].

The behavioral pattern of social field reveals the general law of human behavior to a certain extent and makes the most basic induction and division of various factors affecting behavior. Its conclusion has a high degree of generality and wide applicability. Therefore it has been highly praised by other scholars and has received universal attention and recognition.

Social field theory is not only widely used in sociology but also applied to other fields. For example, pedestrian dynamics modeling of Helbing and Molnár is based on Levin's social field theory [6].

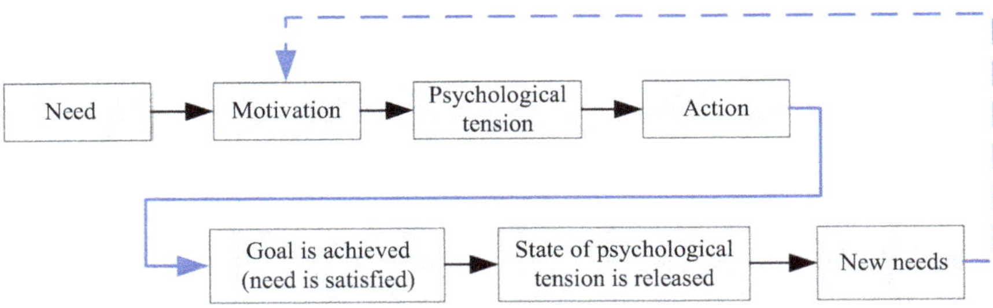

Figure 5.3 Behavioral pattern of social field theory (Lewin's model)

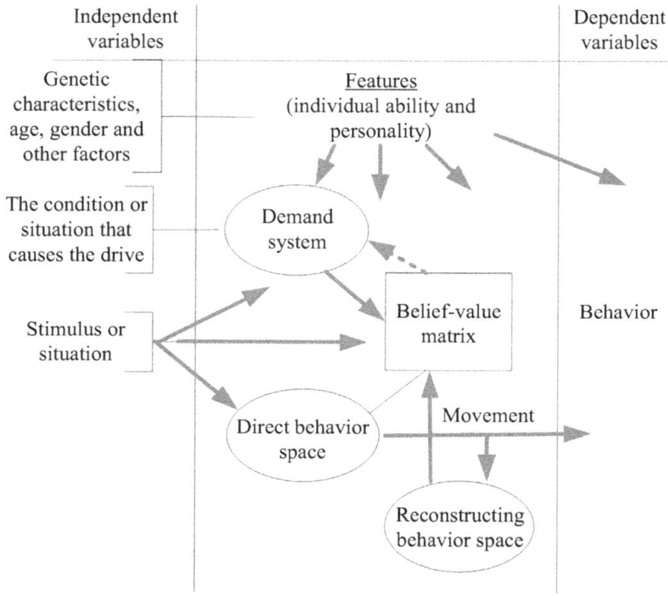

Figure 5.4 Behavioral pattern of social field theory (Tolman's model)

5.4 Behavior analysis patterns in economics

People are the main constituents of the economic system. In terms of behavior research, microeconomic theory emphasizes to analyze how the pursuit of material interests into different behavioral preferences.

Undoubtedly, microeconomic theory studies the complex behaviors that include complex psychological and intellectual activities such as selection, judgment, estimation and decision-making but do not include simple reaction behavior.

Microeconomics-related human behavior research includes consumer choice behavior research, decision-making behavior and gaming behavior.

5.4.1 Consumer choice behavior research

The generalization of *neoclassical economics* on the mechanism of consumer choice behavior is that consumers maximize their utility under income constraints according to their own preferences and the market's established price.

The criterion of choice is the maximization of utility. Therefore, the analysis of consumer choice behavior begins with the utility of consumers. The satisfaction (utility) brought by consumption is a psychological abstract concept. It is not an objective product like product value.

In the eighteenth century, mathematician D. Bernoulli first proposed the *utility theory*. He realized that the amount of wealth a person possessed does not represent the "true value" of this wealth for this person. He proposed using the basic idea of "moral value" (i.e., the concept of utility value now discussed) to show what he called "true value." He believes that the true value people feel is "moral value."

Based on this idea, Bernoulli advocates the establishment of utility functions, using the expected utility value as a priority for the metrics measurement scheme, and using uncertainty to reflect uncertainty. This is the so-called cardinal utility theory.

Later, scholars realized that when they study consumer behavior, they only need to compare consumption more or less. As for how much more or how much less, it does not matter.

Based on this idea, in 1881 F.Y. Edgeworth advocated the use of equivalent curves (surfaces) to reflect the prioritization of the scheme, using the indiscriminate utility curve to evaluate and select the scheme, i.e., the so-called ordinal utility theory. The ordinal utility theory greatly relaxes the hypothesis of the *cardinal utility theory* and makes a new interpretation of the consumer's behavior on the basis of consumer preferences. And it is quickly proved that this does not detract from the systematic and systematic study of consumer behavior.

Therefore, the *ordinal utility theory* has long been the standard utility theory. In 1931, P.P. Ramsey proposed the theorem of utility value operations. In 1944, von Neumann–Morgenstern further developed the utility theory based on Bernoulli's theory of expected utility values. This theory can not only express the spiritual value of wealth under certain conditions, but also express the spiritual value of wealth under the risk situation, and make the calculation of the expected utility value possible, so that the theory of base utility can rise again [9].

In consumer choice behavior study, the *logit model* is the most widely used and typical model.

5.4.1.1 Logit model

The most widely used study of transport selection behavior is the logit model.

The theoretical basis is the logit model theory of consumer choice behavior based on random utility function in microeconomics. The core content of the theory is that consumers choose under the constraint of income or other constraints to meet their maximum personal utility. *Rational man* and *utility maximization* are the key points of this theory, where *rational people* are the cornerstone of theory.

The so-called rational person must have the following two characteristics:

1. Have a well-defined utility function (preference system).
2. Maximize his/her utility under certain constraints.

The general form of the multinomial logit (MNL) model is as follows:

$$P_{it}^{*} = \frac{\exp(V_{it}^{*})}{\sum_{i=1}^{I} \exp(V_{it})} \tag{5.3}$$

where P_{it}^* is the probability value selected by the consumer for the ith scheme; V_{it} is the benefit of the scheme i for the consumer t. When there are only two alternatives to choose from, it is called the binary logit model.

The utility function V_{it} in the model is generally specified as a linear additivity form, namely,

$$V_{it} = \beta' x_{it} \tag{5.4}$$

where the parameter vector β' in the formula is the coefficient of each attribute item, and its value can be estimated by the maximum likelihood method. The likelihood function is as follows:

$$L(P) = \prod_{t=1}^{T} \prod_{i=1}^{I} P_{it}^{f_{it}} \tag{5.5}$$

$$f_{it} = \begin{cases} 1, & \text{if consumer } t \text{ choose the scheme} \\ 0, & \text{else} \end{cases} \tag{5.6}$$

where T observes the number of samples, and I is the number of alternatives.

Among the characteristics of the logit model, the most well-known subject is the *independence of irrelevant alternatives* (IIA).

Due to this feature, the logit model has the biggest drawback: it is assumed that all alternatives are completely independent. However, in many empirical analysis studies, it is found that it is unlikely that all alternatives will be completely independent. Therefore, the *IIA* characteristics have always been a defect of the logit model.

A general solution to this problem is to use the market separation method or the nested MNL model.

The main principle of the nested logit model is to separate the relevant schemes one by one to form a nested selection pattern. At this time, the schemes of each level can be regarded as having *IIA* characteristics, and the selected probability of each scheme can be derived from the logit model.

When we use the microeconomic model of consumer choice behavior to model people's choice behavior, we must have the following assumptions:

1. It is necessary to know the consumer's preference system, that is, to be able to determine the consumer's utility function.
2. The *merchandise bundle collection* (i.e., the *set of alternatives*) faced by the consumer is determined.
3. Consumers can accurately measure variables related to their preferences and circumstances.
4. Consumers can provide a detailed description of all variables related to their preferences and research questions.
5. Consumers have the necessary concepts and tools related to problem solving.

6. Consumers can choose the optimal measurement and calculation method, have the powerful ability to process various information and can explain the problem from various dimensions of the problem.

However, in the real world, the previous assumptions are difficult to hold. As various factors continue to change, people's preferences are also constantly changing, and there is no way to have a stable preference system.

In addition, due to the limited human ability and the high uncertainty of the environment, people's measures on their preferences and circumstances inevitably have errors, and the quality of the results is completely limited by the error.

Moreover, some variables such as psychological and institutional factors are difficult to measure: because the human brain cannot deal with all the elements of taste and preference at the same time, some typical elements occupy a dominant position, while others are difficult to be described.

Since people have limited ability to master abstract concepts, it is impossible to fully understand the concepts, algorithms and methods for problem-solving.

Due to the uncertainty of world change, it is impossible to describe the various possible states of the future world, that is, it is impossible for people to explain the current problems from various dimensions.

In short, because people's choice behavior is a self-organizing one that constantly adapts to the external environment, and people are emotional, it is impossible to deal with problems completely rationally.

Therefore, the microeconomic consumer choice behavior theory based on the above-mentioned perfect assumptions makes it difficult to make a perfect interpretation of the people's choice behavior in reality.

Still it provides a way to describe people's choice behavior and provides a more feasible method for us to establish a model on people's choice of behavior. But it cannot be used as a panacea for solving problems.

In summary, because the hypothesis of *rational man* is irrational, we must seek a more realistic theoretical tool to explain people's choice behavior.

5.4.2 Decision-making behavior

Behavioral decision theory mainly studies people's decision-making behavior from the perspective of empirical, focuses on describing and understanding how people make decisions and belongs to the descriptive decision-making scientific paradigm. The theory holds that decision-making is a series of actions that meet people's certain goals. It can also be said to be a theory to study people's choice behavior [10].

Behavioral decision theory originated from the subjective expectation utility model established by Edwards on the basis of the maximum expected value theory in 1954. The model is as follows:

$$SEU = p_1 U_1 + p_2 U_2 \tag{5.7}$$

where SEU is the subjective expected utility, p_1 and p_2 are the probabilities of occurrence of the two events; U_1 and U_2 are the expected effects of the two events, respectively. The criterion for decision-making is to maximize the SEU.

Edwards believes that to make a good decision requires knowing the effectiveness of the event or the individual's judgment of the value of the event, as well as the probability of occurrence of various events; more importantly, the utility and probability are independent of each other, that is, people are right The evaluation of event utility does not depend on the likelihood of their emergence.

The subjective expectation utility theory belongs to the structural rational model. This model only focuses on the final decision result and does not consider the decision process.

The subsequent development of behavioral decision theory focuses on the study of procedural behavioral models, which consider both the outcome of decision-making and the process of decision-making, especially the psychological characteristics of the decision-making process.

When people make decisions, they are often influenced by various behavioral and psychological factors.

For example,

1. People do not consider all possible options when making decisions. For example, a person does not consider thousands of models when buying a car but only considers a few possible solutions based on his own experience and hobbies.
2. People do not consider all probability distributions when making decisions but also consider the main possibilities.
3. Because of the attention to an event (such as buying a car), and ignoring other possible events (such as buying other things more cost-effective), that is, the values have also changed.
4. Due to a large number of investigations and studies, with the necessary data, the actual selection calculation takes very little time.

Because of various behavioral factors, Simon proposed the principle of *bounded rationality*. Therefore, the theory of behavioral decision analysis first assumes that the decision-making subject is a real person with bounded rationality.

Real people have the following characteristics: (1) real people cannot have complete knowledge; (2) real people cannot always maintain their own orderly and stable preference system: (3) real people have only limited rationality.

The so-called *bounded rationality* means that real people's decision-making has a finite goal, has a moderate level of desire, finds a satisfactory answer according to a certain search method and follows the appropriate benefit principle.

Real-world decision-making must be influenced by behavioral factors, so behavioral models should be used, but only limited rationality can be achieved.

Behavioral decision theory believes that a good decision-making process should consist of four parts:

1. establishing a decision-making framework;
2. collecting information and concentrating wisdom;

3. drawing conclusions through judgment and choice;
4. learning from the decision-making process to accumulate experience as a reference for future decisions.

In the development of behavioral decision theory, many theoretical methods have been produced. In the development of behavioral decision theory, many theoretical methods have been produced. Interested readers can read books on decision theory.

5.4.3 Gaming behavior

5.4.3.1 Overview

The meaning of the game is that under certain conditions, obey certain rules, one or more people or teams with absolute rational thinking, choose and implement the behaviors or strategies that they are allowed to choose, and obtaining the corresponding results or benefits.

The gaming behavior here refers specifically to the process of implementing the selected behavior or strategy.

A complete game should include five aspects:

First, the participants of the game, which is, individuals and organizations that independently make decisions and independently bear the consequences during the game.

Second, the game information, that is, the information that the participants have in his grasp of the selection strategy.

Third, the set of all actions or strategies that the game participants can choose.

Fourth, the order of the game, that is, the order in which the game participants make strategic choices.

Fifth, the gains of the players, that is, the gains and losses of the players after making the decision choices [11].

Researching the theory of people in the game is called *game theory*. It is a new branch of modern mathematics and an important subject of operations research. Game theory considers the predictive behavior and actual behavior of participants in the game and studies their optimization strategies.

This theory is that the two people make their own confrontation strategies according to the opponent's strategies in an equal game to win. The game theory thought has existed in ancient times. The ancient Chinese book "Sunzi Art of War" is not only a military work, but also the earliest game theory book.

Game theory initially focused on the problems of chess, bridge and gambling. People's grasp of the game situation only stayed on experience and did not develop into theory.

Modern research on game theory began in Ernst Zermelo (1913), Borel (1921) and von Neumann (1928).

Modern game theory began with the idea regarding the existence of mixed-strategy equilibria in two-person zero-sum games and its proof by John von Neumann. von Neumann's original proof used the Brouwer fixed-point theorem on continuous mappings into compact convex sets, which became a standard method in game theory and mathematical economics. His paper was followed by the 1944 book *Theory of Games and Economic Behavior*, cowritten with Oskar Morgenstern, which considered cooperative games of several players. The second edition of this book provided an axiomatic theory of expected utility, which allowed mathematical statisticians and economists to treat decision-making under uncertainty.

Game theory was developed extensively in the 1950s by many scholars. It was later explicitly applied to biology in the 1970s, although similar developments go back at least as far as the 1930s. Game theory has been widely recognized as an important tool in many fields. As of 2014, with the Nobel Memorial Prize in Economic Sciences going to game theorist Jean Tirole, 11 game theorists have won the economics Nobel Prize.

5.4.3.2 Game types

The classification of the game also has different classifications according to different benchmarks.

It is generally believed that games can be mainly divided into *cooperative games* and *noncooperative games*. The difference between a cooperative game and a noncooperative game is whether there is a binding agreement between the parties that interact with each other. If there is, it is a cooperative game. If not, it is a noncooperative game.

From the time series of behaviors, game theory is further divided into two categories: *simultaneous games* and *sequential games*.

Simultaneous games are games where both players move simultaneously, or if they do not move simultaneously, the later players are unaware of the earlier players' actions (making them effectively simultaneous). *Sequential games* (or dynamic games) are games where later players have some knowledge about earlier actions. This need not be perfect information about every action of earlier players; it might be very little knowledge. For instance, a player may know that an earlier player did not perform one particular action, while she/he does not know which of the other available actions the first player actually performed.

According to the degree of understanding of the participants to other participants, they are divided into *perfect information games* and *imperfect information games*. Perfect game means that in the process of the game, each participant has accurate information about the characteristics, strategy space and income function of other participants. Examples of perfect information games include tic-tac-toe, checkers, infinite chess and Go.

Imperfect information game means that if the participants do not know accurately about the characteristics, strategy space and income function information of other participants, or do not have accurate information on the characteristics, strategy space and income function of all participants. Many card games are games of imperfect information, such as poker and bridge.

The game theory discussed by economists generally refers to noncooperative games. As cooperative game theory is more complex than noncooperative game theory, the theoretical maturity is far less than noncooperative game theory. The noncooperative game is further divided into: a complete information static game, a complete information dynamic game, an incomplete information static game and an incomplete information dynamic game. The equilibrium concepts corresponding to the previous four games are Nash equilibrium, subgame perfect Nash equilibrium, Bayesian Nash equilibrium and perfect Bayesian Nash equilibrium.

There are still many classifications of game theory. For example, according to the number of games or the length of the game can be divided into *finite games* and *infinitely long games*. According to the form of expression, it can also be divided into *general (strategic) games* and *unfolded games*. According to the logical basis, it can be divided into *traditional games* and *evolutionary games*.

5.4.3.3 Nash equilibrium

In a strategic combination, all agents face such a situation, and when other agents do not change the strategy, his current strategy is the best. That is to say, at this time, if he changes the strategy, his payment will be reduced. At the Nash equilibrium point, every rational agent does not have the impulse to change the strategy alone. The premise of the existence proof of *Nash equilibrium point* is the concept of *game equilibrium pairs*. The so-called game equilibrium pairs is in the two-person zero-sum game. When agent A adopts its optimal strategy a^*, the player B also adopts its optimal strategy b^*, if the agent still adopts b^*, and agent A takes another strategy a, then the payment of agent A will not exceed the payment he took the original policy a^*. This result is also true for agent B.

Thus, the definition of *equalization pairs* is as follows: a pair of strategies a^* (belonging to policy set A) and strategy b^* (belonging to policy set B) are called *equalization pairs*, for any strategy a (belonging to policy set A) and strategy b (belongs to policy set B), the following equation is always fulfilled:

$$(a, b^*) \leq (a^*, b) \leq (a^*, b^*) \tag{5.8}$$

For nonzero sum games, there are also the following definitions: a pair of strategies a^* (belonging to policy set A) and strategy b^* (belonging to policy set B) are called *nonzero game equilibrium pairs*, for any strategy a (belonging to policy set A) and strategy b (belonging to policy set B), the following equation is always fulfilled:

$$\begin{aligned} &(a, b^*) \leq (a^*, b^*) \text{for agent A,} \\ &\text{and } (a^*, b) \leq (a^*, b^*) \text{ for agent B} \end{aligned} \tag{5.9}$$

With the previous definition, the Nash theorem is immediately obtained as follows.

Any two-player game with a limited pure strategy has at least one *equilibrium pair*. This equilibrium pair is called the *Nash equilibrium point*.

The strict proof of Nash's theorem uses the fixed point theory, and the fixed point theory is the main tool for economic equilibrium research. Simply put, finding the existence of an equilibrium point is equivalent to finding the fixed point of the game.

Table 5.2 Payment matrix of prisoner's dilemma game

Prisoner A	**Prisoner B**	
	Prisoner B betrays	**Prisoner B stays silent**
Prisoner A betrays	8, 8	0, 10
Prisoner A stays silent	10, 0	1, 1

The Nash equilibrium point concept provides a very important analytical tool that allows game theory research to find meaningful results in a game structure.

However, the Nash equilibrium point definition is limited to any player who does not want to unilaterally change the strategy but ignores the possibility of other players changing the strategy. Therefore, in many cases, the conclusion of the Nash equilibrium point is not convincing.

Case 1: Prisoner's dilemma

The "prisoner's dilemma" is a standard example in game theory that shows why two completely rational individuals might not cooperate, even if it appears that it is in their best interests to do so. It was originally framed by Merrill Flood and Melvin Dresher while working at RAND in 1950. Albert W. Tucker formalized the game with prison sentence rewards and named it "prisoner's dilemma," [11] presenting it as follows.

Two members of a criminal gang are arrested and imprisoned. Each prisoner is in solitary confinement with no means of communicating with the other. The prosecutors lack sufficient evidence to convict the pair on the principal charge, but they have enough to convict both on a lesser charge. Simultaneously, the prosecutors offer each prisoner a bargain. Each prisoner is given the opportunity either to betray the other by testifying that the other committed the crime, or to cooperate with the other by remaining silent. The offer is as follows:

If A and B each betrays the other, each of them serves 8 years in prison.

If A betrays B but B remains silent, A will be set free and B will serve 10 years in prison (and vice versa).

If A and B both remain silent, both of them will serve only 1 year in prison (on the lesser charge).

Table 5.2 shows the payment matrix for this game.

For prisoner A, although he did not know what choice B would made, he knows that no matter what B chose, his choice of "betray" is always optimal. Obviously, according to the symmetry, B will also choose "betrays," and the result is that both were sentenced to 8 years. However, if they both choose to "stay silent," they will only be sentenced to 1 year. In the four combinations of action choices in Table 5.2 (*prisoner A stays silent, prisoner B stays silent*) is a Nash equilibrium, because any other combination of actions that deviates from this combination of actions selects at least one person's situation. It is not difficult to see that "betray" is the dominant strategy of any criminal suspect, and *(prisoner A betrays, prisoner B betrays)* is a dominant strategic equilibrium.

The dilemma, then, is that mutual cooperation yields a better outcome than mutual defection but is not the rational outcome because the choice to cooperate, from a self-interested perspective, is irrational.

From a mathematical point of view, this is reasonable, but it is obviously inappropriate in the field of sociology where such multidimensional information works together. This is because the social system's binding effect on human behavior changes people's decision-making. For example, from a psychological point of view, the cost of choosing "betray" will be greater. Then the revenge afterward and the label of the **Traitor** will cost him more.

We are now in the era of *Big Data*, and we can grasp as much information as possible and reasonably weight the analysis. The motivation of human activity is complicated, so the prisoner's dilemma can only be used as a reference for simplified models, and specific decisions must be specifically analyzed.

Case 2: Beauty's coin game

A strange beauty came over to talk to you and asked to play a game with you.

Beauty proposes: "Let us each show the side of the coin, either front or back. If we are all front, then I will give you 3 pounds. If we are all back, I will give you 1 pound. The rest of the situation you will give me 2 pounds." That sounds good! If I am a man, I will play anyway, but economic considerations are another matter. Is this game fair enough?

Suppose we have a front probability of x and a back probability of $1-x$ (Table 5.3). In order to maximize the benefits, our income should be equal when the opponent is front or back. Otherwise, the opponent can always change the probability of the front and back to make our total income decrease, thus the equation is as follows:

$$3x + (-2) \times (1 - x) = (-2) \times x + 1 \times (1 - x) \tag{5.10}$$

To put it simply, this equation represents that the benefits you get from the opponent are always the same and the biggest, when your opponent always show the front or always show the back.

The solution of (5.9) is $x = 3/8$, which means that our optimal strategy is to show three times front and the five backs every eight times.

Table 5.3 Beauty's coin game payment matrix

Gentleman	Beauty	
	Front	**Back**
Front	3, −3	−2, +2
Back	−2, +2	1, −1

Substituting $x = 3/8$ into the income expression $3 \times x + (-2) \times (1-x)$ can obtain the expected benefit for each time, and the calculation result is $-1/8$ lb.

Similarly, the probability that the beauty showing front of the coin is y, the probability of showing the back is $1-y$, and the equation would be as follows:

$$-3y + 2(1 - y) = 2y + (-1) \times (1 - y) \tag{5.11}$$

The solution of (5.10) is also $y = 3/8$, and the expected benefit of the beauty every time is $2(1-y)-3y = 1/8$ lb. This tells us that in the case when both sides adopting the optimal strategy, the beauty averagely wins 1/8 lb each time. In fact, as long as the beauty has adopted the (3/8, 5/8) plan, no matter what plan you use, you cannot change the situation. If you show all fronts of the coin, the expected benefit per time is $(3 + 3 + 3-2-2-2-2-2)/8 = -1/8$ lb.

If you all show back, the expected benefit is also $(-2-2-2 + 1 + 1 + 1 + 1 + 1)/8 = -1/8$ lb.

And any strategy is nothing more than a linear combination of the previous two strategies, so it is expected to be $-1/8$ lb. But when you also adopt the best strategy, you can at least guarantee that you lose the least. Otherwise, you will definitely be targeted by the strategy adopted by the beauty, thus losing more.

It seems that this game model seems to be of no use, but in fact this may involve the most important model of financial market pricing—Pricing model.

In general, the essence of "game theory" is to express the contradictions in daily life in the form of games and use mathematical and logical methods to analyze the law of the operation of things.

Since there are players in the game, there must be a game rule maker.

A deep understanding of the nature of competitive behavior helps us to analyze and grasp the relationship between things in competition, and it is more convenient for us to formulate and adjust the rules to finally operate according to the purpose we expected.

5.5 Behavior analysis patterns in traffic engineering

5.5.1 Overview

The main research contents of traffic participants in the field of traffic engineering include strategic and tactical level of choice behavior and microscopic driving (riding/walking) behavior.

The study of traffic travel choices at the strategic and tactical levels began in the early 1960s. This part of the behavior mainly includes travel mode, travel destination, travel frequency, car ownership, travel time, travel frequency and path selection.

Most of the analysis of such selection behaviors previously is based on the *discrete selection theory*, and a small part is based on the theory of utility maximization in continuous time and space.

Many classical traffic planning textbooks have been introduced for the analysis of this type of behavior and will not be repeated here [12].

The microscopic driving (riding/walking) behavior is the core behavior model of the microscopic traffic flow simulation model. It mainly studies the specific microscopic operational behaviors of traffic participants (drivers, cyclists, pedestrians) under specific traffic situations, such as: driver's car-following behavior, overtaking behavior, lane changing behavior, turning and braking behavior; cyclist's overtaking behavior, crossing conflicting traffic flow at intersections, pedestrian routing, crossing streets, crowding, conflict avoidance behavior.

The research and modeling of the microscopic driving (riding/walking) behavior of travelers in traffic engineering is mostly in the application of traffic engineering—for example, the core behavioral model of the microscopic traffic flow simulation model, or providing basic theoretical basis and parameter data for traffic safety analysis, transportation infrastructure design analysis, optimization and evaluation.

This makes the microscopic behavior model of traffic participants more focused on engineering application aspects.

Therefore, it is significantly different from the behavioral analysis model of basic theoretical research done in basic disciplines such as psychology and sociology: the research results of micro-behavior in the field of traffic engineering may not be too complete and deep in theory. However, its advantage is its practicability. The behavior analysis and modeling process pays great attention to the aspects of field data collection, model parameter calibration, model verification and applications of the model.

In addition, attention to the microscopic driving (riding/walking) behavior of most traffic participants concerned by traffic engineering is in the category of reactive behavior. And more specifically, most of them are learned behaviors. A behavior type that is different from the previous type of selection behavior.

For example, the driving behavior of the driver, the cycling behavior of the cyclist and the conflict avoidance behavior that occurs naturally when the pedestrian walks in the crowd, all of which are subconscious behaviors and do not require subjective deliberate judgment and thinking. The pattern of behavior of this type of behavior is a simple *stimulus→organ→reaction* pattern. Similar to simple reaction behavior, when subjected to external stimuli or encountering a particular situation, the traffic participant will behavior automatically responds to his/her fixed, predefined, learned rules.

Therefore, in the field of transportation, the microscopic driving (traveling) behavior of traffic participants is building a relatively simple behavioral model framework, and then using a large amount of field data for model parameter calibration and validation. We also refer to this behavioral analysis modeling method as *semi-black box* (also called *gray box*).

The following is an introduction to the "semi-black box" mode of traffic engineering research in the traffic engineering (Figure 5.5). First, the laws and characteristics of the research system or subject are obtained through observation or other means (the basis of previous research or the theory of other disciplines). Then find the appropriate theoretical method or model as the basic framework based on these regularities and characteristics.

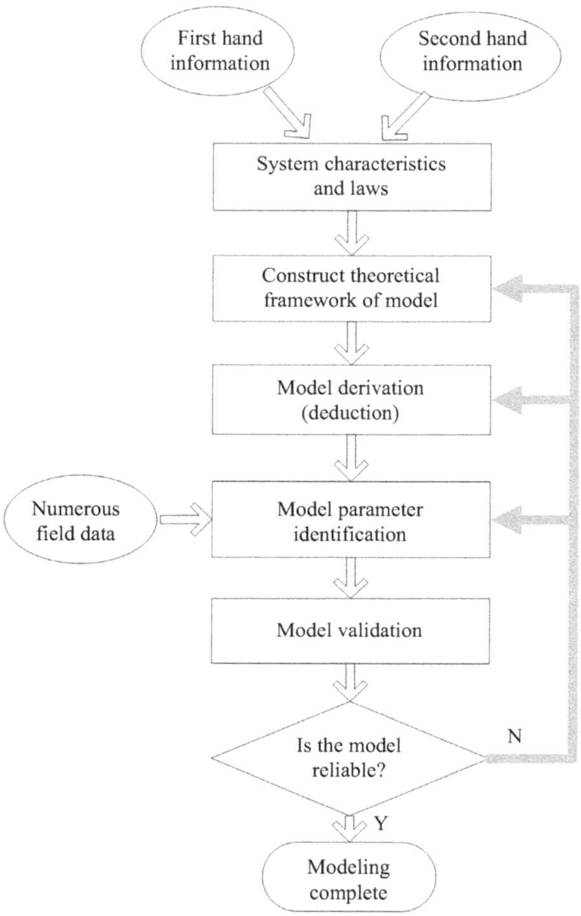

Figure 5.5 Flow chart of behavioral analysis and modeling in traffic engineering

Then some derivation and deduction are carried out under the framework of these theoretical methods and models. This process may require specific assumptions or boundary conditions based on the characteristics of the transportation system and the object to obtain the target behavior model formula.

After obtaining the model formula, the model is parameter identified (calibration) using field data.

An important point after parameter identification is to verify the model, that is, using other field data (data not used for parameter identification) to compare with the model output. The comparison result is usually used as an important index to evaluate the reliability and effectiveness of the behavioral model.

This is a relatively general modeling process for behavioral modeling of traffic participants in traffic engineering.

5.5.2 Car-following behavior model

Taking the earliest and most in-depth linear car-following model as an example of traffic engineering microscopic behavior research, we describe the behavior analysis modeling process in traffic engineering.

A platoon of cars runs in a single lane. In the case of a relatively large traffic density and relatively small vehicle spacing, in order not to cause collisions, each driver in the platoon strives to maintain a certain safe distance.

The driver adopts the corresponding operation according to the information of the preceding vehicle. Therefore, the driving speed and the headway are restricted by the preceding vehicle and are in a non-free flow state. However, the inherent reaction characteristics of the driver determine that the changes in motion between adjacent vehicles are not synchronized. The following car driver needs a reaction time to recognize the state change of the leading car such as acceleration and deceleration.

If the preceding vehicle has a state change at time t, then the rear car can make a corresponding action reaction at time $t + T$ (T refers to the driver's reaction time), thus causing "lag" in time. In order to maintain the safety distance, the driver of the following car is constantly conducting operations, and this operation would pass backwards until the end of the car platoon. This is the so-called car-following behavior.

Models of this phenomenon, which describe the interaction between (typically) adjacent vehicles in the same lane, now form the cornerstone for many important areas of research including (1) simulation modeling, where the car-following model (amongst others) controls the motion of the vehicles in the network, and (2) the functional definition of advanced vehicle control and safety systems, which are being introduced as a driver safety aid in an effort to mimic driver behavior but remove human error.

Therefore, many scholars have carried out a lot of in-depth research on the car-following behaviors and put forward a number of theoretical models. Specifically, there are stimulus–response models, safety distance models, physiological–psychological models, behavioral threshold models, fuzzy logic–based models and cellular automata models, etc.

As an example, we only introduce the linear car-following model. For those interested in other car-following models, please refer to the related literature on car-following models in [13].

In the linear car-following model, the car-following behavior is regarded as the "stimulus–response" pattern. The model is built according to the driver's response after receiving some kind of stimulus of the leading car, formulated as

$$Reaction = Sensitivity \times Stimulus \tag{5.12}$$

The basic assumptions of the vehicle following model are as follows:

1. The performances of the leading and following vehicles are the same.
2. The speed of the follow car will not change in a short time.

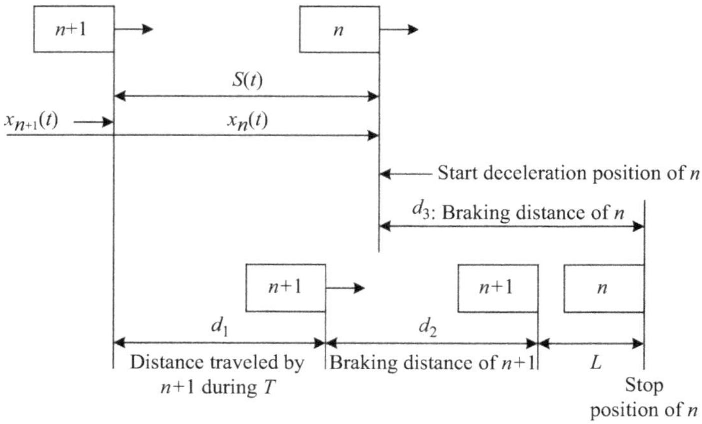

Figure 5.6 Car-following model diagram

In Figure 5.6, the position of the leading car n be x_n (t) at time t, and the position of the following car $n + 1$ be $x_{n+1}(t)$. Then the headway $S(t)$ between the leading car n and following car $n + 1$ at time t should be

$$S(t) = x_n(t) - x_{n+1}(t) \tag{5.13}$$

By the physical dynamics formula, the speeds of the leading and following car are both

$$v(t) = \frac{dx(t)}{dt} = x'(t) \tag{5.14}$$

Accelerations of the leading and following car are both

$$a(t) = \frac{d^2x(t)}{dt^2} = x''(t) \tag{5.15}$$

According to the vehicle's safe driving requirements in car following, in order to ensure that the adjacent two cars do not collide in a sudden braking event, the headway of the leading and following cars should meet the following conditions:

$$S(t) + d_3 = d_1 + d_2 + L \tag{5.16}$$

where L stands for the safety headway between the two cars when stopped; d_1 stands for the running distance during the "reaction" time T of the driver of the following car $n + 1$. From the basic assumptions (2) we can calculate:

$$d_1 = T \times x'_{n+1}(t) \tag{5.17}$$

From the basic assumption (1), it can be seen that the leading and the following car have the same emergency braking distance, that is, $d_2 = d_3$; thus, the $S(t)$ can be written as

$$S(t) = d_1 + d_2 + L - d_3 = d_1 + L = x_n(t) - x_{n+1}(t) \tag{5.18}$$

Find differential in t we get

$$x''_{n+1}(t+T) = \frac{1}{T}\left[x'_n(t) - x'_{n+1}(t)\right] \tag{5.19}$$

where $x''_{n+1}(t+T)$ is the acceleration of the following car at time $(t+T)$, which is called the reaction of the following car; $1/T$ stands for the *sensitivity*, $x'_n(t) - x'_{n+1}(t)$ is the *stimulus* at time t.

It can be seen that the linear car-following model formula and the learned behavior of psychology are consistent in the form of the model.

5.6 Summary

This chapter first analyzes the general pattern of people's behavior analysis, then summary the behavior characteristics of people. We introduce the behavior analysis patterns of the fields of psychology, social field, economics and traffic engineering. It provides a theoretical framework for the construction of the microscopic behavior analysis framework for bicycles at un-signalized intersections in Chapter 6.

References

[1] H. Haken. Principles of Brain Functioning: A Synergetic Approach to Brain Activity, Behavior, and Cognition. Springer Publishing Company, Berlin, 2012.

[2] K. Mainzer and R. Landauer. Thinking in Complexity: The Complex Dynamics of Matter, Mind, and Mankind. Computers in Physics, 1995, 63(8):766–767.

[3] B. Chen (Ed.). Incentive: Introduction to Behavioral Science. Wuhan, Hubei People's Publishing House, 1987, pp. 10 *(in Chinese)*.

[4] J. Wang (Ed.). Traffic Safety Psychology. Beijing, Science and Technology Literature Press Chongqing Branch, 1988 *(in Chinese)*.

[5] K. Lewin. Field Theory in Social Science, New York, Harper, 1951.

[6] D. Helbing and P. Molnár: Social force model for pedestrian dynamics. Physical Review E, 1995, 51:4282–4286.

[7] B. Schmidt. Modeling of Human Behavior – the PECS Reference Model. (1st edition), SCS-Europe BVBA, Ghent, 2000.

[8] R.J. Rummel. Understanding Conflict and War, Vol. 1: The Dynamic Psychological Field. Beverly Hills, CA, Sage Publications, 1975.

[9] L. Silverston. The Data Model Resource Book, Vol. 1: A Library of Universal Data Models for All Enterprises. Wiley Publishing, 2001.

[10] J.E. Russo, A. Baosheng, and X. Liancang. Behavioral Decision Making. Beijing, Beijing Normal University Press, 1998.
[11] D. Fudenberg and J. Tirole. Game Theory. SERBIULA (sistema Librum 2.0),Massachusetts, MIT press, 2019.
[12] J.F.jr. Nash. Equilibrium Points in n-Person Games. Proceedings of the National Academy of Sciences of the United States of America, 1950, 36(1):48–49.
[13] M. Brackstone and M. McDonald Car Following: A Historical Review. Transportation Research Part F: Traffic Psychology and Behaviour, 1999, 2(4):181–196. (doi:10.1016/S1369-8478(00)00005-X).

Chapter 6

Analysis of bicycle microscopic behavior at un-signalized intersections

In Chapter 5, we discussed the analysis of human behavior patterns in various disciplines. Drawing on the methods and research results of behavior analysis in various disciplines, this chapter attempts to construct a theoretical framework for cyclist's microscopic behavior analysis of crossing un-signalized intersections.

We first describe and define the problem that is a detailed description of the behavior of bicycles crossing un-signalized intersections. At the same time, we analyze the psychological and behavioral processes of the cyclist's crossing behavior, and what types of behavior patterns are included.

Then, based on the results of behavioral research in various subject areas (psychology, behavioral science, microeconomics, traffic engineering, etc.), we propose a two-layer hierarchical microscopic behavior analysis framework structure for cyclist's crossing un-signalized intersection behavior. The basic structure and input and output relationship of the framework are introduced.

In the last two sections, we discuss and compare the methods and supporting theories used in the two-layer behavior analysis framework, determine to choose the subjective expectation theory of decision analysis as the tactical-level model support theory, and the psychological field/social field theory as the support theory of the operational-level model.

6.1 Description and definition of the problem

6.1.1 Problem description

Let us imagine a situation where a bicycle rider wants to cross a mixed traffic flow un-signalized intersection.

When the cyclist first arrives at the entrance of the un-signalized intersection, he/she begins to observe and collect the general related information of the intersection, and according to the traffic flow situation (the traffic flow, whether there is a conflict with other motor vehicles, bicycles, pedestrians, etc.), quickly develops a rough path or plan (the decision-making process may take only a few seconds or even less).

In general, if there are small volume of traffic flows at the un-signalized intersection, and there is no conflicting traffic flow, then based on the shortcut characteristics of the human behavior mentioned in the previous chapter, we can speculate that the cyclist is very likely to cross the intersection on the most direct and labor-saving route, that is, from his/her current position x_0, goes straight to the destination D at the other end of the intersection.

If there is a large volume of traffic flow at the un-signalized intersection, the traffic situation becomes complicated; if there are obvious or potential (possible) conflicting traffic vehicles, the situation is completely different.

In this situation, the cyclists need to consider various influencing factors, such as the risk of conflicting vehicles, the urgency of crossing the intersection, the surrounding bicycles and pedestrians in the same direction, etc., so he/she would likely choose the best or satisfactory route after assessing the previous factors.

Then the cyclist rides along the path that he/she just chose. While riding at the intersection, the cyclist will constantly observe the surrounding traffic flow and collect relevant dynamic traffic information (such as the conflicting vehicle's position, speed, the same direction bicycle, pedestrian position, speed, etc.) in the intersection; he/she continuously predicts, measures and judges the position and speed of each impact and conflict vehicle and adjusts the originally established path according to the updated decision.

In addition, similar to the driver who drives the vehicle, the bicycle riding action of the cyclist is a subconscious behavior, that is, a learned behavior. Therefore, during riding along the planned path, the cyclist also makes automatic responses on the basis of the surrounding conditions, such as acceleration and deceleration, avoiding conflicting vehicles, bicycles or pedestrians.

Based on the previous analysis, we can divide the cyclist's crossing behavior at the un-signalized intersection into the following behavioral subprocess: information collection, rough path planning and riding along the planned path, as shown in Figure 6.1.

6.1.2 Problem definition

Based on the previous analysis, we mathematically describe the micro-behavior modeling problem of bicycle crossing un-signalized intersections: the study object is a dynamic traffic system S that moves on a two-dimensional plane. The system S consists of the following elements (Figure 6.2).

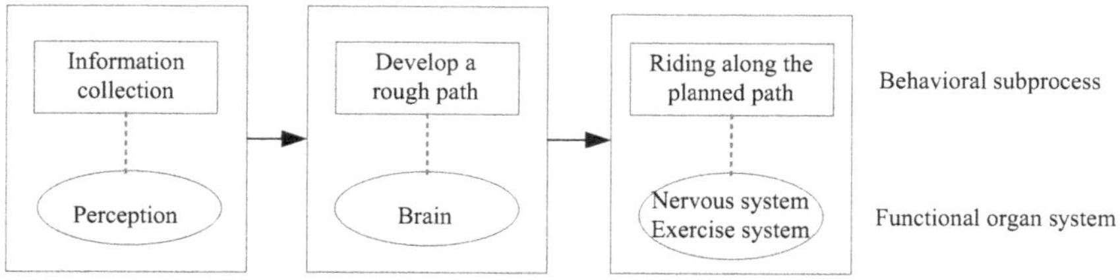

Figure 6.1 Subprocesses of the cyclist's crossing behavior at un-signalized intersection

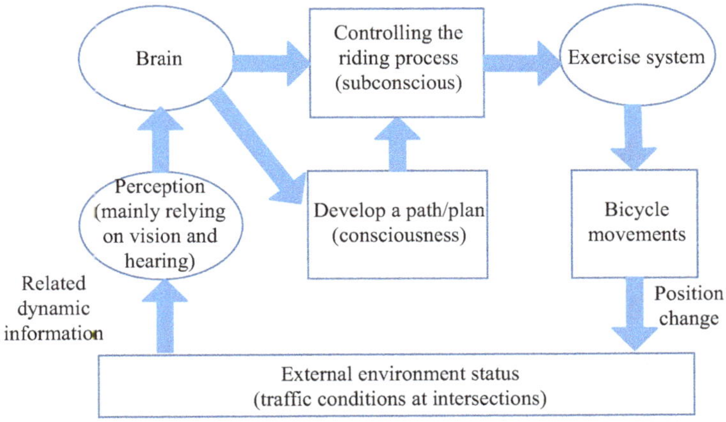

Figure 6.2 Schematic diagram of cyclist's crossing behavior process at un-signalized intersection

6.1.2.1 Transportation infrastructure Ω

Considering the actual situation, the bicycle rider may have collected the intersection traffic information when entering the intersection and made the crossing intersection line, so that the range of the moving vehicle in S (i.e., the infrastructure part in S) is set to Ω, which not only includes the entire un-signalized intersection, the traffic management facilities attached to the intersection, such as safety islands, crosswalks, but also part of the road plane at the intersection entrance. The information of Ω contains all information related to the internal traffic facilities, such as coordinate system, intersection size, crosswalk width and length, safety island location and size, speed limit, management control mode (yield or stopped control).

6.1.2.2 Vehicles V

Traffic vehicles V stands for all moving and stationary vehicles in S, mainly including the following three categories:

B: bicycles within S;
C: motorized vehicles within S;
P: pedestrians within S.

The information of V mainly includes two categories, one is information V^s describing the static characteristics of the individual vehicle, such as the size of the vehicle, the driver (cyclist, pedestrian) information (age, gender, travel purpose, etc.); the other is the information V^d describing that the state of motion of the vehicle is mainly represented by the specific position vector $\vec{x}(t)$ of each vehicle.

The position vector $\vec{x}(t)$ is a two-dimensional vector representing the X, Y coordinates of the vehicle at different times.

By deriving the position vector $\vec{x}(t)$, we can get the velocity and acceleration information of the vehicle:

$$\vec{v}(t) = \frac{d\vec{x}(t)}{dt} = \vec{x}'(t) \tag{6.1}$$

$$\vec{a}(t) = \frac{d\vec{v}(t)}{dt} = \frac{d\vec{x}'(t)}{dt} = \vec{x}''(t) \tag{6.2}$$

By the behavioral model formula by Lewin

$$B = f(P, E) \tag{6.3}$$

We can think of bicycles crossing the un-signalized intersection behavior modeling problem, mainly to study the environmental factors of the un-signalized intersection: **transportation infrastructure** Ω, and other **vehicles** at the intersection V, how to influence the cyclist's crossing **behaviors**.

Therefore, the mathematical description of the problem is that a dynamic traffic system S, within the scope of the Transport infrastructure Ω, the movement trajectory $\vec{x}_{V_{\bar{i}}}(t)$ and state $\vec{x}'_{V_{\bar{i}}}(t)$ and $\vec{x}''_{V_{\bar{i}}}(t)$ of each vehicle $V_{\bar{i}}$ (including B, C, P) except the target bicycle i are known. And the static information V^s of all vehicles V, to find the motion trajectory $\vec{x}_i(t)$ and state $\vec{x}'_i(t)$, $\vec{x}''_i(t)$ of the target bicycle i.

The mathematical formula of the problem is

$$\begin{aligned}\vec{x}_i(t) &= f(\Omega; V^s, V_{\bar{i}}^d | i \in V, V \in S; t)\\ &= f(\Omega; B^s, C^s, P^s, B_{\bar{i}}^d, C^d, P^d | i \in B; B, C, P \subset V; V \in S; t), t \in (t_{O_i}, t_{D_i}]\end{aligned} \tag{6.4}$$

where t_{O_i} and t_{D_i} stand for the time of target bicycle i entering and leaving Ω, O_i stands for the origin of i; D_i stands for the bicycle i's destination crossing the intersection. The problem we have to solve is to find out the mapping f.

6.1.3 Problem scope

From the previous research results of psychology, we inferred that the cyclist's crossing behaviors at the mixed traffic flow un-signalized intersection is a comprehensive and complex behavior. It includes complex behaviors of brain and mental activities, including information collection, judgment and trade-off, as well as simple reaction behaviors or learned behaviors (such as riding along planned path and automatic conflict avoidance behaviors).

In 1985, Ben-Akiva and Lerman proposed in their textbook on travel demand modeling that people's travel decision-making behavior can be regarded as a hierarchical process: they divide the decision-making process when people conduct activities related to travel and other travel-related activities into three levels: long-(strategic level), medium- and short-term [1]. Where the long-term choices include occupation (working place), residence, etc., the mid-term choices include whether or not to own a private car, commute mode selection, etc.; and the specific path choices made by a trip are classified as short-term options. Because Ben-Akiva and

Lerman [1] study the problem of travel demand, it is a problem involving macroscopic-level travel behavior. The division of people's travel behavior is not entirely suitable for our subject problem here. However, they put forward the idea that behavioral hierarchy is worth learning.

In Hoogendoorn and Bovy's model of pedestrian route choice and activity planning behavior, they divide pedestrian travel behavior and choice into the following three levels [2]:

1. strategic-level behaviors: mainly include the choice of travel time and type of activity;
2. tactical-level behaviors: mainly include the sequence arrangement of activities (i.e., activity plan) and route choice behavior of pedestrians;
3. operational-level behaviors: refer to the walking behavior of pedestrians.

As the behavior of cyclists has many similarities with that of pedestrians, we believe that the cyclists' behaviors can also be divided into the following three levels:

1. strategic-level behaviors: mainly include the cyclists' departure time selection and route selection behavior at the road network level;
2. tactical-level behaviors: mainly include the cyclists path planning behavior on a two-dimensional plane (especially within an intersection);
3. operational-level behaviors: refer to cyclists' riding behaviors along a given path.

In this way, the cyclist arrives at the un-signalized intersection, observes the traffic flow at the intersection, receives information (or environmental stimuli), analyzes and judges the information, and selects the path according to his own preferences. Then rides according to the path he/she decides. The whole process involves the latter two levels of behaviors (tactical and operational), so the scope of our research in this book only covers these two levels of behaviors.

6.2 Building a theoretical analysis framework

6.2.1 Behavioral characteristics of cyclist's crossing un-signalized intersections

Based on the previous analysis of the cyclist's crossing behavior through un-signalized intersections, we can derive several characteristics of this behavior:

1. According to the survey and analysis, we found that cyclists tend to make rough path when crossing the un-signalized intersection and then ride according to the established route.

   Thus, the cyclist's crossing behavior at the un-signalized intersection is a comprehensive behavior, including two different types of behaviors at different levels of complex—path planning and riding along a given path.

2. Since there is not much time for a cyclist to choose or make a path to cross an intersection, he/she can only rough path in such a short time. Moreover, in most cases, the cyclist will adjust his/her riding path when the traffic conditions at the intersection change. Therefore, the cyclist's path planning can be seen as a dynamic decision-making process.
3. Due to the small size of most un-signalized intersection, when crossing the cyclists usually have access to all relevant information about the intersection's dynamic transportation system *S*, i.e., the cyclists have access to complete information, meeting the assumptions of being a "rational person."

According to the characteristics (1), the single-model analysis framework cannot fully and effectively analyze the behavior of the bicycle crossing the un-signalized intersection. And different types of behavior patterns should be modeled separately. Therefore, we intend to construct a two-layer model framework to analyze the cyclist's crossing behavior at un-signalized intersections.

6.2.2 Analysis framework structure

The cyclist's crossing behaviors at un-signalized intersections include two different levels of behavior (complex and simple), and the results of complex behavior can be used as input for simple behavior. Therefore, we believe that the cyclist's crossing behaviors at un-signalized intersections behavior model are suitable in the form of a two-layer model framework.

The first level is the *tactical level*, which analyzes the bicycle path planning behavior at mixed traffic flow un-signalized intersections; the next layer is the *operational level*, which analyzes the specific cyclist's riding dynamic movement along a given line. Figure 6.3 shows the basic structure of the analysis framework.

As can be seen from the problem description and Figure 6.3, the input of the analysis framework is the basic information of the infrastructure Ω of the dynamic traffic system *S* and the dynamic information $V_{\bar{i}}^{d}$, static information V^{s} of the vehicles *V* in *S*.

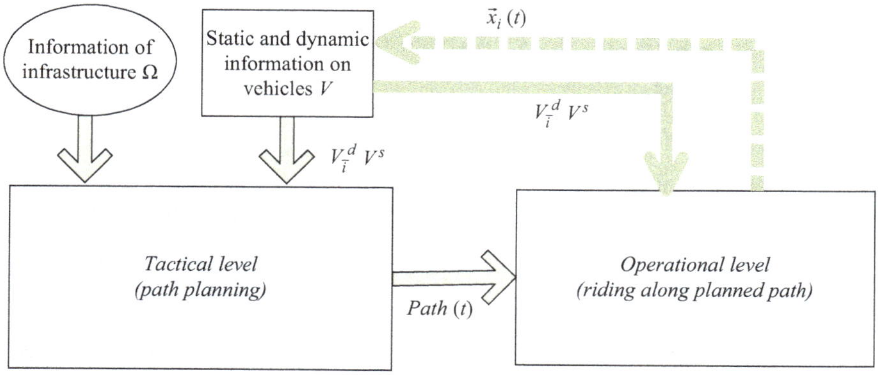

Figure 6.3 Schematic diagram of the analysis framework of cyclist's crossing behavior at un-signalized intersections

This information is used as input to the **tactical-level** module on the one hand as a source of information for decision-making; on the other hand, it is also used as inputs to the **operational-level** module as a source of stimuli for simple reaction behavior. The output of the **tactical-level** module is the established travel line as the input to the **operational-level** module, and the output of the latter is the motion vector of the target bicycle, which is also the output of the entire model.

It is worth noting that the output of the **operational-level** module—the motion vector $\vec{x}_i(t)$ of the target bicycle i will influence the dynamic motion information V_i^d of the vehicles V in the dynamic traffic system S, thereby affecting the input of the **operational-level** module.

Therefore, if the **tactical-level** module of the system framework is continuously updated with a certain time step T, the dynamic state of the target bicycle i and the surrounding vehicle V is constantly updated, and the input of the **tactical-level** model is updated, forming a dynamic path planning process. That is to say, such an analysis framework supports the cyclist's dynamic path planning behavior in the process of crossing the un-signalized intersection.

6.2.3 Support theory for the cyclist's analysis framework

6.2.3.1 Tactical-level model

The goal of **tactical-level** model in the analysis framework is to represent the cyclist's path planning behavior at un-signalized intersections. This path planning behavior is a complex behavior involving psychological and intellectual activities. This behavior can be regarded as a choice behavior or as a kind of decision-making behavior. According to the previous behavior pattern analysis in Chapter 5, the available support theories and methods mainly include the following:

1. *Discrete choice utility theory*
2. *Subjective expected utility theory*
3. Behavioral decision theory, mainly including *prospect theory*

Discrete choice theory originates from microeconomics, and it has a wide range of applications in the field of traffic engineering. It is mainly applied to the modeling of traveler's choice behavior, such as commuter mode and motor vehicle route choice. It has solid theoretical foundation, basically derived from the theoretical assumption of economic benefit maximization.

The advantages of the theoretical method are as follows: the theoretical model is mature; the model parameters are easy to estimate; the discrete modeling method is easy to implement in computer and the operating speed is fast.

However, it is worth noting that in the motor vehicle route choice behavior, because the motor vehicle is running by the lane, its route choice set is usually limited, so it is reasonable to adopt a discrete choice model. According to the previous description of the problem, it is known that the path planning behavior of the bicycle at the un-signalized intersection is based on a two-dimensional plane, so the limited path choice set cannot truly reflect the cyclist's route choice situation at the un-signalized intersections.

Therefore, we think that the **discrete choice theory** is not suitable as a supporting theory for the cyclist's path planning behavior at un-signalized intersections.

Both subjective expectation utility theory and behavioral decision theory belong to the discipline of decision analysis. More specifically, the former belongs to the normative decision analysis theory, and the latter belongs to the descriptive decision analysis theory.

Compared with behavioral decision theory, the theoretical model of subjective expectation utility theory is more mature, the mathematical expression and solution of the model are relatively perfect, and the theoretical framework has good extensibility and supports continuous decision space. Therefore, the previous drawbacks of the discrete choice model are substantially solved here.

Since the continuous space is modeled, theoretically, the decision-making subject has an infinite choice set of path. The selection subject chooses the path according to the maximization of subjective utility, so the decision-making process is concrete.

In the model framework, attributes such as alternatives, surrounding environment and decision-making subject characteristics can be included to complete the model framework. However, compared with the discrete choice model, the subjective utility model also has the defects such as parameters determination, complex calculation and slow computing speed.

Behavioral decision theory is based on the description of the behavioral characteristics of the subject. It fully considers the behavioral characteristics of the subject, that is, different decision-making bodies have decision-making frameworks that lead to different choices.

Therefore, the behavioral decision theory is often difficult to adopt a more unified mathematical form. In particular, the problem of the structure of decision-making can vary from person to person. Theoretical for the modeling of cyclist's path planning behavior, this theoretical approach seems a bit overemphasize on individual differences.

In some respects, the study of bicycle traffic behavior aims to understand the commonality of cyclists' behavior, that is, to explore the rules and characteristics of cyclists' behavior that are meaningful from the overall (or group) of the cyclists. Thus the unique behavior of individual cyclist is not the focus of this book.

Because the behavioral decision theory pays too much attention to the differences, randomness and ambiguity of individual behaviors, we consider that this theory is not suitable for support theory of bicycle path planning behavior at un-signalized intersections.

Therefore, we choose the subjective expectation utility theory as the support theory of the bicycle cyclist's line selection behavior analysis module at the un-signalized intersections. Yet improvements are needed, such as considering the randomness and ambiguity of cyclist's behavior.

6.2.3.2 Operational-level model

The output of *operational-level* module is the riding behavior of cyclist along a given path. This behavior is the dynamic movement behavior of traffic individual.

It is worth noting that bicycles and pedestrians have many similarities in dynamic movements, such as not having protective facilities, sizes are relatively small, low running speed, flexible and driven by human power.

Like pedestrians, bicycles are smarter and more flexible than motor vehicles in steering. They can choose a reasonable speed and direction according to the surrounding environment, instead of the vehicle movement being restricted by the lane; thus, bicycle dynamic movements are more diverse and complicated by the surrounding environment. In addition, when the traffic flow is relatively dense, the cyclists can get off and become a pedestrian pushing the bicycle. These all are the characteristics and modeling difficulties of the bicycle dynamic riding behavior.

According to the previous literature review in Chapter 5 and analysis, the current support theoretical methods for the dynamic microscopic behavior of bicycles at *operational level* mainly include

1. discrete choice utility theory;
2. cellular automata (CAs) model;
3. system dynamic model.

As mentioned earlier, the **discrete choice utility theory** has a solid theoretical foundation and easy estimation of model parameters. The discrete modeling method is easy to implement and the calculation speed is fast, and some random factors are considered in the model. The disadvantage is that because it adopts a discrete modeling method, its choice set is a finite set, which cannot completely reflect the behaviors of the bicycle in the two-dimensional plane at intersection; in addition, there are also subjective influence on the assumption of the random distribution of utility, the choice and disadvantages of the attributes.

In summary, the **discrete choice utility theory** is not a perfect support theory for bicycle dynamic movements at intersection.

A CA is a discrete model studied in computer science, mathematics, physics, complexity science, theoretical biology and microstructure modeling. CAs are also called cellular spaces, tessellation automata, homogeneous structures, cellular structures, tessellation structures and iterative arrays. CA method/model was first proposed in 1983 by Wolfram [3].

The advantages of using CA mainly include the following:

(a) the interaction between individuals in the system is based on intuitive, easy-to-understand rules of behavior, and unlike traditional modeling methods, the individual's behavior or state changes depend on an equation or function;
(b) the CA model has many research results and experience in motor vehicle traffic flow and has a certain research basis in pedestrian traffic flow modeling, which can be applied as a reference;
(c) the CA model has the advantages of easy computer implementation, easy parallel computing, fast running speed and is suitable for large-scale or even ultra-large-scale systems.

The disadvantage of using the CA model approach is that the core of the method—the behavioral rules of the cells—is derived from the traffic flow model.

While the bicycle traffic flow theory model is still in its infancy, it will inevitably contain some subjective factors. In summary, the method is more suitable for simulation implementation and is not suitable for the research and analysis of cyclist's microscopic behavior.

Compared with the CA model and the discrete choice utility theory, the dynamic model drawing on Newton's kinematics has obvious advantages in modeling the bicycle behavior: taking Helbing's "Social Force" model as an example, the theoretical basis of the model is stemmed from the "human territory" and "social field" theory of psychology [4].

In the SF model, he externalized the influence of surrounding peoples on the subject pedestrian psychology as the "force" that affects his movement. At the same time, consider the pedestrians' motivation to reach his/her destination and the psychological motives of attracting and avoiding.

The behavior modeling process is relatively straightforward and reasonable, and the modeling space is a continuous two-dimensional space, which is more inline with the situation where the cyclists arbitrarily select his/her riding path at the intersection plane. However, the Social Force model also has the difficulty of determining the "Social Force," and it is difficult to estimate and implement the model parameters in practice.

Because our research goal in this chapter is to analyze how the environment at the un-signalized intersection (such as conflicting vehicles and obstacles) affects the cyclist's riding behaviors, the "social field" theoretical method based on psychological theory is considered adequate as the analysis method and supporting theory of the **operational-level** module of bicycle microscopic behavior analysis framework.

6.3 Tactical-level model support theory—expected utility theory of decision analysis

Utility theory is a major theoretical basis for microeconomics and decision analysis.

The decision maker (selector) forms a preference for the choice set according to the individual characteristics of the object, the attributes of the choice set and the selection environment at the time. The preference is the ordering of the priority of the decision object by the decision maker (selector). And the quantified value of preference is utility. Thus, utility is the benchmark for decision-making, and any rational person always chooses the option with the largest utility.

The following is a description of the support theory of the tactical level of the cyclist's microscopic behavioral analysis framework—the theory of expected utility.

6.3.1 Basic concepts

6.3.1.1 Natural state, strategy and consequences

Natural state/state

It refers to the uncertain factors that decision makers cannot predict or control in the decision problems they face. The whole state of the natural state is called the *state space.*

Strategy
All of the options available to decision makers are called strategic spaces, and the purpose of the decision is to select an "optimal" solution from the strategic space.

Consequences
After the decision maker adopted policy (option), no matter which state occurs, it will inevitably lead to a certain consequence. All the possible consequences are called the *consequence space*.

6.3.1.2 Subjective probability

Decision makers must make decisions in an uncertain external state, but how do you describe uncertainty? Everyone agrees to use probability to study uncertainties in nature and society, but the connotation of probability concepts must be expanded.

The probability of measuring uncertainty in decision analysis is different from the probability of random event in common probability statistics. Events generally cannot be repeated under the same conditions. It is regarded as the degree of subjective trust of the decision makers under the information conditions they have, based on his level of knowledge, so it is called *subjective probability* or *personal probability*.

6.3.1.3 Preference and utility

Preference
Preference refers to the value of the decision makers in measuring the strengths and weaknesses of the strategy and the consequences. It is described by a binary relationship in mathematics, called the *preference relationship*.

The logical sequence of the decision-making process is the decision maker first compares the *preference relationship* of the decision-making consequences and the *belief relationship* of the state occurrence and then derives the *preference relationship* to the strategy.

Superficially, as long as there is *subjective probability*, that is, the uncertainty of the decision-making problem—the natural state is measured, the decision process can be quantified, because each strategy often has a corresponding benefit (the value of the consequence) and the corresponding possibility of success (probability), it is natural to think that it is appropriate to compare the expectations (average value) of the strategy to show the pros and cons of the strategy.

Utility
In many production and management departments, as well as in general management textbooks, the expected value is commonly used as a criterion for decision-making. However, the famous St. Petersburg paradox shows that it is not enough to use the expected value of the amount as the decision criterion. Therefore, *D.* Bernoulli first proposed to use the "utility" of the consequences relative to the decision maker as a measure of the value of the outcome.

Let D be a non-empty set, $\succ$ be a sequential relationship on D, if the real function u on D satisfies

$$A \succ B \Leftrightarrow u(A) \succ u(B), \quad \forall A, B \in D \tag{6.5}$$

Then u is called the utility function on D. If D is a policy space, u is a utility function for the strategy, which is accustomed to the uppercase letter U; if D is the consequence space, u is a utility function for the consequences, which is customarily represented by the lowercase u.

The utility function not only qualitatively reflects the decision maker's preference for strategy or outcome but also quantitatively depicts the relative strength of the preference.

6.3.1.4 Expected utility

Let S, X, F represent **state space**, **consequence space** and **strategy space**, respectively, u is the utility function on X, P is the probability measure on S, then for strategy f, $U(f) = \int_S u(f(s))dP$ called the expected utility of strategy f, u can be regarded as F utility function.*

Expected utility theory is an important principle for solving risk-based decision-making problems. The main founder of the theory is L. J. Savage. Expected utility decision makers believe that a reasonable strategy is mainly based on two factors: one is the subjective probability of the decision maker's uncertainty about the state of nature; the other is the utility value of the decision makers for each relevant consequence, and the decision makers use the maximum expected utility value as the criterion for evaluating the pros and cons of the strategy. That is to solve

$$\max_{f \in F} U(f) = \max_{f \in F} \int_S u(f(s))dP \tag{6.6}$$

For example, in the St. Petersburg paradox, if the decision maker's utility function for the consequences is u, then the expected utility values for f and $\bar{f}$ are

$$U(f) = \sum_{n=1}^{\infty} u(2^n - 100) \times 2^{-n} \tag{6.7}$$

and

$$U(\bar{f}) = \sum_{n=1}^{\infty} u(0) \times 2^{-n} = u(0) \tag{6.8}$$

In most cases, u is a bounded function, making $U(\bar{f}) > U(f)$, so that not participating in the bet has a greater expected utility. Expected utility is a combination of subjective probability and utility function, so it is still a subjective concept.

*As the spatial structure and the probability measure have different meanings, the meaning of the integral in the expected utility definition is also different. For example, when S is a discrete space, the integral degenerates into a normal summation.

6.3.2 Basic concept of utility function

6.3.2.1 Preference and binary relationship

The basic principle of decision analysis is that decision makers compare the advantages and disadvantages of the consequences and strategies according to their own values.

The values have two aspects of meaning: one is a qualitative comparison, for example, for programs *f*, *g* and *h*, *f* is preferred to *g*, and *g* to *h*; the other is a relatively quantitative comparison, for example, relatively speaking, I like *f* to a greater than *g*.

Habitually, the former is called "**preference**" and is described by a **binary relationship** called "order"; the latter is called "**utility**" and is characterized by a real-valued function or functional called a "utility function," which are closely related.

Let D be a non-empty set, a subset R of the product set $D \times D = \{(x,y)|x,y \in D\}$ is called a *binary relationship* on D. For $x,y \in D$, if $x,y \in R$, then it is said x have a relationship with y, which is recorded as xRy; if $x,y \notin R$, then it is said that x have no relationship with y and is recorded as $x\bar{R}y$. The *binary relationship R* has the following eight common properties that, in turn, can be divided into four pairs:

$$\begin{cases} 1. \quad \text{Reflexivity} : xRy, \\ 2. \quad \text{Inrreflexivity} : x\bar{R}y; \end{cases}$$

$$\begin{cases} 3. \quad \textit{Symmetry} : \text{IF } xRy, \text{ THEN } yRx, \\ 4. \quad \textit{Asymmetry} : \text{IF } xRy, \text{ THEN } y\bar{R}x; \end{cases}$$

$$\begin{cases} 5. \quad \textit{Transitivity} : \text{IF } xRy \text{ AND } yRz, \text{THEN } xRz, \\ 6. \quad \textit{Negative transitivity}; \text{IF } x\bar{R}y \text{ AND } y\bar{R}z, \text{THEN } x\bar{R}z; \end{cases}$$

$$\begin{cases} 3. \quad \text{Connetedness} : xRy \text{ OR } yRx, \\ 4. \quad \textit{Weak connetedness} : \text{IF } x \neq y, \text{THEN } xRy \text{ OR } yRx; \end{cases}$$

A binary relationship cannot have ALL the previous properties at the same time, rather, has some of the properties. It often refers to the transmitted binary relationship as an order and can also define various orders. For binary relationships with asymmetry, the common notation "$\succ$" or "$\prec$" is indicated.

Definition 6.1. *Let R be the binary relationship on set D*

1. *if R has asymmetry and negative transitivity, then R is said to be weak ordering;*
2. *if R be a weak ordering with weak connectedness, then R is said to be strict weak ordering;*
3. *if R has reflexivity, symmetry, and transitivity, then R is said to be an equivalence relation.*

For example, let D be the real number set, then the "greater than" relationship "$>$" is a weak ordering, which is also strict weak ordering, and the "equal" relationship "$=$" is an equivalence relation.

The more common practice in decision analysis is to describe the basic preferences of decision makers in weak order. Let $\succ$ be a weak ordering. If $x \succ y$, say x is better than y or y is worse than x; if neither $x \succ y$ nor $y \succ x$, then there is no difference between x and y, denoted as $x \sim y$; in addition, if $x \succ y$ or $x \sim y$, then x is not inferior to y or y is not better than x, noted as $x \geq y$.

It is natural to use weak ordering as the starting point of comparative consequence and strategy. The requirement of asymmetry indicates the coordination of the decision maker's choice. If you think that y is better than x, you cannot think that x is better than y at the same time. The weak ordering defined by asymmetry and negative transitivity also contains many properties such as transitivity.

Proposition 6.1. Set $\succ$ be the weak ordering on D, then

1. *For any* $x, y \in D$, *three relationships:* $x \succ y, y \succ x$ *and* $x \sim y$, *there is one and only one is established;*
2. $\succ$ *is transitive;*
3. $\sim$ *is an equivalence relationship;*
4. *by* $x \succ y, y \sim z \Rightarrow x \succ z$; *and by* $x \sim y, y \succ z \Rightarrow x \succ z$;
5. $\geq$ *is transitive and connected;*
6. *the relationship* $\succ$ *defined on the equivalence class* $D/\sim \{a|a \subset D, \forall x, y \in a, x \sim y\}$ *is as follows: set to a certain* $x \in a$ *and* $y \in b$, *have* $a \succ b \Leftrightarrow x \succ y$, *then* $\succ$ *is strictly weak order.*

The proof of the proposition is straightforward, so we will not prove it here. The meaning of an equivalence class in Property (6) is to treat all undifferentiated strategies or consequences as the same element.

6.3.2.2 Basic axiom and theorem of utility function

The utility theory is characterized by first establishing a set of strategic preference axioms, which have a certain practical background. Then from this set of axioms, the decision maker's preferences about the strategy can be derived, which can be quantitatively expressed with the expected utility value of the consequences.

In 1944, von Neumann and Morgenstern first established a set of axioms for classical utility theory in their famous book *The Theory of Games and Economic Behavior*. These axioms are constantly refined by scholars. And the basic idea is simplified to the following three axioms:

A1 (***weak order axiom***) $\succ$ is defined as weak order in the strategy space F;

A2 (***Archimede's axiom***) if $P \succ Q, Q \succ R$, then exist $\alpha, \beta \in (0, 1)$, which makes $Q \succ \beta P + (1 - \beta)R$ and $\alpha P + (1 - \alpha)R \succ Q$;

A3 ***(independence axiom***) if $P \succ Q$, then for any $\alpha \in (0, 1)$ and R, there is $\alpha P + (1 - \alpha)R \succ \alpha Q + (1 - \alpha)R$.

It is not difficult to understand the nature of weak order—*weak order axiom* is a preference hypothesis that combines the properties of transfer and asymmetry.

Archimede's axiom states that people's preferences are not absolute and can be compromised. As long as you think that P is better than Q and Q is better than R, no matter how strong your preference for P, you can always incorporate enough R to make the resulting mixture (draw lots) worse than Q; conversely, no matter how much R is. Unpopular, it is always possible to incorporate enough P to make the resulting mixture better than Q.

The *independence axiom* is the core of classical utility theory, although its meaning is very intuitive, that is, the relationship between P and Q is not affected by the mixing of other elements (in the same proportion), but the independence (and the inclusion by weak-order transmitting) has always been the focus of controversy.

The following preference properties implied by **A1**, **A2** and **A3** are the key properties that prove the basic theorem of utility functions (i.e., the Savage theorem that will be introduced later), and they can also be used independently as axioms.

Proposition 6.2. Let F be a mixed set, $\succ$ defines on F and satisfies **A1**, **A2** and **A3**, then for any $P, Q, R, T \in F$, there is

C1 (monotonic)$(P \succ Q, 0 \leq \alpha \leq \beta \leq 1) \Rightarrow \beta P + (1-\beta)Q \succ \alpha P + (1-\alpha)Q$;

C2 (continuity)$(P \geq Q, Q \geq R, P \succ R) \Rightarrow$, *there is a unique* $\alpha \in [0,1]$, *which makes* $Q \sim \alpha P + (1-\alpha)R$;

C3 (mixed dominance)$(P \succ Q, R \succ T, 0 \leq \alpha \leq 1) \Rightarrow \alpha P + (1-\alpha)R \succ \alpha Q + (1-\alpha)T$;

C4 (non-specificity)$(P \sim Q, 0 \leq \alpha \leq 1) \Rightarrow \alpha P + (1-\alpha)Q \sim P$;

C5 (replacement)$(P \sim Q, 0 \leq \alpha \leq 1) \Rightarrow \alpha P + (1-\alpha)R \sim \alpha Q + (1-\alpha)R$.

Theorem 6.1. Set F a mixed set, and relationship $\succ$ is defined on F, the necessary and sufficient conditions for $\succ$ satisfying **A1**, **A2** and **A3**:

1. *there exists a real-value function* U *on* F, *for any* $P, Q \in F$, *there is* $P \succ Q \Rightarrow U(P) > U(Q)$;
2. *for any* $P, Q \in F$ *and* $\alpha \in [0,1]$, *there is* $U(\alpha P + (1-\alpha)Q) = \alpha U(P) + (1-\alpha)U(Q)$.

Furthermore, U is unique in the sense of positive linear transformation: if the function V on F also satisfies (1) and (2), then there exist constants $a > 0$ and b, make $V(P) = aU(P) + b, \forall P \in F$.

We do not prove the previous theorem. The proof process is referred to [5].

6.3.3 Savage theorem (expected utility theorem)

The basic feature of Savage's theory of expected utility is that the existence of the utility function and the subjective probability are based on the decision maker's preference hypothesis about the strategy (option). Thus, Savage's theory is often referred to as *subjective expected utility theory*.

Let S be the state space, X be the consequence space, $F = \{f : S \to X\}$ be the policy space, $\succ$ is a binary relationship on F, indicating the decision maker's preference for the strategy, no difference relationship $\sim$ and $\geq$ as defined earlier. Each consequence $x \in X$ is seen as a special degradation strategy $f : S \to X$, where

$f(s) = x, \forall s \in S$ or $x(s) = x, \forall s \in S$. Therefore, the same symbol will be used to indicate the decision maker's preference for the strategy and the consequences.

A strategy that only corresponds to a limited number of consequences is called a simple strategy (or a step function), and $(x_1, A_1; x_2, A_2; \ldots; x_n, A_n)$ represents a simple strategy *f*, where $f(s) = x, \quad \forall s \in A_i, 1 \leq i \leq n$.

For $f, g \in F$ and $A \subset S$, define the mapping $f|_A^g$ as $f|_A^g = \begin{cases} g(s), s \in A \\ f(s), s \in A^c \end{cases}$, then $f|_A^g \in F$, which is called a hybrid strategy of *f* and *g*, which can be understood as when event *A* occurs *g* works, while event *A* does not occur, *f* works.

Definition 6.2. *For $f, g \in F$ and $A \subset S$, if there is a certain $h \in F$, make $f|_{A^c}^h \succ g|_{A^c}^h$, then we call f is superior to g with respect to event A, denoted as $f \overset{A}{\succ} g$, also called $\overset{A}{\succ}$ as conditional preference.*

The other two conditional preference relationships $\overset{A}{\sim}$ and $\overset{A}{\geq}$ can be defined accordingly.

Definition 6.3. *For $A \subset S$, if for any $f, g \in F$, there is $f \overset{A}{\geq} g$, then A is called invalid.*

It is easy to see that if *A* is an invalid set, then for any $s \in A^c$, there is $f, g \in F$, making $f(s) = g(s)$.

Savage's expected utility theorem relies on the following seven axioms:

P1 $\succ$ is weakly ordered on *F*;
P2 $\forall f, g, h_1, h_2 \in F$ and $A \subset S, f|_{A^c}^{h_1} \succ g|_{A^c}^{h_1} \Leftrightarrow f|_{A^c}^{h_2} \succ g|_{A^c}^{h_2}$;
P3 if *A* is not an invalid set, then for any $f \in F$ and $x, y \in X$, there is $f|_A^x \succ f|_A^y \Leftrightarrow x \succ y$;
P4 $\forall x_1, y_1, x_2, y_2 \in X, x_1 \succ y_1, x_2 \succ y_2$ and $A, B \subset S$, $(x_1, A; y_1, A^c) \succ (x_1, B; y_1, B^c) \Leftrightarrow (x_2, A; y_2, A^c) \succ (x_2, B; y_2, B^c)$;
P5 at least exist $x^*, x_* \in X$, make $x^* \succ x_*$;
P6 $\forall f, g, h \in F$ and $f \succ g$, there is a finite partition $\{A_1, A_2, \ldots, A_n\}$ of *S*, which makes $\forall i, f|_{A_i}^h \succ g$ and $f \succ g|_{A_i}^h$;
P7 $\forall s \in S, f \overset{A}{\geq} g(s) \Rightarrow f \overset{A}{\geq} g$.

P2 is called the *sure-thing principle*, and its implication is that the superiority and inferiority of the strategy are not affected by events that can cause the same consequences. The *sure-thing principle* is equivalent to the independence axiom A3, which is the core assumption of the Savage theory.

P3 guarantees the correspondence between consequence preferences and strategic preferences. P4 indicates that preferences should be consistent.

P5 is a structural assumption, explaining that the decision makers are not fully able to distinguish the pros and cons of the consequences.

P6 is a strong structural condition, which implies that the state space is an *uncountable infinite set*, which makes the subjective probability in the theorem conclusion have some continuity.

P7 is the dominance axiom.

Theorem 6.2 ***(Savage).*** Set $\succ$ on F meets axioms P1–P7, then

1. *there is a unique (finitely additive) probability P defined on the entire subset of S, P has the properties* $\forall B \subset S$ *and* $\alpha \subset (0, 1]$, *there is a subset C of B, such that* $P(C) = \alpha P(B)$;
2. *there is a bounded real-valued function u on X, u is unique in the sense of a positive linear transformation;*
3. $\forall f, g \in F, f \succ g \Leftrightarrow E\{u[f(s)], P\} > E\{u[g(s)], P\}$.

6.3.4 Mathematical model of utility function

6.3.4.1 Subject risk attitude and utility function [6]

In practical applications, it is not enough to measure the utility of some points in the simple preference object set X. It is also necessary to know the utility of all points in X. This raises the question of establishing a mathematical model of the utility function. The utility function is a quantitative description of the preferences of the decision maker. In theory, as long as this preference satisfies the rational behavior axiom and has the necessary preference structure characteristics, then the mathematical model of the utility function of the decision subject can always be established within the scope of the elementary function.

Of course, when using the elementary function as a model of the utility function, it is necessary to first identify certain properties of the set of utility values (utility functions) of the preferred object described by the model, such as continuity, monotonicity, boundedness and so on. In practice, a number of object attributes are often limited to the extent that the actual object can reach, and most of them are bounded.

For example, the sales volume of a certain period of time, the market share of a certain product and the energy consumption index of the equipment.

Therefore, all attributes with the best and worst indicators can be considered as bounded attributes. People's preferences for bounded attributes are always limited. Therefore, as the utility of quantitative description of preferences, there must be an upper or lower bound.

In this sense, the utility function is bounded. Deriving the mathematical model of the utility function on the set of nondeterministic preference objects, in addition to verifying the continuity, monotony and boundedness of the function, there is also an extremely important problem that is the subjective risk attitude.

The subjective risk attitude determines the basic characteristics of the utility function of the subject's preference. The trait of the utility function is mainly determined by the risk attitude of the subject.

Risk is used to describe the structural characteristics of any object that contains uncertainties. The so-called risk refers to the possibility that some undesired result appears in the object set of uncertainty preference. Since the worse consequences make people feel more dangerous, the risk refers to the harm that may be caused by some undesired results.

In practical problems, no matter what kind of subject, there may be different preference characteristics on the specific nondeterministic preference object set, and even the changing preference characteristics may appear.

These preference characteristics of the subject are collectively referred to as the subject's attitude toward the object, for short also called *subject's risk attitude*.

Let ω denote a risk with expected value $E(\omega)$. Let τ denote the risk ω when the subject's property level is x. In order to define various risk attitudes, the concept of risk compensation is introduced first. The so-called risk reward is the difference between the expected value of a certain risk and the determined equivalent of the risk at the subject's property level of $x \in X$. Let the risk reward as π, we have

$$\pi(x, \omega) = E(\omega) - \tau(x, \omega) \tag{6.9}$$

Using risk reward $\pi(x, \omega)$, we can divide the attitude of the subject toward risk, that is, the preference model in value judgment, into three basic types of risk: *conservative*, *adventurous* and *intermediate.* When expressed in terms of norms, they are called *risk aversion*, *risk seeking* and *risk neutral.*

1. **Risk aversion:** If and only if $\pi(x, \omega) > 0$, then the subject is said to be risk averse. At this time, the subject's utility function $u(x, \tau)$ is convex.
2. **Risk seeking:** If and only if $\pi(x, \omega) < 0$, then the subject is said to be risk seeking. At this time, the subject's utility function $u(x, \tau)$ is concave.
3. **Risk neutral:** If and only if $\pi(x, \omega) = 0$ then the subject is said to be risk neutral. At this time, the subject's utility function $u(x, \tau)$ is linear.

It can be seen that the basic trait of the utility function u defined on x depends on the risk attitude of the subject.

According to the actual survey, *risk aversion* is the most common kind of subject attitudes. In major economic activity choices or social activity choices, the attitudes taken by the subjects are often *risk aversion*.

In general management activities or recurring decisions, it is common to be *risk neutral. Risk seeking* is rarely seen in the normal favorable situation. Only when the loss is involved, will the situation of *risk seeking* be seen.

6.3.4.2 Utility function under different changing risk attitudes

The subjective risk attitudes can change. Since *risk aversion* is one of the most prevalent risk attitudes, *risk seeking* can often be dealt with in the dual case of *risk aversion*. Therefore, we take the risk aversion situation as an example to illustrate the risk attitude of change. Based on the different changes in risk reward $\pi(x, \omega)$ vs. x, there are three kinds of *changing risk aversion* attitudes.

1. The subjective *decreasing risk aversion*: if and only if $\pi(x, \omega) > 0$, and for all ω, $\pi(x, \omega)$ is a decreasing function of x.
2. The subjective *increasing risk aversion*: if and only if $\pi(x, \omega) > 0$, and for all ω, $\pi(x, \omega)$ is an increasing function of x.
3. The subjective *constant risk aversion*: if and only if $\pi(x, \omega) > 0$, and for all ω, $\pi(x, \omega)$ remains unchanged on x.

The following briefly introduces the utility functions corresponding to these three kinds of changing risk attitudes.

Risk aversion function

Let the subject's property level be $x \in X$. The subject's risk t is ω. Let ω have a certain distribution, and it is assumed that the possible values of ω are small compared with the property level x owned by the subject, so the variance σ is also small compared with x. The determined equivalent of ω has been assessed as $\tau(x, \omega)$. When u is a utility function defined on X, the corresponding utility equation would be

$$\begin{aligned} u[x + \tau(x, \omega)] &= u\{[p(\omega_1), x + \omega_1; \ldots; p(\omega_n), x + \omega_n\} \\ &= E[u(x + \omega)] \end{aligned} \tag{6.10}$$

Define

$$r(x) = -\frac{u''(x)}{u'(x)} \tag{6.11}$$

Let $r(x)$ be the *risk aversion function*. After introducing the risk aversion function, the risk reward π can be expressed as

$$\pi(x, \omega) = -\frac{1}{2}\sigma^2 r(x) \tag{6.12}$$

To determine the form of the utility function, by (6.12), there is

$$r(x) = -\frac{d}{dx} \ln u'(x) \tag{6.13}$$

After sorting out the second integral of the previous formula, we have

$$u(x) = c_1 \int e^{-\int r(x)dx} dx + c_2 \tag{6.14}$$

Therefore, as long as $r(x)$ is an integrable functional form, the form of the utility function $u(x)$ can be determined by (6.14). Thus, on the one hand, the risk aversion function $r(x)$ is considered as the measure of risk reward, on the other hand, it stipulates the traits of various utility functions in quantity.

By the intermediary role of $r(x)$, the various types of subject's risk attitudes are related to the corresponding various types of utility functions.

Utility function under different changing risk attitudes

In the following, various basic utility functions are derived on the basis of the type of subjective risk attitude reflected by $r(x)$.

Constant risk aversive utility function: In the actual economic and social decision-making problems, when the level of property changes within a small scope, the premise of decision-making can generally be regarded as a constant risk aversion.

When the subjective risk attitude is constant risk aversion, $\pi(x,\omega) > 0$, and for all ω, the value of π remains constant. At this time, $u'(x) > 0$, $u''(x) < 0$. The corresponding utility function is convex, i.e., $r(x) = c > 0$, according to formula (6.14), we can get

$$\begin{aligned} u(x) &= c_1 \int e^{-\int r(x)dx} dx + c_2 \\ &= c_1 \int e^{-\int cdx} dx + c_2 \\ &= c_1 \int e^{-cx} dx + c_2 \\ &= \frac{c_1}{c} e^{-cx} + c_2 \end{aligned} \tag{6.15}$$

Change the integral constants to a and b, and (6.15) has the following form:

$$u(x) = a - be^{-cx}, c > 0, b > 0 \tag{6.16}$$

The previous formula is the derived constant risk aversive utility function. Under the set conditions, the larger value of c, the more convex the utility function be.

Increasing risk aversion utility function: Incremental risk aversion refers to the subjective risk attitude that becomes more and more cautious, and risk avoidance when the level of subject's property increases.

When the subject's risk attitude is increasing risk aversion, $\pi(x,\omega) > 0$, and for all ω, the value of π is an increasing function of x. At this time $r(x) > 0$ and is an increasing function of x. At this time $u'(x) > 0$, $u''(x) < 0$, the corresponding utility function is convex.

After analysis and comparison, in the form of $a/(b-x)$, $r(x)$ can be used to derive the form of the incremental risk aversive utility function. When selected $r(x) = a/(b-x)$, $a, b > 0$, $0 < x < b$ and the above $r(x)$ is substituted into the formula (6.14), there is

$$\begin{aligned} u(x) &= c_1 \int e^{-\int (a/b - x)dx} dx + c_2 \\ &= c_1 \int e^{-\int a \ln(b-x)dx} dx + c_2 \\ &= c_1 \int (b-x)^a dx + c_2 \\ &= -\frac{c_1}{a+1}(b-x)^{a+1} + c_2 \end{aligned} \tag{6.17}$$

Change the integral constants to c and d, and (6.17) has the following form:

$$u(x) = c - d(b-x)^{a+1}, 0 < x < b \tag{6.18}$$

The previous formula is the general form of the derived increasing risk aversion utility function. In the formula, when $a = 1$, then there is

$$u(x) = c - d(b - x)^2 \tag{6.19}$$

The utility function shown in (6.19) has the form of a quadratic function, which is a practical utility function in the case of increasing risk aversion situations.

Decreasing risk aversion utility function: Decreasing risk aversion means that as the level of property increases, the subject faces the same risk, and he/she gradually changes from a more cautious attitude (stronger risk aversion) to a less risk that is averse risk (weaker risk aversion).

When the subject's risk attitude is decreasing risk aversion, $\pi(x, \omega) > 0$, and for all ω, the value of π is a decreasing function of x. At this time, $r(x) > 0$ is a decreasing function of x. At this time, $u'(x) > 0$, $u''(x) < 0$, the corresponding utility function is convex. However, unlike the incremental risk aversion utility function, the convexity of $u(x)$ gradually decreases as x increases.

According to the simple and practical requirements put forward by the utility function, after analysis and comparison, when $r(x) = a/x$, $a > 0$ is selected, the previously listed $r(x)$ is substituted into formula (6.14), then there is

$$\begin{aligned} u(x) &= c_1 \int e^{-\int (a/x)dx} dx + c_2 \\ &= c_1 \int e^{\ln x^{-a}} dx + c_2 \\ &= c_1 \int x^{-a} dx + c_2 \end{aligned} \tag{6.20}$$

According to the difference of the parameter a, the following three forms of basic decreasing risk aversion utility function can be derived.

1. when $0 < a < 1$

$$u(x) = \frac{c_1}{1 - a} x^{1-a} + c_2 \tag{6.21}$$

Let $q = 1 - a$ change the integral constant to b and d, then the decreasing risk aversion utility function in general power function form would be

$$u(x) = bx^q + d, \quad 0 < q < 1 \tag{6.22}$$

2. when $a = 1$

$$u(x) = c_1 \ln x + c_2 \tag{6.23}$$

The obtained formula is a decreasing risk aversive utility function in the logarithmic function form.

3. when $a > 1$

$$u(x) = -\frac{c_1}{a-1}x^{1-a} + c_2 \tag{6.24}$$

Let $q = -(1-a)$ change the integral constant to b and d, then the decreasing risk aversion utility function in general power function form would be

$$u(x) = -bx^q + d, \quad q > 0 \tag{6.25}$$

Utility function construction theorem

Here is another way to derive utility functions. It is to start from the basic utility function of the existing one and directly construct a new utility function with good properties and practical value. In this method, constructing various new utility functions is based on the *utility function construction theorem*. The construction theorem of four utility functions is given next. Here we focus on the construction of utility function for *decreasing risk aversion*. The construction of the risk-seeking utility function or the construction of the utility function of the *increasing risk aversion* can be handled accordingly.

Theorem 6.3. If $u(x)$ is a decreasing risk aversion function in interval $[ax_0 + b, ax_1 + b]$, a and b are constants, and $a > 0$, then on interval $[x_0, x_1]$

$$u(x) = u_1(ax + b), \quad a > 0 \tag{6.26}$$

is also a decreasing risk aversion utility function.

Theorem 6.4. If $u_1, u_2, \ldots, u_n$ are constant risk utility functions on interval $[x_0, x_1]$, $r_1, r_2, \ldots, r_n$ are their risk aversion function, respectively; let $a_1, a_2, \ldots, a_n$ be constants and each of them is greater than 0, if at least one pair in $r_1, r_2, \ldots, r_n$ is not equal, then on interval $[x_0, x_1]$

$$u(x) = \sum_{i=1}^{n} a_i u_i \tag{6.27}$$

is a decreasing risk aversion utility function.

Theorem 6.5. If $u_1, u_2, \ldots, u_n$ are decreasing risk aversion utility functions on interval $[x_0, x_1]$, and $a_1, a_2, \ldots, a_n$ are all constants and each of them is greater than 0, then on interval $[x_0, x_1]$

$$u(x) = \sum_{i=1}^{n} a_i u_i \tag{6.28}$$

is a decreasing risk aversion utility function.

Theorem 6.6. If $u_1(x)$ is a decreasing risk aversion utility function on interval $[x_0, x_1]$, $u_2(x)$ is a decreasing risk aversion utility function on interval $[u_1(x_0), u_1(x_1))$, then on interval $[x_0, x_1]$

$$u(x) = u_2[u_1(x)] \tag{6.29}$$

is a decreasing risk aversion utility function.

The basic utility functions such as exponential function, logarithmic function and power function, which have been derived before, can be used as the basic function forms to construct the new utility function.

Based on these basic utility functions, the previous four theorems can be applied flexibly to derive or construct various concise and practical utility functions of decreasing risk aversion.

If a preference structure satisfies the four conditions of completeness, reflexivity, transitivity and continuity, there exists a continuous utility function $u(x)$.

Therefore, we only need to make a few axiomatic assumptions on preferences, and we can get a utility quantity result similar to the *cardinal utility*.

However, it should be noted that the purpose of this quantification is simply to express preferences in a more convenient way. The only function of the utility function is to sort the different consumer bundles, indicating which of the two consumer bundles is more acceptable to the consumer. Therefore, the value of $u(x)$ itself has no economic meaning.

6.3.5 *Multi-attribute utility theory*

The choice of traffic mode is a multi-attribute choice problem, which is the result of the "comprehensive trade-off"ocomparison between the various attributes (speed, cost, comfort, convenience and safety) of the transport mode (alternatives). Regarding the comprehensive evaluation values of the respective traffic mode, each traffic mode is sorted according to the evaluation value and finally chooses the most suitable traffic mode.

In the multi-attribute selection, due to the contradictions and checks and balances between the various attribute indicators, it is often impossible to find a solution that is absolutely optimal for all attributes. For example, the cost of fast and comfortable traffic mode is very high. Each traffic mode has its own merits, and we can only choose a satisfactory one.

6.3.5.1 Standardization of attributes

"Attribute" is the characteristic, quality or performance parameter of alternative.

In order to facilitate the evaluation and comparison of each option, the attributes are generally expressed by quantity. However, since different attributes represent different things, the unit of measure and the size of the dimensions are different, which is not convenient for comprehensive comparison. Therefore, many multi-attribute choice methods require that each attribute value be normalized so that all attribute values are represented by a uniform dimension and unit. It is generally standardized as a dimensionless value (0-1), similar to a utility value.

There are many ways to standardize the attribute [6].

Linear transformation methods

In order to make the standardized attribute values change in the same direction as the utility, the larger the attribute value, the greater the utility.

Therefore, two different transformation methods are adopted for the benefit objective and the cost objective, respectively. The benefit objective adopts the linear transformation in the same direction method, and the cost objective adopts the inverse linear transformation method.

Benefit objective:

$$y_{ij} = \frac{x_{ij}}{\max_i x_{ij}}; \tag{6.30}$$

Cost objective:

$$y_{ij} = 1 - \frac{x_{ij}}{\max_i x_{ij}}, \quad \text{or} \quad y_{ij} = \frac{\min_i x_{ij}}{x_{ij}} \tag{6.31}$$

where x_{ij} is the original value of the jth attribute value of the ith scheme; y_{ij} is the standardized value of the jth attribute value of the ith scheme; $\max_i x_{ij}$ is maximum of the original value of the jth attribute; $\min_i x_{ij}$ is the minimum value of the original value of the jth attribute.

The previous symbols are applicable to the following other standardization methods.

Interval transformation

The standardized attribute values also change in the same direction as the utility values.

Benefit objective:

$$y_{ij} = \frac{x_{ij} - \min_i x_{ij}}{\max_i x_{ij} - \min_i x_{ij}} \tag{6.32}$$

Cost objective:

$$y_{ij} = \frac{\min_i x_{ij} - x_{ij}}{\max_i x_{ij} - \min_i x_{ij}} \tag{6.33}$$

Vector transformation

This is generally only used in TOPSIS (technique for order preference by similarity to an ideal solution). The standardized attribute value changes in the same direction as the original attribute value, so the same standardized method is adopted regardless of benefit objective or cost objective, namely,

$$y_{ij} = \frac{x_{ij}}{\sqrt{\sum_j x_{ij}^2}} \quad (6.34)$$

Different methods of attribute standardization have a certain influence on the choice of methods and the choice of weights. This should be paid attention when selecting multiple attributes alternatives.

6.3.5.2 Determination of attribute utility value

The attributes of the scheme are divided into qualitative and quantitative attributes. For quantitative attributes (such as travel time), the determination of its utility value is actually to convert its original attribute value into utility value according to a certain conversion formula. For example, for traffic mode, the shorter the travel time, the greater the utility value.

The qualitative attribute itself has no quantitative representation (such as the comfort of travel), and its utility is determined by directly estimating its utility value by a certain estimation method. There are many methods for estimating utility, as mentioned next.

Subjective assessment method

The rater directly scores according to the observation, just like the referee directly scores in the competitions such as singing, dancing and gymnastics.

Sorting assessment method

That is, the orders of the attribute of all schemes are first evaluated and then converted into utility values by a certain method. The specific methods are as follows:

Simple coding method: When the ordering of several schemes has been determined, the natural number of 1, 2, 3, ..., is used to assign values to each scheme, the worst scheme is assigned 1 and the optimal scheme is assigned n, and each of the intermediate schemes is assigned to value from 2 to $n-1$.

Then, the utility value is standardized to a value of 0–1 according to the method of attribute value standardization described earlier.

In essence, this simple coding method determines the utility values of each scheme are $1/N$, $2/N$, N/N, respectively, in order from poor to excellent.

Equal ratio coding method: The previous simple coding method has a major disadvantage, that is, the more inferior the scheme, the larger the ratio of the utility values of the two adjacent schemes, the more superior the scheme, the smaller the ratio of the utility values of the two adjacent schemes. The equalization coding method is to overcome this shortcoming, so that the ratio of the utility values of the adjacent two schemes is a constant. If the ratio is set to a, the utility values of the schemes from the inferior to the superior order are $(1/a^{n-1}), (1/a^{n-2}), (1/a^{n-3}), \ldots, 1$, respectively.

Ring ratio method: This method is used under the condition that not only the order of all schemes is known, but also the ratio of the utility values between the two adjacent schemes can be known.

For example, there are four schemes A, B, C and D, the order of inferior to superior is as $B - C - A - D$. If it can be judged that the utility of C is equal to 2.5 times of B, the utility of A is equal to 2 times of C, and the utility of D is equal to 1.6 times of A, then the utility of D is equal to $2.5 \times 2 \times 1.6 = 8$ times of B, and the utility of A is equal to $2.5 \times 2 = 5$ times of B.

Let the maximum utility of the D scheme be 1, then the utilities of the four schemes B, C, A and D are 1/8, 2.5/8, 5/8 and 8/8, i.e., 0.125, 0.3125, 0.625 and 1, respectively.

Stepwise adjustment method: Let us know that the utility order of the four schemes A, B, C and D is as follows: A is superior to B, B is superior to C and C is superior to D.

First, according to this order, the four schemes are given a certain utility value, as long as the utility values are in the range of 0–1, and the order can be matched. Now the utility value of the assignment is $A=1$, $B=0.7$, $C=0.3$, $D=0.1$.

The next step is to ask some additional questions and then see if the assigned utility values meet the answers to these supplementary questions.

If they meet, the assignment is unchanged; if they do not, adjust the utility values to meet the answers. For example, we can ask question: Is the utility of B less than the sum of the utility of C and D? If the answer is yes, since 0.7 is greater than the sum of 0.3 and 0.1, the assignment should be adjusted, say $B=0.4$, $D=0.2$ is just fine.

Then we ask is the utility of A greater than the sum of the utility of B, C and D? If the answer is yes, it is not necessary to adjust as $1 > (0.4 + 0.3 + 0.2)$. Thus, the utilities of A, B, C, D are 1, 0.4, 0.3 and 0.2, respectively.

6.3.5.3 Utility conversion of quantitative attributes

The utility conversion of quantitative attributes is to convert the quantitative attributes by a certain utility function curve. To do this, we must first determine the functional relationship between the original attribute value and the utility value, usually to determine whether the relationship is a linear or a nonlinear relationship, and then separately processed.

Linear relationship

We can use the linear or interval transformation in the attribute value standardization introduced earlier to convert the original attribute value into utility value. There is a special transformation similar to the linear relationship called *three-broken-line curve*. The utility function of *three-broken-line curve* is the conditional equation for the utility line:

$$u(x) = \begin{cases} 0, & (x_{\min} \le x \le x_a) \\ \dfrac{x - x_a}{x_b - x_a}, & (x_a \le x \le x_b) \\ 1, & (x_b \le x \le x_{\max}) \end{cases} \tag{6.35}$$

$$u(x) = \begin{cases} 1, & (x_{\min} \le x \le x_a) \\ \dfrac{x - x_b}{x_b - x_a}, & (x_a \le x \le x_b) \\ 0, & (x_b \le x \le x_{\max}) \end{cases} \tag{6.36}$$

In the previous formula, x_a and x_b are the two turning points of the utility curve (i.e., the *three-broken-line curve*): next x_a and upper x_b, the utility values, are all constant (0 or 1). The utility value between x_a and x_b varies in a straight line.

The *three-broken-line* utility function is applied because it is often the case in reality; on the one hand, the attribute value of the target should have a minimum requirement, and those below this requirement are considered invalid (i.e., the utility value is equal to 0). On the other hand, the target attribute value should not exceed a certain limit. If the limit value exceeds the limit function, the utility value cannot be increased, so the utility value is 1 from the limit.

Nonlinear relationship

Although the nonlinear utility function can have different forms, the most common one is the curve form of decreasing utility change rate, that is, as the value of the attribute increases, the change rate of its utility becomes smaller and smaller.

For this type of utility curve, it is usually described by a negative exponential curve or a logarithmic curve.

The most common method used in the negative exponential curve is based on only two of the attribute values that are the corresponding maximum utility and the minimum utility of the attribute. As these two values are usually given, this method is the easiest. At this time, the maximum utility value is set to 1, but the minimum utility value cannot be set to 0 but can only be a value slightly higher than 0, usually 0.1.

The utility function for the benefit objective is

$$u(x) = 1.1 - e^{-(2.3(x-x_{\min})/x_{\max}-x_{\min})} \tag{6.37}$$

The utility function for the cost objective is

$$u(x) = e^{-(2.3(x-x_{\min})/x_{\max}-x_{\min})} \tag{6.38}$$

where e in the formula is the base of the natural logarithm.

Although the previous negative exponential curve has the advantage of simple calculation, its disadvantage is not accurate enough because it assumes that the rate of change of utility is given.

A more precise method is described by a logarithmic curve:

$$u(x) = A + B\ln(x + C) \tag{6.39}$$

There are three parameters in this formula: A, B and C, so it is determined by the utility value of the three points, that is, in addition to the attribute values corresponding to the known maximum utility value and the minimum utility value (1 and 0, respectively). We should also think of ways to find the value of the attribute corresponding to an intermediate value.

In addition to the previous simple curves, many other forms of curves can be theoretically proposed, but they are often too complicated to use. Therefore, we will not introduce them here.

6.3.5.4 Calculation of the synthetic utility

To synthesize the utility values of each attribute in each scheme into a synthetic utility value, the key is to choose what kind of synthesize method, that is, what kind of reasonable synthesize algorithm is chosen. This depends on the type of relationship between the attributes. Generally, the following synthesize algorithms can be used:

Weighted sum method

That is, the synthetic utility value is equal to the weighted sum of the utility values of the attributes:

$$TU = \sum_{i=1}^{n} \omega_i u_i(x) \tag{6.40}$$

where TU is the synthetic utility value, ω_i is the weight of the ith attribute and u_i is the utility value of the ith attribute. The weighted sum method is a very common method because it is simple and easy to understand.

However, the weighted sum method is only applicable to the case where the attributes are independent of each other, and the synthetic utility can be linearly compensated or replaced.

This is the same as the student test scores. The subjects are basically independent of each other, but the reduction of one subject score can be compensated by the scores of other subjects. That is, just the total score can represent the student's whole performance in learning.

The disadvantage of the weighted sum method is that it cannot be used in situations where the utility of each attribute cannot be compensated by each other, i.e., cannot be replaced.

Multiplication method

In this method, the synthetic utility value is equal to the product of each attribute's utility value:

$$TU = \prod_{i=1}^{n} u_i(x) \tag{6.41}$$

This method emphasizes the irreplaceability of attributes. If the utility value of one attribute is 0, the combined utility value also be 0; the utilities of the remaining attributes make no difference.

For example, when we consider the design of an electronic computer, the calculation speed, storage capacity and reliability of the computer can be used as three important attributes. We consider that the three attributes—reliability, calculation speed and storage—are irreplaceable.

If the reliability is very low, no matter how high is the calculation speed or how large the storage capacity is, we would not choose the computer. In this case, we have to use the multiplication method.

This feature of multiplication is also convenient for those choices that have minimum requirements for the attribute values of each scheme. We only need to specify the minimum requirement as a utility value that is equal to 0.

Since the comprehensive utility value of the serialization result will become very small, although it does not affect the evaluation and selection of each scheme, it is intuitively easy to give an illusion.

Therefore, some people advocate finding the *n*th root after the multiplication, that is, the geometric mean method is adopted:

$$TU = \sqrt[n]{\prod_{i=1}^{n} u_i(x)} \tag{6.42}$$

Multiplication is generally not weighted, indicating that each attribute is equally important. If we feel the need to give different attributes different status in the multiplier, we can also use weighting, that is

$$\ln TU = \sum_{i=1}^{n} \omega_i \ln u_i(x) \tag{6.43}$$

Substitution method

This method applies to the situation that each attribute has great substitutability to the choice maker and in particular requires at least one attribute to reach the highest value.

That is to say, as long as one attribute reaches the highest value, it does not matter if the other attributes are extremely badly implemented.

When the measure value of the utility value of each attribute is 0–1 (i.e., maximum utility 1, minimum utility 0), the calculation formula of the synthetic utility value using the substitution method is

$$TU = 1 - \prod_{i=1}^{n} [1 - u_i(x)] \tag{6.44}$$

The shortcomings of the substitution method are the same as the multiplication method. When the number of items multiplied is large, the right half of the formula becomes very small, and the *TU* becomes very close to 1, and the difference is extremely small, which is inconvenient to compare.

Therefore, this method is also rarely used. Now many people tend to use the following *extreme value method* instead of the substitution method.

Extreme value method

Use the maximum value of an attribute utility value as the synthetic utility value. That is

$$TU = \max_i u_i(x) \tag{6.45}$$

When actually calculating the synthetic utility value, if the number of attributes is large, it is often not only use one type of the previous synthesize algorithm, but several synthesize algorithms are used together. For example, a weighted sum is calculated between several attributes, and several other attributes may be calculated by multiplication or other synthesize methods to form a complex calculation formula.

6.3.5.5 Determination of attribute weight

Many of the methods for calculating integrated utility involve the problem of weighting individual effects with individual attribute weights. Even if the utility value of each attribute of a scheme is unchanged, once the weight of each attribute changed, the synthetic utility value of the scheme may change, thereby affecting the comparison and choice of the schemes.

In fact, not only in the application of multi-attribute utility theory, but also in many other multi-attribute selection methods, there are many related attribute weights, so the determination of attribute weights becomes an important issue in multi-attribute choice problem.

Determining attribute weights in multi-attribute choice is to solve two problems, one is to reflect the relative importance of each attribute, and the other is to reflect the degree of dispersion of each attribute value.

In multi-attribute choice, sometimes only the first kind of weight is used, and sometimes the two kinds of weights are used together. The method for determining the weights is as follows.

Methods of determining the weights reflecting the relative importance of the attribute

There are many such methods, but none of them are recognized as the best and can only be selected depending on the natures of the choices. Here we just introduce two common methods:

Sorting determination method: This kind of methods first determines the order of importance of each attribute and then converts the order into weights.

This is similar to the ranking method used in the utility estimation of qualitative attributes. There are also different methods such as simple coding method, equal ratio coding method, ring ratio assignment method and stepwise adjustment method. The calculation methods are very similar.

The only difference is that the minimum value of the utility value can be zero, while the minimum weight cannot be zero, otherwise it is equivalent to canceling the effect of attribute with the minimum weight.

Subjective estimation method: Such methods rely on experienced experts who are familiar with the choice objects to directly estimate the weights of all attributes.

To avoid the subjective one-sidedness of individual expert, it is usually through the investigation of a certain number of experts. Let the experts make personal estimates and then use the statistical method to synthesize the average to obtain the weight, so this method is also called the *expert investigation method.*

There is no fixed model for the investigation methods of experts. The *Delphi method* is the most widely used.

It should be noted out here that the weights obtained by the expert method are the average results estimated by the them. Whether these weight values are usable depends on whether the weights estimated by the experts are relatively close.

If the estimates differ greatly from each other, such weight estimates lack reliability and require additional ideas to correct (such as whether the selected experts should adjust? Does the survey method switch to the *feedback Delphi method*?).

Methods of determining the weight reflecting the attribute value dispersion degree

If an attribute has the same value between various schemes, it is not necessary to consider this attribute when evaluating the schemes, because this is equivalent to the weight of this attribute which is zero.

If the value of an attribute has only a small difference between various schemes, then this attribute should be placed in a less important position in the comprehensive evaluation of the scheme, that is, the weight should be set smaller. On the contrary, if the difference is large, the weight should be set larger. Therefore, there should be a class of weights to reflect the extent of this difference in the distribution of attribute values.

According to the nature of such weights, it is most appropriate to use "entropy" to represent it. This is called the *entropy method* of determining the weight, because entropy is the distribution difference. The formula for calculating entropy is

$$S(P_1, P_1, \ldots, P_n) = -k\sum_{i=1}^{n} P_i \ln P_i \tag{6.46}$$

where P_i originally means probabilities and can also indicate the proportion of each part. It can be seen from the previous equation that only when all P_is are the same, that is, $P_i = 1/n$, $S = k \ln n$ reaches the maximum value. Now the attribute value x_{ij} of each scheme (i.e., the jth attribute value of the ith scheme) is converted into a PU value, the condition is

$$\sum_{i=1}^{n} P_{ij} = 1 \tag{6.47}$$

Then get

$$P_{ij} = \frac{x_{ij}}{\sum_{i=1}^{n} x_{ij}} \tag{6.48}$$

Then the entropy of the jth attribute is

$$S_j = -k\sum_{i=1}^{n} P_{ij} \ln P_{ij} \tag{6.49}$$

In order to make S_j only change between 0 and 1, the maximum value of S_j should be 1. That should be $k = 1/\ln n$, then there are

$$S_j = -\frac{1}{\ln n}\sum_{i=1}^{n} P_{ij} \ln P_{ij} \tag{6.50}$$

Considering that the larger the entropy, the smaller the weight should be, so it should be

$$d_j = 1 - S_j \tag{6.51}$$

Therefore, the weight reflecting the difference in the distribution of attribute values should be

$$W_j = \frac{d_j}{\sum_{j=1}^{m} d_j} \tag{6.52}$$

Here is a question that should be handled with great care in multi-objective selection, that is, what kind of situation is only necessary to use the weights that reflect the importance of the attributes and under what circumstances both kinds of weights should be used (the weights and reflections reflecting the importance of the attributes, and the weight reflecting the attribute value dispersion degree, by the product of these two kings of weights)?

The answer to this question is we compare the original value of the attribute with the converted utility value.

If the dispersion of utility value of each attribute is the same as the original attribute value, then we can only use the importance weight as the weight. Conversely, if the degree of dispersion of the utility value does not reflect the degree of dispersion of the original attribute value, then we should use the product of two kinds of weights.

6.4 Operational-level model support theory—Social Force model

6.4.1 Basic concepts

6.4.1.1 Psychological field

The *psychological field* is a commonly used concept in the **gestalt psychology** [7].

This school cites theories and concepts of "field" in modern physics and directly uses them to demonstrate their theoretical views. Yet, they do not have a clear explanation of the terms they have created. They think that the *physical field* is the field of the geographical environment; the *behavior field* and the *physiological field* are the connections between the conscious physiological processes; the *psychophysical field* is organized, it first expresses the polarity of itself and the environment, and second, it shows that each part of the two polarity has its own structure. Gestalt theorists remind everyone that our heart physics field exists within a real organism.

To summarize the so-called *physical field* refers to the field of geography; the so-called *behavior field* refers to the field of consciousness or direct experience; the so-called *psychophysical field* is a broader field that covers the field of behavior and the physiological field, while Koffka. K. will refer to all the abovementioned fields except the *physical field* collectively referred to as the *psychological field*.

Psychologist Levin K. developed field theory. In his *field theory* proposed in 1951, he emphasized the following points [8]:

1. Use a method of construction or occurrence.
2. Field theory is a kind of motivational research.
3. Field theory emphasizes the study of psychological processes.
4. Field theory is to analyze the environment as a whole.
5. Field theory is a systematic study that is different from traditional psychology research.
6. Field theory describes the psychological environment in mathematical forms.

Based on the previous points of *field theory*, Lewin put forward a series of basic concepts: living space, behavior and movement, force and force field, tension and tension system and other unique psychological concepts.

Living space: That is, *Field*, which includes people and their related psychological environment, refers to the overall impact of individual psychological factors in a specific time.

Behavior and movement: Behavior refers to the movement of people in the *living space*, and psychological movement refers to the approaching or avoiding of a certain goal.

Force and force field: In 1936, Levin called force the cause of change. It has various properties of direction, strength and fulcrum. As far as *living space* is concerned, *force* represents the direction and tendency of change.

Tension and tension system: Any *living space* or *field* has opposite forces. When two or more positive forces are applied to different goals, or positive forces and negative forces are applied to the same target area, a state of tension will arise as a result.

6.4.1.2 Understanding of psychological field theory

Lewin's dynamic field theory is closely related to the field theory of physics. In defining his concept of field, Lewin quoted Einstein's definition: "Field is the whole of interdependent facts" [8].

But Levin's field mainly refers to people's psychological and behavioral field. Lewin borrowed the concept of field theory in physics, that is to say, he used field theory as a method of analyzing and solving problems.

Lewin's psychological research and theoretical construction under the guidance of this concept and method is a brand-new creation.

There are two basic concepts in Levin's theory of psychological field: the **psychological tension system** and **living space**. The former shows a dynamic view, while the latter highlights the significance of the integrity in dynamics.

Dynamics and integrity are the two basic points of Levin's dynamic field theory.

Tension refers to the relationship between the state of a system and the state of its surrounding systems, and it already reflects the dynamic characteristics of the field. *Living space* is the psychological field in which human behavior occurs. People and the environment are seen as a common dynamic integrity.

Combining the concept of dynamic with the holistic view is the basic form of Levin's psychology field theory.

Levin said, "Any kind of behavior arises from the whole of various interdependent facts, and these interdependent facts have the characteristics of a dynamic field. This is the basic claim of field theory."

Both the tension system and the living space have a characteristic that highlights the psychological significance of the subject's cognition. Levin believes it is important to study the psychological process when explaining a psychological law, that is, to study the perception of psychological events rather than the events themselves, and to study the behavior of human beings rather than the results of behavior. According to M. Deutsch, Levin provided a psychological interpretation for behavioral process, assuming that all behaviors are directional. Yet, the physiological mechanism of the behavior or the physiological process itself is not too much investigated.

This kind of psychological orientation is one of the characteristics of Levin's dynamic field theory, and it is also a viewpoint of Levin's research on psychology.

In addition to the concept of force, holistic view and psychological orientation, integration is also a major feature of field theory. To exert this integration effect of field theory, Levin proposed a specific theoretical construction method. The method advocates studying the nature of things in the mutual relationship of objects, and by building the elements or constructs (concepts such as tension, force, etc.) to conduct conceptual analysis and theoretical generalization, thereby achieving an integrated theoretical construction.

Levin uses a precise mathematical language to express his theory. He believes that mathematics can not only express quantitative relationships but also can express qualitative relationships. He chose topology to describe and explain the psychological field, and thus mathematics also became a view of its field theory.

In summary, Levin's dynamic field theory has the following characteristics: dynamic view, holistic view, system view, psychological orientation, integration tendency, construction method and mathematical expression.

Among them, "dynamics" is the basis and the most prominent feature of Levin's field theory. "Dynamics" is not only the foundation of all human activities, but also the core and theoretical construction of Levin's psychological field theory.

The next section will introduce the dynamics of the social field theory.

6.4.2 Social field dynamics

6.4.2.1 Levin's dynamic concept

In psychology, the term "dynamic/dynamics" has meanings such as force, energy and activity, and it can include all intrinsic or underlying factors that determine the behavior of an organism.

For different psychologists, "dynamic/dynamics" may have different meanings; the concepts they use are not the same, and the views about it are also quite different. For example, Freud's psychoanalysis is a dynamic psychology, and his focus is on the role of subconsciousness or libido. W. McDougall is also a dynamic psychologist whose dynamic concept mainly refers to instincts and its instigation of behaviors.

Levin's dynamic concept (or dynamic view) can be summarized as a psychological tension system. He not only pays attention to the meaning of need and dynamic energy but also emphasizes the interaction of various psychological dynamics in a system, and the relationship between human internal dynamics and environmental factors.

Psychological tension system

The "psychological tension system" is a dynamic analysis of the root causes of human behavior from the conceptual level, where tension and need are the two most basic terms.

Levin believes that "as long as there is a psychological need within a person, there will be a tension system." The release of tension can provide dynamics and energy for mental activity and behavior. Therefore, it constitutes a potential factor in determining people's psychological activities and behaviors.

In order to prove this theoretical idea, Levin directed his students to conduct a series of experimental studies. For example, in Zeigarnik's experiment, she asked the subjects in the experimental group to do 18–20 simple tasks, but only half of the subjects completed the tasks, and the other half of the subjects were stopped halfway, and then she asked all the subjects to recall the work they have done.

According to Levin's theoretical assumptions, it can be predicted that (1) a work to be completed is equal to a quasi-need, and it will produce a corresponding psychological tension; (2) if the work is completed, the tension will be released; (3) if the work is blocked and uncompleted, the psychological tension will continue to exist and will affect the behavior and psychological activities of the participants.

The results of Zeigarnik's experiment fully confirm Levin's theoretical assumptions. The average recall of the unfinished work was 68%, and the average recall of completed work was 43%.

This is the famous "Zeigarnik effect" in the history of psychology [9].

In the dynamic theory of Levin's psychological field theory, the goal of behavioral or psychological activity also has a kind of force nature, which is called "valence." Positive valence has a force of attraction, and negative valence has a force of rejection.

Therefore, human behaviors are not only driven by internal needs and tension but also by the attraction (or rejection) of the target itself. As the intensity of need increases, the valence of the target associated with the need increases, while the increase in valence will, in turn, affect the strength of need.

This is like the situation: the more hungry a person is, the more attractive the food is to him; the more attractive the food, the stronger the inner desire for food.

Therefore, in Levin's dynamic theory, people are closely related to the environment. His psychological tension system includes the relationship between people and the environment, including the interaction between tension and goals.

Levin's dynamic research is based on the new physics worldview. For this, we can further examine another important concept of Levin: the living space or the dynamic field.

Tension system and life space

In Levin's view, "tension" is always in a "system." Therefore, in order to truly understand "tension," it is necessary to understand the meaning of such a system, and to understand the needs and tensions, tensions and goals, the attraction and rejection of goals and the interaction of various forces.

Or, if you want to understand the dynamic in Levin psychology, you must understand his dynamic field at the same time, and this dynamic field is also called "life space" by Levin.

The *life space* includes all the facts that determine individual behavior and mental activity at a certain time.

Levin said, "In order to understand or predict behavior, people and their environment must be regarded as a collection of interdependent factors. We refer to the whole of these factors as the life space of the individual and use $B = f(P, E) = f(LS)$ to express" (B represents behavior, P represents the subject of the behavior, E represents the environment, LS is the abbreviation of the living space). *Life space* includes people and their environment. And behavior occurs within this life space—it is a function of people and the environment, and also a function of living space.

Levin uses the topology to describe the dynamic structure in the life space. With the help of vector science, the dynamics in the life space and the direction of dynamic and behavior in the life space are analyzed. It is called "topological psychology" and "vector psychology." The combination of topological psychology and vector psychology, or the combination of dynamic research and holistic research, that is, to study human behavior and psychological activities within the life space or dynamic field, has formed the psychological field theory of Levin.

6.4.2.2 Levin's dynamic model

Extending the dynamic concept of Levin's psychological field theory, it is expressed as a basic dynamic mode. Levin calls it the "quasi-homeostasis equilibrium process."

Just as Levin's dynamic concept and experimental research have caused revolution in dynamic psychology, his dynamic model also provides a new method for the development of dynamic psychology.

The concept of *homeostasis* or steady state was introduced by the physiologist Walter B. Cannon in 1929. In 1932, in his study "The Wisdom of the Body" [10], he made a more specific exposition of the significance of *homeostasis*. Cannon believe that "the word (*homeostasis*) does not imply something set and immobile, a stagnation. It means a condition – a condition which may vary, but which is relatively constant."

The imbalance of homeostasis will result in the tension of organisms and urge organisms to acquire new balance through adaptive behavior. Lewin's psychological tension system follows this dynamic model of homeostasis theory.

In describing his system of psychological tension, Levin pointed out that only when the need breaks the original psychological balance, will it cause internal tension. The behavior motivated by this tension is not to avoid hardships but to gain a new internal balance.

Levin said, psychological processes usually came from a tendency to balance, just like universal biological processes, as well as physical, economic or other processes. From a steady state to a process, the changes occur in the process of stability. It could be derived from the fact that at a certain point, the balance was broken, and the process of moving toward a new equilibrium state begins.

Levin sees balance as a dynamic concept and sees tension as a process of quasi-homeostasis equilibrium. Levin's dynamic theory not only follows the basic model of the homeostasis theory but also emphasizes the characteristics of organic theory such as the holistic structure and system.

6.4.2.3 Summary of Lewin's field theory

In his book R. J. Rummel summarized Lewin's field theory [11]:

> *In summary, what is Lewin's field theory? First, it is an emphasis on a person's subjective perspective. Second, it incorporates the whole that is subjectively relevant to a person and that organizes behavior, goals, needs desires, intentions, tensions, forces, and cognitive processes into one system. Third, the elements composing this whole are interdependent and stand in dynamic mutual relationship. Fourth, the key to the dynamic nature of this subjective whole is the idea of tension (energy) systems created by needs and discharged by achieving associated goals. Fifth, the dynamic psychological construct is that of inner-personal forces, which result from the intensity of personal needs and the valence of associated goals. Sixth, blocked goals can lead to increase in tension and a variety of behavioral and psychological consequences. And finally, inner-personal conflict is the result of opposing psychological forces.*

The following is the individual behavior mathematical model in the social field theory.

6.4.3 Behavioral mathematical model of social field theory

6.4.3.1 Model formula

The method of statistical physics has achieved rich results in the field of physics. In the last 20 years, statistical physics has become increasingly important in interdisciplinary subjects. For example, experts in the field of physics recently used statistical physics to derive mathematical model formulas for individual behavior in the social field.

The model they used was in the form of Boltzmann equations that were derived from the fields of kinetics of gases and chemical reactions.

Boltzmann-like equations also play an important role in the quantitative model of sociology, such as the logistic equation describing the finite growth process, the so-called gravity model used in the application the spatial exchange processes, as well as the game dynamic equation describing the competition and cooperation process. These are all special cases of the Boltzmann equation [4].

In addition, recent scholars have analyzed the pair interactions and derived that the Boltzmann equation can be used to describe pedestrian conflict avoidance behavior and attitude formation process [4].

The following is a brief introduction to the mathematical model formula for individual behavior under the influence of social fields. This formula is called the Boltzmann–Fokker–Planck model.

Helbing *et al.* analogize the behavior of subjects under the influence of social fields with the physical model of the behavior of charged particles (such as protons) in electromagnetic fields [12,13].

The mathematical model formula of individual behavior in the social field is as follows:

$$\frac{dx}{dt} \approx F_\alpha(x,t) + \xi \tag{6.53}$$

where

$$F_\alpha(x,t) := \begin{pmatrix} F_{\alpha 1}(x,t) \\ \vdots \\ F_{\alpha n}(x,t) \end{pmatrix} \tag{6.54}$$

$$F_\alpha(x,t) = -\nabla V_\alpha(x,t) \tag{6.55}$$

that is

$$F_{\alpha i}(x,t) = -\frac{\partial}{\partial x_i} V_\alpha(x,t) \tag{6.56}$$

In (6.56), dx/dt represents the change rate of individual's behavior x, and ξ represents a random interference term indicating the difference in behavior between individuals; vector $F_\alpha(x,t)$ is the individual behavioral dynamics that drives the individual α's behavior $x(t)$ to undergo a systemic change. At the same time, $F_\alpha(x,t)$ is the force of the social field to the individual α. The vector $V_\alpha(x,t)$ can be understood as the potential energy function of the social field, and the dynamic force $F_\alpha(x,t)$ generated by the social field to the individual α can be obtained by finding the partial derivative (i.e., the gradient ∇ of the social field potential energy).

Obviously, the force $F_\alpha(x,t)$ must be representative of the impact of the environment on the behaviors of individual α.

However, the force of the social field to the individual is not a physical effect on the individual but an amount describing the specific *motivation* of the action. In the case of pedestrians walking, this motivation causes the individual's physical movements, such as acceleration and deceleration, as pedestrian's behavioral reactions to the perceived information about their surroundings.

Scholars apply the previous social field motion model to pedestrian walking behavior, thinking of the impact of other pedestrians and obstacles on pedestrians as a social field. The model has achieved good effects on the description and simulation of pedestrian behavior and interaction behavior [13].

6.4.3.2 Discussion on the dynamics

In the previous model, the dynamics for behavior change is the Social Force.

Obviously, the Social Force does not follow Newton's law of physical mechanical kinematics. On the contrary, the Social Force $F_\alpha(x,t)$ is a vector with the following properties.

$F_\alpha(x,t)$ is driving force that causes another vector of the individual α—the behavior $x(t)$—to change dx/dt over time.

Ignoring the influence of random interference, if $F_\alpha(x,t)$ disappears, the individual's behavior $x(t)$ will not change, i.e., $F_\alpha(x,t)=0$, corresponding to an extreme value of social field $V_a(x,t)$, because this indicates

$$\nabla V_a(x,t)=0 \tag{6.57}$$

Here, we can compare and analyze the results and Levin's social field theory. From the social psychology introduced earlier, the individual's behavior is determined by the influence of the environment and the individual personality.

Inspired by the theory of electromagnetic fields, social field theory believes that the impact of the environment on individuals can be regarded as a dynamic field, thereby it can be described in mathematical form. Temporary changes in the social field will cause the individual's mental tension, causing an individual's behavior to release this tension.

It is assumed that the individual's goal is to behave in an optimal way according to the social field intensity $V_a(x,t)$, that is, the individual always tends to act in a manner where the social field intensity is minimal.

If the individual's behavior $x(t)$ is inconsistent with the minimum value of the social field intensity $V_a(x,t)$, this creates a force:

$$F_a(x,t) = -\nabla V_a(x,t) \tag{6.58}$$

The magnitude and direction of the force is determined by the gradient of the social field intensity $V_a(x,t)$, and the force points to the direction in which the social field intensity decreases the fastest. The power $F_a(x,t)$ acts as a mental tension, and this mental tension causes a change in behavior:

$$\frac{dx}{dt} \approx F_a(x,t) \tag{6.59}$$

And the change of behavior dx/dt promotes the individual's behavior $x(t)$ close to the social field intensity $V_a(x,t)$'s minimum value x_a, when the individual behavior $x(t)$ reaches the social field intensity minimum value x_a, there are

$$\nabla V_a(x,t)=0$$

In this way, the dynamic force $F_a(x,t) = 0$, indicating that the individual releases mental stress (mental tension) through previous behavior changes. When the mental tension $F_a(x,t) = 0$ is eliminated, the individual's behavior will not change except for the effects of random disturbances, that is, the individual reaches a balanced state in the social field.

It is noted that for different types of individuals α, the social field $V_a(x,t)$ has a different minimum value x_a, which indicates that different individuals have different mental tensions for the same social field. In other words, α represents the personality traits of different individuals.

6.5 Summary

In this chapter, we first analyze the human behavior analysis patterns and research results in various related disciplines. On this basis, we construct a two-layer analysis framework suitable for the cyclist's microscopic behaviors at mixed traffic flow un-signalized intersections.

Then, we introduce the structure of the framework and determine the supporting theories for each layer of the framework.

Finally, we introduce the supporting theory of the analysis framework—the expected utility theory of decision analysis for the tactical-level model, and the Social Force model for the operational-level model.

References

[1] M. Ben-Akiva and S. Lerman. Discrete Choice Analysis: Theory and Application to Travel Demand. Cambridge, MA, The MIT Press, 1985.

[2] S. Hoogendoorn and P. Bovy. Pedestrian Route-Choice and Activity Scheduling Theory and Models. Transportation Research Part B, 2004, 38:169–190.

[3] S. Wolfram. Statistical Mechanics of Cellular Automata. Reviews of Modern Physics, 1983, 55(3): 601–644.

[4] D. Helbing and M. Treiber. Gas-Kinetic-Based Traffic Model Explaining Observed Hysteretic Phase Transition. Physical Review Letters, 1998, 81:3042–3045.

[5] Z.H. Ma, Q.L. Liu, X. Lu *et al.* (Eds.). Editorial Board of Modern Applied Mathematics Handbook, Operational Research and Optimization Theory Volume. Beijing, Tsinghua University Press, 1998.

[6] Y. Zhang and H. Chen (Eds.). Utility Function and Optimization. Beijing, Science Press, 2000.

[7] V.J. Nordby and C.S. Hall. A Guide to Psychologists and Their Concepts. San Francisco, CA, W. H. Freeman, 1974.

[8] K. Lewin. Field Theory in Social Science, New York, Harper, 1951.

[9] R.F. Baumeister and E.J. Finkel. Advanced Social Psychology: The State of the Science. In Advanced Social Psychology: The State of the Science. New York, Oxford University Press, 2010.
[10] W.B. Cannon. The Wisdom of The Body. Revised and Enlarged Edition. The Journal of Nervous and Mental Disease, 1940, 92(2):262.
[11] R.J. Rummel. Chapter 3 Psychological Field Theories. In Understanding Conflict and War, Vol. 1: The Dynamic Psychological Field. Beverly Hills, CA, Sage Publications, 1975.
[12] D. Helbing. Interrelations Between Stochastic Equations for Systems With Pair Interactions. Physica A, 1992, 181:29–52.
[13] D. Helbing, L. Buzna, A. Johansson, and T. Werner. Self-Organized Pedestrian Crowd Dynamics: Experiments, Simulations, and Design Solutions. Transportation Science, 2005, 39(1):1–24.

Chapter 7
Microscopic behavior model of bicycle crossing un-signalized intersections

Based on the framework of the cyclist's microscopic behavior analysis at un-signalized intersections in Chapter 6, in this chapter, we specifically model the microscopic behavior model of bicycle crossing un-signalized intersections.

In Section 7.1, we propose a normative cyclist behavior (NCB) theory and model and introduce the basic premise assumptions of the model, then we introduce the overall framework of the theoretical model, expound and discuss the basic concepts of the theory, and summarize its innovation points and characteristics.

In Section 7.2, we model the tactical and operational-level behavior modules theoretically. We specifically model the general formula of the theoretical model for the microscopic behavior characteristics of bicycles at un-signalized intersections.

7.1 Normative cyclist behavior theory and model

Based on the microscopic cyclist's behavior characteristics and the microscopic behavior analysis framework of bicycle crossing un-signalized intersections, we proposed the normative cyclist behavior theory and model (short for NCB model). The theoretical model is introduced systematically. The first are the basic theoretical assumptions for NCB model.

7.1.1 Theoretical assumptions

The assumptions of the NCB theory mainly include the following two aspects:

Assumption 7.1. *The cyclist's microscopic behaviors are a comprehensive behavior, including different levels of complexity and different types of behaviors.*

According to the analysis of the cyclist's microscopic behavior patterns in Chapter 5, we found that cyclist's travel behaviors (including crossing the un-signalized intersections) are comprehensive behaviors.

This cyclist's behaviors include different levels of complexity and different types of behaviors. They include long-term and medium decision-making behaviors (strategic level), short-term choice behaviors (tactical levels) and subconscious action response behaviors (operational levels).

From the degree of complexity, cyclist's behaviors include complex behaviors involving complex mental and brain intelligence activities such as information collection, information processing, subjective prediction and evaluation, judgment, trade-off, selection and simple reaction behavior involving only neural reflex.

The NCB theory fully recognizes that different levels of complexity and types of behavior have different characteristics and modeling methods, and these various levels of behaviors are hierarchical.

That is to say, there is an inherent interrelationship between the different levels of cyclist's behaviors.

Generally speaking, the higher level behavior module involves a wider range, and the lower level behavior module involves a relatively narrow scope.

The inputs for high-level behavior are less specific and less accurate, and the corresponding output is also more macroscopic.

The inputs for low-level behavior module require more specificity and accuracy, and the output is more specific and accurate correspondingly.

And the output of the high-level behavior module is often the input to the next-level behavior module.

The cyclist's behaviors hierarchy is described in detail in Section 7.1.2.

Assumption 7.2. *All the behavior of the cyclist, whether it is performing an activity or riding along a certain route, will have an effect on the cyclist, or a corresponding cost. The behavior of the cyclist will be determined on the basis of the outcome of the prediction and optimization of this utility or cost, that is, the behavior of the cyclist is determined by maximizing profit or minimizing costs.*

The core idea of Assumption 7.2 is to maximize subjective utilities, and its supporting theory is *subjective expectation utility theory*, also known as *normative rational decision theory*.

The advantages and disadvantages of this theory have been discussed in detail in Chapter 5 and will not be repeated here.

Although the *normative rational decision theory* cannot completely reflect people's choice behavior in daily life, it does provide a very practical theoretical model framework for human decision-making behavior modeling [1].

In addition, field data from some recent studies indicate that modeling methods based on utility maximum theory are quite feasible for pedestrian route choice behavior [2,3].

Since the cyclist's behaviors have many similarities with those of pedestrians, the authors believe that the theory is more feasible in the application of bicycle complex choice behavior.

The *subjective expectation utility theory* is applicable as a theoretical basis for the complex behavior modeling of cyclists. But does it also apply to the modeling of cyclists' low-level behaviors (such as riding behavior, conflict avoidance behavior, etc.)? The answer is yes.

Human acquisition behavior is essentially an act based on utility maximization rules.

In his article on the modeling of pedestrian behavior, Helbing pointed out that the behavior of pedestrians is also based on the utility maximization: pedestrians always move along the most convenient route when they arrive at a destination, trying to minimize their delays and avoid obstacles, tend to choose the best route and save energy consumption [4].

The acquisition of these behaviors is a subconscious learning process in which pedestrians use the principle of utility maximization and automatically obtain the best walking behavior rules after a large number of walking experiences through trial and error.

Explain in psychological terms, that is, through continuous stimuli, the best reaction mode is solidified into a certain neural circuit. When the corresponding situation or stimuli is reproduced, the neural circuit is excited, and the person will naturally make the corresponding action without the participation of subjective consciousness. Therefore, the authors believe that the idea of utility maximization is also applicable as a modeling support theory for simple reactive behavior.

In summary, the normative rational decision theory of utility maximization can be used as the basic support theory of cyclist's microscopic travel behavior, which is also the origin of the name of NCB theory.

7.1.2 Theoretical framework of NCB model

7.1.2.1 The hierarchical cyclist's travel behavior

Scholars have previously realized that people's travel behavior is a multilevel integrated behavior.

For example, Ben-Akiva and Lerman considered people's travel decision-making behavior as a multilevel process in their textbook on traffic demand modeling in 1985.

They divide people's decision-making process on travel and other activities related to travel into three levels: long- (strategic level), medium- and short-term [5]. In addition, Hoogendoor also divided pedestrian traffic behavior into three different levels in his research [6].

They put people in activities related to travel and other travel-related activities. The decision-making process is divided into three levels: long- (strategic level), medium- and short-term [5]. In addition, Hoogendoor [6] also divided pedestrian traffic behavior into three different levels in his research.

Based on the analysis of cyclist's microscopic travel behavior, the *NCB theory* proposed in this book divides cyclist's travel behavior into the following three levels:

1. *Strategic-level behavior*: It mainly includes the cyclist's departure time, travel planning activities and route choice behavior at the road network level.
2. *Tactical-level behavior*: It mainly includes the activities and path planning behavior of cyclists on two-dimensional space in different transportation facilities (including road sections, intersections, squares and stations).
3. *Operational-level behavior*: It refers to the cyclist's specific riding behavior along the planned path from the tactical-level module, including collision avoidance behavior, reaction to the signal light and so on.

Through the previous hierarchical division of the cyclist's microscopic travel behaviors, the cyclist's microscopic traffic behaviors of all levels and types are included in the NCB theoretical model framework: from the choice of travel activities to the specific riding behavior along the planned path.

The authors consider that the choice of traffic mode is a kind of macroscopic travel behavior and does not fall within the scope of the theoretical framework.

7.1.2.2 NCB theoretical model framework

The framework structure of the NCB theoretical model is shown in Figure 7.1.

7.1.2.3 Basic components of the NCB model

The dynamic transport system S

The NCB theoretical model framework is a dynamic traffic system S in a two-dimensional plane space. The system S is mainly composed of the following elements:

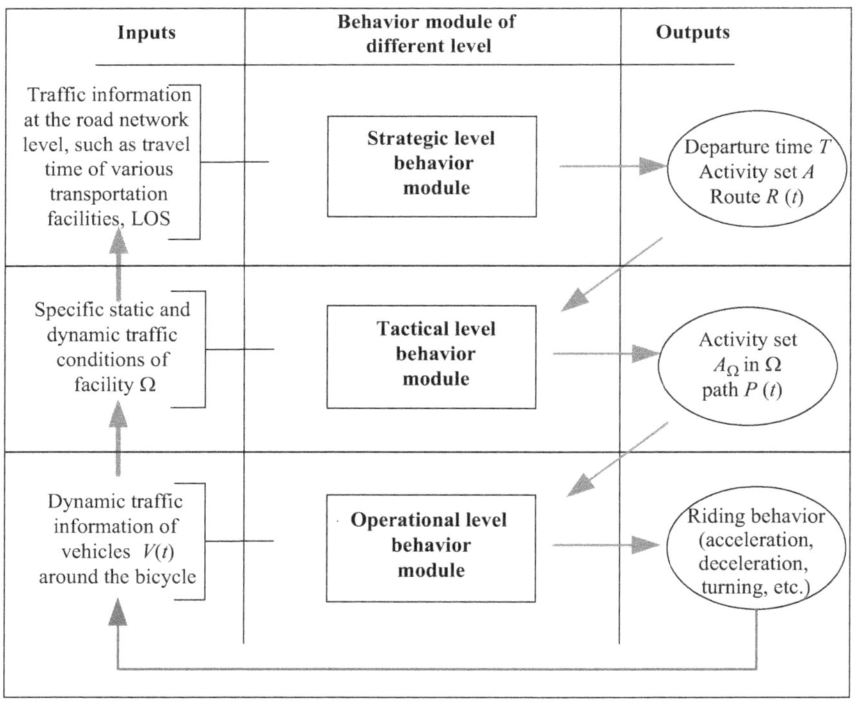

Figure 7.1 Schematic diagram of NCB model framework. Note: The solid line indicates the input of the information and the resulting result, and the dashed arrow indicates the transfer and feedback loop of the influence

Road network N and transportation facility Ω: Let the road network N be the overall transportation facilities of the dynamic transport system S, that is, N is the aggregate of the specific transportation facilities Ω in S:

$$N = \bigcup_i \Omega_i \tag{7.1}$$

Ω can be all kinds of transportation facilities in the road network, including road sections, intersections (including signalized and un-signalized intersections, plane intersections and interchanges), stations, squares, as well as traffic management facilities attached to various transportation infrastructures such as safe islands, crosswalks and pedestrian bridges.

The information of Ω contains all the information related to transportation facilities Ω, such as coordinate system, facility scale, passing capacity, plane slope, management control mode (speed limit of road section, signal control scheme of signalized intersection, etc.).

The vehicles V: The vehicles V representatives of all moving and stationary vehicles in S, including the following three categories:

B: bicycles in the road network N;

C: motor cars in the road network N;

P: pedestrians in the road network N.

The information of V mainly includes two categories, one is information V^s describing the static characteristics of the individual vehicle, such as the size of the vehicle, the driver (cyclist, pedestrian) information (age, gender, travel purpose, etc.); the other information V^d describing the state of motion of the vehicle, mainly represented by the specific position vector $\vec{x}(t)$ of each vehicle.

The position vector $\vec{x}(t)$ is a two-dimensional vector representing the X, Y coordinates of the vehicle at different times. By deriving the position vector $\vec{x}(t)$, the speed and acceleration information of the vehicle can be obtained, namely,

$$\vec{v}(t) = \frac{d\vec{x}(t)}{dt} = \vec{x}\,'(t) \tag{7.2}$$

$$\vec{a}(t) = \frac{d\vec{v}(t)}{dt} = \frac{d\vec{x}\,'(t)}{dt} = \vec{x}\,''(t) \tag{7.3}$$

Main variable of the theoretical model

Activity: Combining the activities of the cyclist with the traffic behavior to consider modeling is a major innovation and advantage of the theory.

Transportation is an additional requirement, that is, people usually do not travel for travel but travel for a certain activity. In this way, people often have more than one destination and activity when they travel.

For example, if a person goes home from work and rides a bicycle, he may not go home directly but go to the bank; then go to the kindergarten to pick up his/her children, then go to the supermarket for shopping, and finally go home.

In this way, his trip is a *trip chain*—consisting of four head-to-tail sub-trips: from working place to bank→bank to kindergarten→kindergarten to supermarket→supermarket to home, and the corresponding activities set $A = \{$financial affairs; pick up children; shopping$\}$.

The choice of departure time t_O and trip plan activity A and route R is often considered together, and any change in one variable will affect the decision of other variables.

For example, in the previous example, if the cyclist is late for work, he/she may give up the activities of going to the bank and go directly to the kindergarten to pick up the child. Then his/her plan activity set A changes because of the late departure time t_O, so the travel route R also changes accordingly.

It can be seen that the framework structure that combines people's activities and traffic behaviors can reflect people's travel behavior more flexibly and realistically. Therefore, *activity* is an important basic concept of the NCB theoretical model.

In the NCB theoretical model, activity is a collection of behaviors, a vague and broad concept. It refers to all the actions directly carried out by the rider (traveler) during the trip.

It is noted that the activities here are not direct travel behaviors such as route selection, riding, walking, but acts that directly affect these traffic behaviors, such as shopping, doing things. Corresponding to bicycle travel behavior, the activities of the cyclist are also divided into the following three levels and types:

Planned activity (PA): Trip plan activity corresponds to the activities set A developed at the strategic level.

The so-called *plan activity* refers to the activities that the rider plans to carry out during the trip before or during the trip, such as shopping and work. The components of a strategic-level trip *plan activity* set A are all activities of this type.

Expected temporary activity (ETA): The so-called *expected temporary activity* means that the activity is not the activity planned by the agent before the trip, but the specific dynamic traffic information at the time of the traffic facility Ω makes the behavior subject expected temporary events (also known as activities) that can occur.

For example, at an un-signalized intersection, the cyclist expects that the vehicle c will collide with himself/herself at some place and that collision is an *temporary activity* (or event) a that the subject expects to occur on Ω due to dynamic traffic flow information. Event a is considered as an activity when he/she is planning the path.

The tactical-level activity set A_Ω includes not only the *expected temporary activities* (ETA) but also the travel planned activities (PA) described earlier. Therefore, the activity set A of the strategic-level behavior module is a set indicating the expected activities of the traveler before the trip, and the activity set A_Ω at the tactical level is a set of activities specific to a certain traffic facility Ω.

A_Ω includes not only the expected activities A planned by the traveler before the trip, such as buying tickets for a station a_1 and buying newspaper a_2, but also the actual dynamic traffic flow on the traffic facilities at the time that the traveler did not anticipate. Incidents, such as avoiding a potential conflict a_3, bypassing obstacles (or traffic accidents) a_4, and even encountering acquaintances to go to say hello, are called *expected activities*.

So $A_{station} = \{a_1, a_2, a_3, a_4\}$, where a_1 and a_2 are expected activities, also belong to the activity set A of the strategic-level module. a_3 and a_4 belong only to the active set A_Ω of the tactical-level module and not to the activity set A of the strategic-level module.

In addition, NCB theory sets the strategic level of trip *planning activity* set A not to be an empty set (at least one activity a_d at the trip destination), while the tactical-level activity set A_Ω can be an empty set. That is, the cyclist can have no planned activities on a certain traffic facility. At the same time, due to the small traffic flow on the traffic facilities at the time, the dynamic traffic flow did not produce any *expected temporary activities*.

Unexpected temporary activity (UTA): *Unexpected temporary activity* is a specific type of activity at the operational level.

Similar to the *expected temporary activities*, this type of activity was also temporarily generated by the specific dynamic traffic conditions on the transportation facilities at that time. The occurrence of such activities is "unexpected." For example, the behavior of other vehicles is inconsistent with the expectations of the cyclist when planning path P at the tactical level, so the resulting activity is unexpected.

Such activities are sudden and occur in the consciousness of the subject (cyclist), so they cannot be incorporated into the tactical level for complex thinking and response. Moreover, the occurrence of such situations is often sudden, and the cyclist can only determine the activity or behavior to be dealt with through instinct or nerve reactions.

These characteristics of *unintended temporary activity* (UTA) make it impossible to describe it specifically, and thus we use the dynamic information $V(t)$ of the vehicle around the cyclist on the traffic facility Ω to be uniformly represented.

Note that the *activities* defined here and the traffic *incident* are two closely related yet different concepts. Usually, an *activity* is caused by a specific dynamic traffic *incident* on the transport facility Ω.

The concept of *activity* in NCB theory emphasizes the actions of the actors, while the *incident* emphasizes an objective situation. Following, we use a case to illustrate. For example, when bicycle i crosses an un-signalized intersection, he/she found (or expected) that a vehicle c traveling laterally will collide with itself at time t_1 and area A_1 (also understood as a collision), as shown in Figure 7.2.

Then, the expected traffic *incident* triggers the cyclist's *expected temporary activity* of a_1={avoiding a potentially conflicting car c}.

It is expected that the area where the conflict traffic incident may occur is A_1, but the place where the evasion activity A_1 may occur is at any location within the physical limit speed of the bicycle within the traffic facility Ω.

Route, path and riding behavior: Route, *path* and *riding behavior* are also basic variables of the NCB theoretical model.

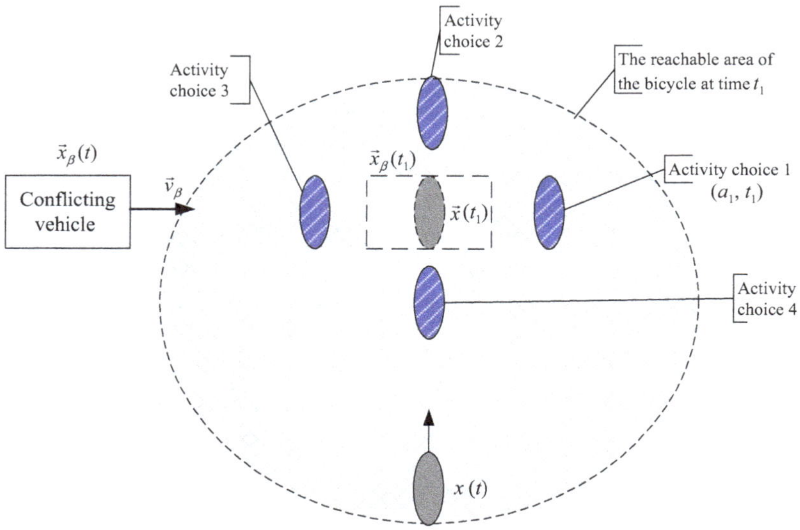

Figure 7.2 Illustration of expected temporary activities

Route: A *route* is an ordered set of links from the origin to the destination of the cyclist before or during the trip, and the corresponding junctions that join the links together.

Therefore, *route R* is an ordered collection of specific traffic infrastructures Ω in *S*, usually expressed as

$$R(t) = \{\Omega_{L1}, \Omega_{J1}, \ldots, \Omega_{Lm}, \Omega_{Jm}\}, \;\; \Omega_{Li}, \Omega_{Ji} \in N, i = 1, 2, \ldots, m \tag{7.4}$$

where Ω_L represents the links in the road network N, and Ω_J represents the junctions in the road network N. Considering the dynamic path selection behavior of the bicycle, that is, R is not a static set but will change with time t, so it is considered as a function of time t, expressed as $R(t)$.

Path: The *path* refers to the specific trajectory line planned by the cyclist in a certain traffic facility Ω. In theory, it should be a dynamic curve equation based on the coordinate system of the traffic facility Ω.

Taking into account the basic characteristics of the cyclist's traffic behavior, the *path* is simplified here.

Since cyclists plan the *path* during the dynamic process of cycling, the path planning process is often very short (usually less than a few seconds). Due to the short time, the surrounding traffic dynamic information collected by the cyclist has great ambiguity. Therefore, we believe that the path planned by the cyclist at this time is only a rough plan, a polygon formed by a number of "temporal destinations" (TDs). Cyclists may have an idea of when they arrive at the "TDs," but how to reach these TDs, that is, the specific riding behaviors are given to the subconscious, the next level of behavioral modules.

Therefore, $Path(t)$ is an ordered set of specific plane points D_j within the traffic facility Ω_i and the time t_j at which it arrives, it is usually expressed as

$$Path(t) = \{(D_j, t_j)\}, \quad D_j \in \Omega_i, \qquad j = 1, 2, \ldots, k, \quad t_j \in (t_o, t_d] \tag{7.5}$$

where D_j represents the positional coordinates of the TD and time t_j represents the time the cyclist is expected to reach the TD, which should within the range the time the cyclist enters and leaves the transportation infrastructure, represented by t_o and t_d. Similarly, recognizing that the path made by the cyclist is not a static variable, but a dynamic variable that changes over time t, it is expressed as $Path(t)$.

Riding behavior: Riding behavior refers to the specific dynamic movements of a cyclist riding along a planned *path*. Therefore, it is the most basic behavior, and its behavior pattern is the *Stimulus→Reaction* mode.

The corresponding stimulus is the UTA defined earlier. Depending on the type of situation or UTA, the cyclist often has a corresponding fixed behavior response rule.

The representation of riding behavior is mainly expressed by the dynamic behavior and trajectory of the bicycle. That is, the bicycle's acceleration vector $\vec{a}(T)$, or the velocity vector $\vec{v}(T)$, or the trajectory vector $\vec{x}(T)$ in a very short future time T (may be the next few seconds, or the next second, or even half a second, etc.).

In two-dimensional space, acceleration $\vec{a}(T)$, velocity $\vec{v}(T)$ and position $\vec{x}(T)$ are two-dimensional vectors that can be characterized by the coordinate system $(x, y)_\Omega$ of the traffic infrastructure Ω in which they are located:

$$\vec{a}(T) = [a_x(T), a_y(T)], a_x(T), a_y(T) \in [a_{\min}, a_{\max}] \tag{7.6}$$

$$\vec{v}(T) = [v_x(T), v_y(T)], v_x(T), v_y(T) \in [v_{\min}, v_{\max}] \tag{7.7}$$

$$\vec{x}(T) = [x(T), y(T)], x(T), y(T) \in \Omega \tag{7.8}$$

where $a_x(T)$ and $a_y(T)$ are the components of the acceleration vector $\vec{a}(T)$ in the x and y directions of the coordinate system $(x, y)_\Omega$, respectively. $a_{\min}$ and $a_{\max}$ are the upper and lower limits of the bicycle acceleration, respectively.

$v_x(T)$ and $v_y(T)$ are the components of the velocity vector $\vec{v}(T)$ in the x and y directions of the coordinate system $(x, y)_\Omega$, respectively. $v_{\min}$ and $v_{\max}$ are the upper and lower limits of the bicycle velocity, respectively. $x(T)$ and $y(T)$ are the coordinate values of the x and y directions of the bicycle position.

7.1.2.4 Basic process of the NCB model framework

The basic process of the NCB theoretical model framework is as follows:

1. First, the input information required by the *strategic-level behavior module* is the traffic information at the road network level, such as the travel time of each transportation facility, the level of service (LOS) and the like. The output of this level module is the departure time T_O, the trip plan activity set A and the *route* $R(t)$.

2. The outputs of the *strategic level*—the departure time T_O and the *route* $R(t)$—are the inputs to the *tactical-level behavior module*.

   The $R(t)$ determines the specific traffic facilities Ω_i corresponding to the *tactical-level behavior module*, and all relevant information about the traffic infrastructure included in Ω_i, such as the coordinate system $(x, y)_\Omega$ of the traffic infrastructure Ω_i, the rider at Ω_i final destination D_f, speed limits $V_i^{\min}$ and $V_i^{\max}$, etc. The *tactical-level behavior module* also requires the dynamic and static information V^d and V^s of the vehicle V in Ω_i.
3. Then, the *tactical-level behavior module* judges the previous various types of information and outputs the cyclist's initial *planned activity* A_Ω and the planned path $P(t)$ in Ω_i. $P(t)$ contains the rough location of the "TD" of the path and the corresponding approximate time to reach the TDs.
4. The output of the *tactical-level behavior module—the planned path* $P(t)$—is used as the input of the operation-level behavior module. The cyclist will ride to the nearest TD D_i and make a corresponding response $\vec{a}_i(T)$ (or $\vec{v}_i(T)$, or $\vec{x}_i(T)$) according to the actual surrounding vehicle dynamic traffic conditions $V^d(t)$.
5. At certain time interval ΔT, the *tactical-level behavior module* judges and updates the activity plan $A_\Omega(t + \Delta T)$ and the planned *path* $P(t + \Delta T)$ in the traffic facility according to the new dynamic information $V^d(t + \Delta T)$ in the traffic facility Ω_i. The updated $P(t + \Delta T)$ is inputted to the operational-level behavior module, which reflects the cyclist's dynamic path planning behavior when riding.
6. At a certain time interval ΔT_s, the strategic-level behavior module will judge and formulate a new travel plan activity set $A(t + \Delta T_s)$ and route $R(t + \Delta T_s)$ according to the dynamic information of each traffic infrastructure Ω at the road network level, such as travel time T_Ω, service level LOS_Ω. Then the updated output is inputted to the tactical-level behavior module. This step reflects the cyclist's dynamic route planning behavior during the trip.

It can be seen from the previous process that the NCB theory model framework not only supports the dynamic path planning behavior of bicycles in tactical level travel behavior but also supports the dynamic route choice behavior of strategic-level travel behavior.

7.1.2.5 Outputs of each level of behavior module

Strategic-level behavior module

The set of trip activities output by the strategic-level behavior module is an ordered collection of activities performed by the cyclist on the trip, indicating the planned activities (PA) and the order of activities of the trip, usually expressed as

$$A(t) = \{a_0, \ldots, a_d\}, a_i \in \Sigma, i = 0, 1, \ldots, k \tag{7.9}$$

where Σ is the overall space of the activity. a_d is the activity of the cyclist at the travel destination. Since the cyclist may only have one activity a_d in one trip, let a_0 indicate that there is no activity, that is, $a_0 = Null$. And the number of activities planned by the rider in one trip is $k + 1$. Since the determination of the activity set changes over time, the activity set is considered to be a function of time A, expressed as $A(t)$.

Another output path $R(t)$ of the strategic-level behavior module is an ordered set Ω of specific traffic infrastructures in S, usually expressed as

$$R(t) = \{\Omega_{L1}, \Omega_{J1}, \ldots, \Omega_{Lm}, \Omega_{Jm}\}, \Omega_{Li}, \Omega_{Ji} \in N, i = 1, 2, \ldots, m \tag{7.10}$$

where Ω_L is the links in the road network N and Ω_J is the junction in the road network N. Considering the dynamic route choice behavior of the cyclist, that is, R is not a static set, but will change with time t, so R is considered to be a function of time t, expressed as $R(t)$.

Tactical-level behavior module

The tactical level of output *activity set* A_Ω includes strategic-level travel planned activities (PA) and tactical-level expected temporary activities (ETAs).

The cyclist judges, chooses and sorts the two types of activities, forming the final activity set A_{Ω_i}, which contains not only the activity a_j but also the order of the a_j occurrence and the approximate location (or area) P_{aj} and time T_{aj}. Therefore, A_{Ω_i} is an ordered collection of events and event occurrences and event end time pairs.

Among them, Σ_Ω is the overall space of Ω activities on the transportation facilities Ω. The activity plan made by the rider A_{Ω_i} may not be a static variable, but a dynamic variable that changes with time t and is therefore expressed as $A_{\Omega_i}(t)$.

Note that the position D_l of the TD in path $P(t)$ is usually coincident with the area P_{aj} of an active a_j in A_{Ω_i}. In fact, the development of the TDs in the path $P(t)$ is closely related to the planned activity A_{Ω_i}. The location of the activity a_j causes the cyclist to pay attention to it and is included in the path planning behavior of the tactical level as a TD. Otherwise, it falls into the simple reaction behavior of the operational level.

Therefore, in the NCB theoretical model, it is considered that each TD in the tactical level path $P(t)$ corresponds to an active a_j (or vice versa). And the coordinate position D_j of the TD is the position (or area) of the corresponding active a_j, that is

$$D_j = P_{aj}, D_j, P_{aj} \in \Omega_i, \quad j = 0, 1, \ldots, k \tag{7.11}$$

The corresponding activity end time T_{aj} has a consistent relationship with the arrival time T_j of D_j of the path $P(t)$. That is

$$T_j = T_{aj}, T_j, T_{aj} \in (T_o, T_d], \quad j = 0, 1, \ldots, k \tag{7.12}$$

In the following theoretical model explanation, it is considered that the time–space consistency relationship between the TDs of the path $P(t)$ and the $A_{\Omega_i}(t)$ is established, that is, (7.11) and (7.12) are established unless otherwise specified.

Operational-level behavior module

The input of the operational-level behavior module includes the output of the tactical-level behavior module path $P(t)$ and the dynamic and static information

$V^d(t)$ and V^s of the vehicle V in the traffic facility Ω_i. $V^d(t)$ represents the dynamic information of all vehicles V in the traffic facility Ω_i, which is a time variable, and V^s represents static information, such as the size and type of the vehicle (including motor vehicle, bicycle and pedestrians), and the characteristics of the driver.

The output of the operational-level behavior module is the dynamic motion behavior of the target bicycle i at a certain moment T in the future. The variable can use acceleration $\vec{a}_i(T)$, or variables such as speed $\vec{v}_i(T)$ and position $\vec{x}_i(T)$ to specifically represent the dynamic information of the bicycle.

In addition, $V_i(T)$ is used to generalize the dynamic information of vehicle i in some places in the theoretical model framework.

It should be noted that *route* and *path* are two different concepts.

A route is an ordered collection of traffic facilities that a rider passes between an origin and a destination and belongs to a category at the *road network* level.

The path refers to the planning of a specific riding trajectory of a cyclist in a two-dimensional plane in a traffic facility, which is an ordered collection of plane coordinate points within the traffic facility and the planned time of arrival at that point. The path belongs to a category of a specific transportation facility.

7.1.3 Features of the NCB theoretical framework

In summary, the NCB theoretical model framework has the following main features.

7.1.3.1 Hierarchical behavioral framework

The hierarchical travel behavior assumption is one of the premise hypotheses of the NCB theory: *the cyclist's microscopic traffic behavior combines various levels, different levels of complexity and different types of behaviors.*

Bicycle travels include the long- /medium-term decision-making behaviors (strategic level), short-term choice behaviors (tactical level) and subconscious action response behaviors (operational level).

From the degree of complexity, bicycle behavior includes complex behaviors involving complex mental and brain intelligence activities such as information collection and processing, subjective evaluation, judgment, trade-off conflict, choice and simple reaction behavior involving only neural reflex.

The theoretical framework fully recognizes the characteristics and the interrelationships between different levels of behaviors.

Generally speaking, the high-level behavior module involves a wide range, and the low-level behavior module involves a relatively narrow scope. The level of input and accuracy required for high-level behavior is lower, and the corresponding output is more macroscopic. The low-level behavior module requires a relatively high degree of specificity and accuracy, and the output is correspondingly more specific and accurate. The output of the high-level behavior module is usually the input of the next-level behavior module, and the output of the low-level behavior module also affects the upper-level behavior module.

7.1.3.2 Utility maximization

One of the assumptions of the NCB theory is that all the behavior of the cyclist, whether it is performing an activity or riding along a certain path, will have a utility on the cyclist, or a corresponding cost. The behavior of the cyclist will be based on predicting and optimizing this utility or cost (i.e., maximizing utility and minimizing cost).

Previous studies have proved that the normative rational decision theory based on subjective utility maximization theory provides a very practical theoretical model framework for human decision-making behavior modeling.

In addition, Helbing and Treiber also discussed low-level simple behaviors, such as pedestrian walking and conflict avoidance behavior, the essence of which is also a result of the law of maximizing utilities: pedestrians always move along the most convenient route when they arrive at a destination, trying to minimize their delays and avoid obstacles, tend to choose the best route and save physical exertion [4].

Therefore, the idea of maximizing benefits can also be used as a modeling support theory for simple reaction behavior.

In summary, the normative rational decision theory of utility maximization can be used as the basic support theory of cyclist microscopic traffic behavior, which is also the origin of the theoretical name—the NCB.

7.1.3.3 Combine the cyclist's activities with traffic behavior

The NCB theory combines the activities of the rider with the traffic behavior, which is more in line with people's real travel behavior and supports the "trip chain" behavior modeling.

At the tactical level, expected temporary activities (ETA) and unexpected temporary activities (UTA) are both considered. The temporal activities are temporary events caused by dynamic traffic flow, such as conflicts. According to the expectation of the cyclists, these temporary activities are divided into two types: expected and unexpected temporary activities. The former is included in the path planning behavior of the tactical level as the cyclist realizes it. The latter is included in the operational-level module because it is unexpected, relying on the simple response of the cyclist.

Detailed classification of activities can better realistically model the cyclist's behavior.

7.1.3.4 Extensive and flexible framework

Since the cyclist's path planning/selection behavior at un-signalized intersections is based on two-dimensional planes, the limited choice set cannot truly represent the cyclist's path planning/selection at un-signalized intersections.

The NCB model theory has an extensive and flexible theoretical framework structure, which not only supports the continuous space–time choice problem but also discrete choice problem by discretizing the continuous space–time according to the user's accuracy requirements and computer computing power.

In addition, the proposed activities in the model are more extensive and universal for the cyclist behavior model. By different definitions of activities, the NCB model is applied to different application areas, such as evaluation models.

In addition, subjective and subjective prediction models can introduce subjective randomness and fuzziness according to the actual modeling situation; therefore, the NCB model realizes the modeling of random and fuzzy behavior.

7.1.3.5 Support for dynamic behavior modeling

The operational-level behavior module of the NCB theory affects the dynamic information $V^d(t)$ of all vehicles in the traffic facility Ω_i by changing its own dynamic state $V_i(T)$, thereby generating feedback effects on the output of the hierarchical behavior module by affecting the input information of the operational-level behavior module.

Thus, if the tactical-level module of system S is continuously updated with a certain time step ΔT, a dynamic path planning process at the tactical level is formed.

Similarly, the tactical-level behavior module affects the travel time T_Ω of the transportation facility Ω_i and the service level LOS_Ω by the path planning, thereby generating feedback effects on the output of the hierarchical behavior module by influencing the information input to the strategic-level behavior module.

If the strategic-level behavior module of system S is continuously updated with a certain time step ΔT_s, a dynamic route choice process at the strategic level is formed.

In this way, the NCB theory model framework not only supports the dynamic path planning behavior of bicycles at tactical-level traffic behavior but also supports the dynamic route choice behavior of strategic-level behavior (Figure 7.3).

7.1.4 NCB tactical-level behavioral theory model

This section will introduce the general NCB tactical-level theory model, which is applicable to all types of transportation infrastructure Ω in road network N, such as

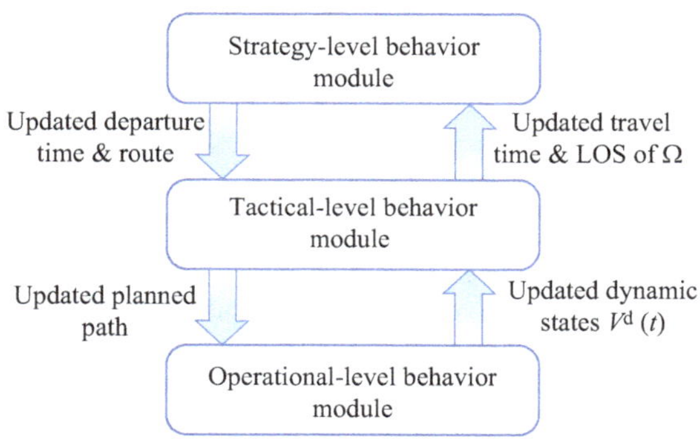

Figure 7.3 Interactions between different-level behavior modules

road sections, signal control intersections or un-signalized intersections. Therefore, many parts can only give generalized formulas of theoretical models rather than concrete ones, such as the internal (subjective) motion model of the behavior subject (cyclist), the introduction of the randomness of motion or behavior, the specific form of the utility function and the corresponding optimization algorithm.

The NCB-specific modeling of the microscopic behavior of bicycles crossing un-signalized intersections will be systematically elaborated in Section 7.2.

7.1.4.1 Decision variable of tactical-level behavior module

Before the modeling, we will first introduce the decision variables of the NCB tactical-level behavior module. In the tactical-level behavior module, the main decision variables and intermediate variables faced by the behavior subject (cyclist) mainly include the following:

1. Cyclist's own and related vehicles dynamic information $V^d(t)$.
   $V^d(t)$ includes the cyclist's own speed $\vec{v}(\bullet)$ and the corresponding trajectory $\vec{x}(\bullet)$ in the traffic facility Ω, and the speed $\vec{v}_j(\bullet)$ and trajectory $\vec{x}(\bullet)$ of other related vehicles V_j in the traffic facility Ω. These variables are predicted by the cyclist's internal (subjective) motion model $V(\tau) = f_{sub}(V(t), \tau)$, $\tau > t$.
2. The cyclist's activity space Σ_Ω within the transport facility Ω.
   Σ_Ω includes the planning activities (PA) and expected temporary activities (ETA) within Ω; these are also based on internal subjective motion model $V(\tau) = f_{sub}(V(t), \tau)$.
3. Activity plan A_Ω within Ω.
   The cyclist judges, selects and sorts all the activities in Σ_Ω and forms the activity plan $A_\Omega(t) = \{(a_i, P_{ai}, T_{ai})\}$, $a_i \in \Sigma_\Omega$, $P_{ai} \in \Omega$, $T_{ai} \in (T_o, T_d]$. A_Ω is the collection of the activity, sequence and location (or area) P_{ai} and event end time T_{ai} that the cyclist is going to perform in Ω.
4. Temporary destination D_i of planned path P.
   The temporary destination D_i of the planned path P within facility Ω indicates the position or area of the ith temporary destination of P, and the time T_i to reach D_i, such that $\vec{x}(T_i) \in D_i$.
   The so-called cyclist's internal motion model $V(\tau) = f_{sub}(V(t), \tau)$, $\tau > t$ means that at the initial time t_o, the behavioral subject (i.e., the cyclist), first obtains the dynamic traffic information $V^d(t_o)$ of each vehicle in the $\Omega \in \mathbb{R}^2$ when it arrives at a certain traffic facility Ω.

Then, based on the information and the subjective internal motion model, the future motion state of the vehicle in Ω is predicted, that is, $V^d(\tau)$, $\tau > t$. Meanwhile, the UTA is also determined according to the expected dynamic information, combined with the travel planning activity (PA). Thus, the activity space Σ_Ω within the transportation facility Ω is formed.

Then the cyclist judges, selects and sorts all the activities in Σ_Ω and forms the activity plan $A_\Omega(t) = \{(a_i, P_{ai}, T_{ai})\}$, $a_i \in \Sigma_\Omega$, $P_{ai} \in \Omega$, $T_{ai} \in (T_o, T_d]$, and then the temporary destination D_i of the planned path P, and the time T_i to reach D_i, such that $\vec{x}(T_i) \in D_i$.

If we set the *final destination* of the cyclist in Ω is D_f, the corresponding D_f is also the end point (last temporary destination) of the planned path P in Ω.

In the previous analysis, the relationship between the temporary destination D_i of the planned path $P(t)$ and the location of the activity (region) P_{ai} in the activity plan $A_\Omega(t) = \{(a_i, P_{ai}, T_{ai})\}$ in the traffic facility Ω is considered to be consistent, which is

$$D_i = P_{ai}, D_i, \qquad P_{ai} \in \Omega, \quad i = 0, 1 \ldots k \tag{7.13}$$

The corresponding T_{ai} has a consistent relationship with the time T_i to reach D_i of the planned path $P(t)$, which is

$$T_i = T_{ai}, T_i, \qquad T_{ai} \in (T_o, T_d], \quad i = 0, 1, \ldots, k \tag{7.14}$$

Therefore, in the later modeling process, a_i is reduced to i, then $A_\Omega(t) = \{(i, D_i, T_i)\}$, $i \in \Sigma_\Omega$, $D_i \in \Omega$, $T_i \in (T_o, T_d]$, unless the definition of the planned activity variable is specified.

Note that for activity i in A_Ω, the rider has a certain freedom to choose its occurrence zone D_i. The NCB model uses different activity occurrence regions (which can also be regarded as candidate TD points) D_{ij} to represent, $D_{ij} \subset \Omega$, $j = 1, 2, \ldots, J_i$, and the subscript i represents activity i. j represents the jth candidate area where activity i occurs, and J_i represents the total number of optional areas of activity i.

For example, to bypass an obstacle, the cyclist can produce different D_{ij} in many different places, such as the left and right sides of the obstacle, a little further or closer to the obstacle. The NCB theory framework supports the choice of continuous space, i.e., D_{ij} can be a continuous variable, so that J_i can be infinite. While in practical applications, the discretization process is usually performed according to the actual situation. Thus, J_i is a finite positive integer.

Since in the cyclist's planned path T_i represents the time to reach D_i, $(T_i - T_{i-1})$ is the expected riding time from D_{i-1} to D_i. From the subjective self-motion trajectory $\vec{x}(\bullet)$ and $(T_i - T_{i-1})$ of the cyclist, the position of the TD can be obtained from the time T_i reaching the TD, i.e., $x(T_i) \in D_{ij}$ (assuming the rider does not move while performing activity i).

From the physics kinematics, the rider's motion trajectory $\vec{x}(\bullet)$ is a continuous equation determined by the dynamic velocity $\vec{v}(\bullet)$:

$$\vec{x}(\cdot) := \left\{\vec{x}(\tau) \in \Omega \middle| t_0 \leq \tau \leq t_1 s.t. \vec{x}'(\tau) = \vec{v}(\tau) \text{ and } \vec{x}(t_0) = \vec{x}_0\right\} \tag{7.15}$$

where $\vec{x}'(\tau) = \vec{v}(\tau)$ is the formula of velocity and trajectory relationship, and $\vec{x}(t_0) = \vec{x}_0$ is the initial position and time relationship of the rider. These two formulas are the boundary conditions in formula (7.15).

7.1.4.2 Subjective utilities optimization at the tactical level

According to the NCB theory, at time t_0 the cyclist is located at $\vec{x}_0$ in the traffic facility Ω. Based on the dynamic traffic information within Ω, the cyclist optimizes the subjective expected utility (or cost) C, determines the activity plan and gets the initial planned path $P(t_0)$ in Ω.

The decision process includes all of the decision variables described earlier: the activity space Σ_Ω within the traffic facility Ω; the planned activity $A_\Omega(t_0)$; the D_i in planned path $P(t_0)$; and the cyclist's own expected speed $\vec{v}(\bullet)$ and the corresponding expected trajectory $\vec{x}(\bullet)$.

At this time, the subjective expected utility or cost C of the rider is mainly determined by the following parts: (1) activity content and sequence in activity plan $A_\Omega(t_0)$; (2) locations of D_i where the activities are performed; (3) two adjacent. The location of the activity, that is, the riding cost between two adjacent TDs is $P(t_0)$.

Thus, the total utility/cost C of the cyclist can be written as

$$C\left(A_\Omega, \{T_i\}_{i\in A_\Omega}, \vec{v}(\bullet)\middle| t_0, \vec{x}_0\right) = \sum_{i\in A_\Omega} C_i(T_i, \vec{v}(\bullet)|T_{i-1}, \vec{x}(T_{i-1})) + \psi(A_\Omega) \tag{7.16}$$

where C_i represents the sum of the riding cost between the location $\vec{x}(T_{i-1}) \in D_{i-1j}$ of the activity $i-1$ and the location $\vec{x}(T_i) \in D_{ij}$ of the activity i, and the cost (or utility) U_{ij} of the activity at the activity site D_{ij}.

By optimizing the subjective expected utility/cost C_i of the rider, an optimal overall selection result is obtained:

$$(A_\Omega*, \{T_i\}*_{i\in A_\Omega}, v*(\bullet)) = \arg\min C(A_\Omega, \{T_i\}_{i\in A_\Omega}, \vec{v}(\bullet)|t_0, \vec{x}_0) \tag{7.17}$$

Since the event location of activity i can be selected between a number of different locations D_{ij} in Ω, activity i can have different utility or cost corresponding to different locations D_{ij}.

These different utilities or costs U_{ij} reflect some of the rider's preferences for the choice of locations D_{ij}, such as the *obstacles avoidance* activity, some riders prefer to circumvent from the left side, while others ride likes to circumvent from the right side.

In addition, there are some activities in activity plan A_Ω that are optional (discretionary), that is, the cyclist may or may not perform the activity, while the other activities are mandatory, meaning that the cyclist must perform such activities.

The distinction between these two types of activities is expressed by the size of the utility/cost—for the required activities, a high penalty factor of ϕ_i is used to indicate that if the required activity i is not completed, it can cause great utility loss to the rider or generate a lot of cost.

The order of activities in activity plan A_Ω is determined primarily by the "shortcut" law in behavioral science and the selection process in the tactical-level behavior module [7].

In general, the "shortcut" rule of the activity sequence is mainly reflected in the riding cost formed by the cyclist's trajectory $\vec{x}(\bullet)$. The introduction of the variable $\psi(A_\Omega)$ is to describe the situation related to the sequential order. That is, the previous activity must be completed before the next activity, for example, activity $i-1$—"enter the queue" and activity i—"leave the queue."

In the NCB tactical-level behavior module, all the factors that may affect the cyclist's subjective riding cost between the two areas of activity are grouped into the following three categories:

1. *Intrinsic characteristics of the planned path*: Such factors refer to the attributes related to the planned path, mainly, including the riding distance, the riding time and the related attributes of the directness of the line (the number of path turns).
2. *Traffic conditions factors*: Such factors mainly refer to the attributes related to traffic conditions, such as the distance from other moving vehicles in the traffic facility Ω (mainly potential conflicting vehicles or other influencing vehicles), and physical obstacles in Ω (such as barrier and safety island), as well as the number of possible conflicts with other traffic vehicles.
3. *Individual characteristics of the rider*: Such factors mainly include all the characteristics associated with the individual of the bicycle and cyclist, such as gender, age, vehicle type (ordinary bicycle or tricycle, etc.), riding status (riding alone or in a bicycle group).

In the specific modeling of the microscopic behavior of bicycle crossing un-signalized intersections in Section 7.2, small-scale interviews and questionnaire surveys were conducted for Beijing cyclists (see Section 8.1 in Chapter 8) to analyze the previous various influencing factors. The effects of these factors on the cyclist's crossing behavior at un-signalized intersection were analyzed, and the microscopic behavior was specifically modeled on the basis of the interview analysis results.

7.1.5 NCB operational-level behavioral theory model

This section will introduce the generally applicable NCB operational-level behavioral theory model. Like the tactical-level behavioral theory model, the model formula for this section applies to all types of traffic infrastructure in traffic networks, such as road segments, signal-controlled intersections or un-signalized intersections. Therefore, many parts can only give generalized formulas of theoretical models rather than specific formulas, such as the form of the psychological field, the expression of dynamics and so on.

The specific NCB operational-level behavioral model formula for bicycles at un-signalized intersections will be systematically explained in Section 7.2.2.

As mentioned earlier, both bicycle and pedestrian traffic belong to the non-motorized traffic mode and have many similarities in travel behavior. The bicycle riding at the un-signalized intersection plane is quite similar to the pedestrians walking in the square.

The current "social field" dynamics model is mainly aimed at pedestrian/crowd modeling. The model primarily focuses on pedestrian walking behavior, the interaction among pedestrians, between pedestrians and fixed obstacles, and how these interactions affect the pedestrian's waling behavior.

Here, we are modeling the dynamic cycling behavior. It is necessary to examine the interaction between bicycles and motor vehicles that are in motion and stationary, the interaction among bicycles, the interaction between bicycles and pedestrians affects, and how these interactions affect the cyclist's riding behavior.

7.1.5.1 Variables of the operational-level behavioral model

Cyclist α rides in the traffic facility Ω and we use his/her trajectory $\vec{x}(\bullet)$ to indicate his/her dynamic riding behavior. According to the "social field" theory, it is considered that the cyclist α has a psychological field $\vec{w}_\alpha(\vec{x},t)$, which varies with time and place, and is therefore recorded as a function of time and position.

From the Levin's "social field" theory, the individual's behavior is determined by the influence of the environment in which he/she is located, E and individual personality characteristics P.

Inspired by the theory of physics, social field theory believes that the impact of the environment on individuals can be regarded as a dynamic field, so it can be described in mathematical forms. Temporary changes in the social field will cause the individual's *mental tension*, causing an individual's behavior to release this tension.

We assumed that the individual's goal is to *behave in an optimal way* according to the social field strength, that is, the individual always tends to act in a way that minimizes social field strength.

If the individual's behavior $\vec{x}(\bullet)$ is inconsistent with the minimum value of the social field strength $\vec{w}_\alpha(\vec{x},t)$, which creates a mental tension $\vec{f}_\alpha(\vec{x},t)$ that wants to change state $\vec{x}(\bullet)$:

$$\vec{f}_\alpha(\vec{x},t) = -\nabla\vec{w}_\alpha(\vec{x},t) \tag{7.18}$$

The magnitude and size of the tension is determined by the gradient of the social field strength $\vec{w}_\alpha(\vec{x},t)$, pointing to the direction in which the social field strength $\vec{w}_\alpha(\vec{x},t)$ drops the fastest. The *force* $\vec{f}_\alpha(\vec{x},t)$ acts as a mental tension, and this mental tension F eventually forms the driving force of the subject's behavior $\vec{F}_\alpha(\vec{x},t)$:

$$\vec{F}_\alpha(\vec{x},t) = F(\vec{f}_\alpha(\vec{x},t)) \tag{7.19}$$

where $\vec{F}_\alpha(\vec{x},t)$ causes a behavior change:

$$\frac{d\vec{x}}{dt} \approx \vec{F}_\alpha(\vec{x},t) + \xi \tag{7.20}$$

and $\vec{F}_\alpha(\vec{x}, t)$ is a behavioral dynamic vector:

$$\vec{F}_\alpha(\vec{x}, t) := \begin{pmatrix} \vec{F}_{\alpha 1}(x, t) \\ \vdots \\ \vec{F}_{\alpha n}(x, t) \end{pmatrix} \tag{7.21}$$

Then

$$\vec{F}_{\alpha i}(\vec{x}, t) \approx \frac{d\vec{x}_i}{dt} + \xi \tag{7.22}$$

where $d\vec{x}/dt$ represents the change rate of individual behavior $\vec{x}(t)$, and ξ represents a random interference term, indicating the behavioral differences between individuals; vector $F_\alpha(x, t)$ is the motivation of individual behavior, driving individual α behavior $\vec{x}(t)$ to undergo system changes. At the same time, $F_\alpha(x, t)$ is the force of social field to individual α.

The vector $\vec{w}_\alpha(\vec{x}, t)$ can be understood as the potential energy function of the social field, and the tension $\vec{f}_\alpha(\vec{x}, t)$ generated by the social field to the individual can be obtained by finding the partial derivative (i.e., the gradient ∇ of the social field potential energy). And the behavioral change $d\vec{x}/dt$ promotes the individual's behavior $\vec{x}(t)$ toward the minimum strength value $\vec{x}_\alpha$ of social field $\vec{w}_\alpha(\vec{x}, t)$, when the individual behavior reaches the social field strength minimum $\vec{x}_\alpha$, there would be

$$\nabla \vec{w}_\alpha(\vec{x}, t) = 0$$

Thus, the mental tension $\vec{f}_\alpha(\vec{x}, t) = 0$ indicates that the individual releases mental tension through changes in the previous behavior. When the mental tension $\vec{f}_\alpha(\vec{x}, t)$ is eliminated, the behavior of the individual will not change except for the influence of random disturbances. That is to say, the individual reaches a state of balance in the social field, and at this time the behavioral power $\vec{F}_\alpha(\vec{x}, t) = 0$.

It is worth noting that for different cyclists α, the social field $\vec{w}_\alpha(\vec{x}, t)$ has a different minimum value $\vec{x}_\alpha$, in other words, α represents the personality traits of different individuals.

From the previous model description, the main variables in the "social field" dynamics theoretical model mainly include the *social field* $\vec{w}_\alpha(\vec{x}, t)$ of the cyclist α (also known as the psychological field); the *mental tension* $\vec{f}(\vec{x}, t)$ that causes the rider α to change his/her motivation; and the force $\vec{F}_\alpha(\vec{x}, t)$ that causes the cyclist α to change his/her behavior.

The three variables are described separately next.

The cyclist's social field: The psychological field is a commonly used concept in the Gustav psychology [8]. This school introduces theories and concepts of "field" in modern physics and directly uses them to demonstrate their theoretical views. However, they do not have a clear explanation of the terms they have created. They think that the physical field is the field of the geographical environment; the behavioral field and the physiological field are the connections between conscious physiological processes. It can be described as follows: the so-called physical field refers to the field of geography; the so-called behavior field refers to the field of consciousness or direct experience; and the so-called psychophysical field is a broader field that covers the field of behavior and the physiological fields, while some scholars also collectively referred to all "fields" but "physical field" of the previous fields, as the "psychological field."

Psychologist Levin developed the field theory in psychology. He proposed a series of basic concepts: life space, behavior and movement, force and force field, tension and tension system and so on. In the NCB theory, these psychological concepts are applied and limited to the scope of traffic psychology and traffic behavior.

In the NCB theory, it is considered that the "social field" (i.e., "psychological field") of the rider refers to the sum of various factors that affect the cyclist α's riding behavior in the time and space within the traffic facility Ω. It includes two characteristics of the cyclist's personality characteristics P and external environment E.

The personality characteristics P of the cyclist α, including the factors affecting the riding behavior, such as gender, age and travel destination, are included in the function of the field strength, that is, the subscript α of the *social field* has included personality characteristics P, so the *social field* of the cyclist can be expressed as

$$\vec{w}_\alpha(\vec{x},t) = W_\alpha(E,\vec{x},t) \tag{7.23}$$

where W_α represents the *social field* strength formula of the cyclist α, and E represents various factors in the external environment that may affect the cyclist's riding behavior, mainly, including static traffic facilities Ω and other dynamic moving vehicles V_β.

Later we discuss several properties of the cyclist's *social field* in the NCB theory.

Superposition: Since the cyclist's *social field* refers to the sum of various factors affecting the cyclist α's riding behavior in the time and space scope within the traffic facility Ω, referring to the knowledge of the physical field, we consider that the cyclist's *social field* also satisfies the *superposition* principle, which is

$$\vec{w}_\alpha(\vec{x},t) = W_\alpha(E,\vec{x},t) = \sum_{\beta\in E} W_\alpha(\beta,\vec{x},t) \tag{7.24}$$

where β represents the various environmental factors affecting the cyclist's *social field* $\vec{w}_\alpha(\vec{x},t)$. The *social field* superposition function $\sum_{\beta\in E}$ represents the superposition of the social field $W_\alpha(\beta,\vec{x},t)$ generated by each environmental factor β. It is noted that the superposition function $\sum_{\beta\in E}$ may not be a simple linear superposition. In the process of concrete modeling, the *social field* superposition function should be embodied according to the actual situation.

Subjective perception: It is noted that the cyclist α's *social field* is a psychological concept, which is a subjective *psychological field* rather than an objectively measurable *physical field* such as a magnetic field. Similar to the subjective state prediction model of the cyclist in the tactical-level behavior model, each cyclist α has its own *social field* $\vec{w}_\alpha(\vec{x}, t)$.

This explains why different cyclist has different behavioral responses to the same environmental stimulus. Because the same stimulus is NOT the same in different cyclists' *social field,* it produces different mental tensions, thus leading to different behavioral responses.

Another need to emphasize is the impact of the cyclist's perception on his/her *social field.* The external environment E's stimulation needs to be perceived by the cyclist in order to be included in his/her *social field.* Here, we use $P(E)$ to indicate the perception of the external environment E by the cyclist α. Then the formula (7.24) becomes

$$\vec{w}_\alpha(\vec{x}, t) = \sum_{\beta \in E} W_\alpha(P(\beta), \vec{x}, t) \tag{7.25}$$

In the process of specific modeling, the cyclist's perception function should be embodied according to the actual situation.

The cyclist's mental tension: The *psychological tension system* is a dynamic analysis of the root causes of human behavior from the conceptual level, where tension and need are the two most basic terms.

Levin believes that "as long as there is a psychological need within a person, there will be a tension system." The release of tension can provide dynamics and energy for mental activity and behavior. Therefore, it constitutes a potential factor in determining people's psychological activities and behaviors.

According to Levin's theory of *psychological field*, the key to the dynamic nature of individual behavior is the *tension system*. The *mental tension* of the system is generated by the *need* and released by achieving the relevant goal. In addition, the magnitude of mental tension is related to the intensity of the individual's *need* and the *valence* of the relevant *goal*.

In the dynamic theory of Levin's psychological field theory, the goal of behavioral or psychological activity also has a kind of force nature, which is called *valence*. Positive valence has a force of attraction, and negative valence has a force of rejection.

Therefore, human behaviors are not only driven by internal needs and tension, but also by the attraction (or rejection) of the target itself. As the intensity of need increases, the **valence** of the target associated with the need increases, while the increase in **valence** will, in turn, affect the strength of need.

This is just as the more urgent a person is, the more attractive the destination D_f is to him; the greater the attraction of the destination D_f, the stronger the inner desire to move for the person.

The NCB operational-level behavior model considers that there are two types of sources that cause the cyclist α's mental tension: one is the output of the previous tactical-level behavior module—the TDs in the planned path $P(t)$ and the expected position and arrival time (D_i, T_i) of the TD, the other type is the perceived surrounding influencing vehicles V_β in traffic facility Ω. The mental tension produced by the former is called the goal tension; the mental tension produced by the latter type is called the effect tension.

Goal tension: In the NCB theoretical model, it is considered that the goal tension of the cyclist α is only the TD in the planned path $P(t)$ at current time t. We think that the goal tension to the cyclist α is only from the TD (D_i, T_i) in the planned path $P(t)$ that is closest to the current time t in time. As for the other TDs, they has no effect on the cyclist α's psychology for the time being.

The effect of TD (D_i, T_i) on the cyclist α is through the goal of *social field*, that is, the TD produces a corresponding "goal social field" $\vec{w}^0{}_\alpha(\vec{x}, t)$:

$$\vec{w}^0{}_\alpha(\vec{x}, t) = W^0_\alpha(D_i, T_i, \vec{x}, t) \tag{7.26}$$

And $\vec{w}^0{}_\alpha(\vec{x}, t)$ corresponds to a goal tension $\vec{f}^0(\vec{x}, t)$, according to the *social field* model, there are

$$\vec{f}^0(\vec{x}, t) = -\nabla\vec{w}^0{}_\alpha(\vec{x}, t) = -\nabla W^0_\alpha(D_i, T_i, \vec{x}, t) \tag{7.27}$$

Considering that in traffic behavior, the traveler usually has a destination (or temporary destination), so the goal tension of the cyclist can be used as a basic drive for generating riding behavior.

That is to say, when there is no other factor, at least the goal tension has an influence on the cyclist's riding behavior. At this time, some behavior characteristics of the cyclist such as the shortcut rule may work.

Effect tension: The effect tension is the mental tension "valence" generated by the perceived traffic facility Ω and the other dynamic vehicles V_β around the cyclist α.

The external influence factor $\beta \in E$ is first perceived by the cyclist α, and the resulting *subsocial field* $\vec{w}^i{}_\alpha(\beta, \vec{x}, t)$ can produce the corresponding mental tension $\vec{f}_\beta(\vec{x}, t)$:

$$\vec{w}^\beta{}_\alpha(\vec{x}, t) = W^\beta_\alpha(P(\beta), \vec{x}, t) \tag{7.28}$$

$$\vec{f}_\beta(\vec{x}, t) = -\nabla\vec{w}^\beta{}_\alpha(\vec{x}, t) = -\nabla W^0_\alpha(P(\beta), \vec{x}, t) \tag{7.29}$$

Considering some influence of factors $\beta \in E$ on cyclist α may be both repulsive and attractive, so it is recorded as $\vec{f}^r_\beta(\vec{x}, t)$ and $\vec{f}^a_\beta(\vec{x}, t)$, respectively. And the corresponding attraction and repulsion have different field strengths:

$$\vec{w}^{\beta^r}_\alpha(\vec{x}, t) = W^{\beta^r}_\alpha(P(\beta), \vec{x}, t), \quad \vec{f}^a_\beta = -\nabla\vec{w}^{\beta^a}_\alpha(\vec{x}, t) = -\nabla W^{\beta^a}_\alpha(P(\beta), \vec{x}, t) \tag{7.30}$$

$$\vec{w}^{\beta^a}_\alpha(\vec{x}, t) = W^{\beta^a}_\alpha(P(\beta), \vec{x}, t), \quad \vec{f}^r_\beta = -\nabla\vec{w}^{\beta r}_\alpha(\vec{x}, t) = -\nabla W^{\beta^r}_\alpha(P(\beta), \vec{x}, t) \tag{7.31}$$

In the NCB theoretical model, the sources of effect tension mainly include related facilities in the traffic facilities Ω, such as safety islands, railings, as well as surrounding related dynamic vehicles, such as vehicles, bicycles and pedestrians traveling in the same direction or opposite to the cyclist α.

Cyclist's behavioral force: The result of the combined action of the goal tension and various effect tensions in the *social field* of the cyclist α forms the final behavioral force $\vec{F}_\alpha(\vec{x}, t)$ of α. Recorded as

$$\vec{F}_\alpha(\vec{x}, t) = F_\alpha\left(\vec{f}^0, \vec{f}^r_\beta, \vec{f}^a_\beta\right), \quad \beta \in E \tag{7.32}$$

Unlike the mental tensions, behavioral force is not only an internal psychological force, but also the driving force of the ultimate external behavior of the behavioral subject. Therefore, the behavioral force $\vec{F}_\alpha(\vec{x}, t)$ is objective and measurable, while the mental tension $\vec{f}_\beta(\vec{x}, t)$ is unmeasurable.

7.1.5.2 Basic process of operational-level behavior module

The basic structure of the operational-level behavior module of NCB theory is as follows.

As can be seen from Figure 7.4, the NCB operation-level behavior module (social field model) is mainly composed of four functional sub-modules: the perception module $P(\beta)$, the goal social field $\vec{w}^0_\alpha(\vec{x}, t)$, the effect social field $\vec{w}^\beta_\alpha(\vec{x}, t)$ and the integrated social field module $\vec{w}_\alpha(\vec{x}, t)$.

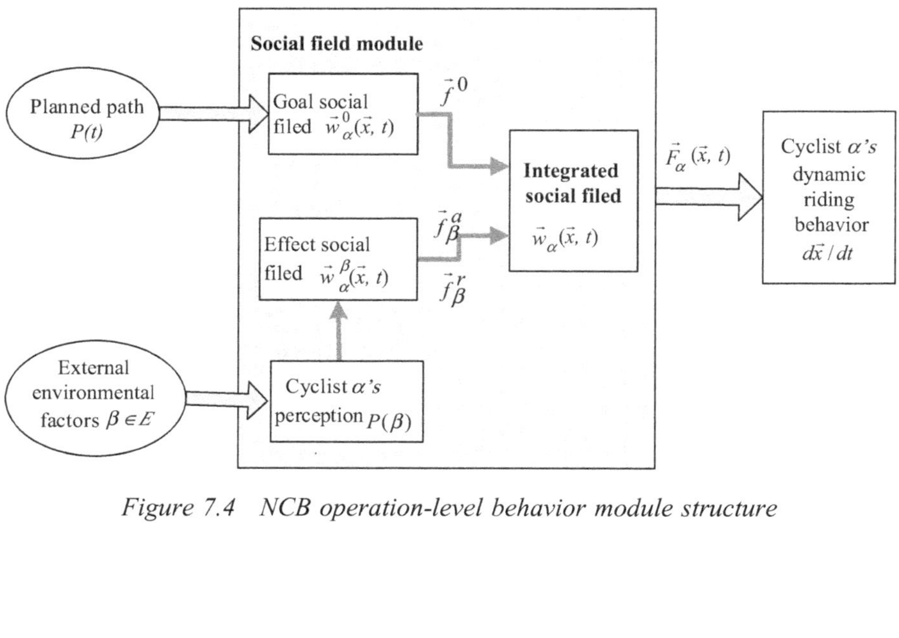

Figure 7.4 NCB operation-level behavior module structure

The personality characteristics of the cyclist α are the main factors affecting these social field model function sub-modules. The basic flow of the NCB operation-level behavior module is described as follows:

1. First, the output of the tactical-level behavior module—the planned path $P(t)$ enters the goal social field module, forms the goal social field $\vec{w}^0{}_\alpha(\vec{x}, t)$, and generates the goal tension $\vec{f}^0(\vec{x}, t)$ for the cyclist α's riding behavior.
2. At the same time, external environmental factors $\beta \in E$ (mainly including transportation facilities and related dynamic vehicles) are perceived by the cyclist α's perception system $P(\beta)$, and the perceived information generates the effect social field $\vec{w}^\beta_\alpha(\vec{x}, t)$.
3. The generated effect social field $\vec{w}^\beta{}_\alpha(\vec{x}, t)$ will have a corresponding mental tension $\vec{f}_\beta(\vec{x}, t)$ on α. The mental tension $\vec{f}_\beta(\vec{x}, t)$ can be divided into two categories: repulsive tension $\vec{f}^r_\beta(\vec{x}, t)$ and attractive tension $\vec{f}^a_\beta(\vec{x}, t)$.
4. The goal tension $\vec{f}^0(\vec{x}, t)$ generated by (1) and the repulsive tension $\vec{f}^r_\beta(\vec{x}, t)$ of each influencing factor $\beta \in E$ of (3) and the attraction tension $\vec{f}^a_\beta(\vec{x}, t)$ are integrated and weighed to obtain the integrated social field $\vec{w}_\alpha(\vec{x}, t)$ and the corresponding riding force $\vec{F}_\alpha(\vec{x}, t)$.
5. Ignoring the random interference factor, we get $\vec{F}_\alpha(\vec{x}, t) = d\vec{x}/dt$, so that the NCB social field model can achieve the function of dynamically changing the cyclist's riding behavior $\vec{x}(t)$ according to the surrounding environment.

Social field model formula summary:

- Goal social field and corresponding goal tension:

$$\vec{w}^0_\alpha(\vec{x}, t) = W^0_\alpha(D_i, T_i, \vec{x}, t) \tag{7.33}$$

$$\vec{f}^0(\vec{x}, t) = -\nabla \vec{w}^0{}_\alpha(\vec{x}, t) \tag{7.34}$$

- Effect social field and the corresponding effect tension:

  Repulsive social field and repulsive tension:

$$\vec{w}^{\beta^r}_\alpha(\vec{x}, t) = W^{\beta^r}_\alpha(P(\beta), \vec{x}, t) \tag{7.35}$$

$$\vec{f}^r_\beta = -\nabla \vec{w}^{\beta r}_\alpha(\vec{x}, t) = -\nabla W^{\beta^r}_\alpha(P(\beta), \vec{x}, t) \tag{7.36}$$

  Attractive social field and attractive tension:

$$\vec{w}^{\beta^a}_\alpha(\vec{x}, t) = W^{\beta^a}_\alpha(P(\beta), \vec{x}, t) \tag{7.37}$$

$$\vec{f}^a_\beta = -\nabla \vec{w}^{\beta^a}_\alpha(\vec{x}, t) = -\nabla W^{\beta^a}_\alpha(P(\beta), \vec{x}, t) \tag{7.38}$$

Finally, the result of the cyclist's mental tension (i.e., behavior change force) can be obtained in two ways. One is to find the cyclist α's integrated social field $\vec{w}_\alpha(\vec{x}, t)$, and then to find the social field gradient ∇, namely,

$$\vec{w}_\alpha(\vec{x},t) = \sum_{\beta \in E} \left\{ W_\alpha^\beta(\beta,\vec{x},t), W_\alpha^0(D_i,T_i,\vec{x},t) \right\} \tag{7.39}$$

$$\vec{F}_\alpha(\vec{x},t) = -\nabla \vec{w}_\alpha(\vec{x},t) \tag{7.40}$$

Or in another way, by integrating α's mental tensions $\vec{f}^0$, $\vec{f}^r_\beta$ and $\vec{f}^a_\beta$, that is

$$\vec{F}_\alpha(\vec{x},t) = F_\alpha(\vec{f}^0,\vec{f}^r_\beta,\vec{f}^a_\beta) \tag{7.41}$$

Finally, according to the social field's behavior change force definition: $(d\vec{x}/dt) \approx \vec{F}_\alpha(\vec{x},t)$ we derive the acceleration $d\vec{x}(t)/dt$ of the cyclist at time t and position $\vec{x}(t)$. Based on the previous formulas, we obtain the basic formula of the NCB operation-level behavior module.

7.2 Microscopic behavior modeling of bicycle crossing un-signalized intersections

The main content of this section is based on the NCB tactical level and operational-level behavior model described in Section 7.1, to specifically model microscopic behavior of bicycle crossing un-signalized intersections (i.e., give specific formulations).

In the tactical-level path planning behavior modeling, we will specify the cyclist's internal traffic state prediction model, the introduction of uncertain factors, the functional form of the subjective expected utility/cost and the determination of each attribute that are required. These should be embodied according to the actual situation of the cyclist crossing the un-signalized intersection.

In the modeling of cyclist's operational-level behavior model, we need the specific form of the four functional sub-modules—cyclist's perception $P(\beta)$, goal social field $\vec{w}^0_\alpha(\vec{x},t)$, effect social field $\vec{w}^\beta_\alpha(\vec{x},t)$ and integrated social field $\vec{w}_\alpha(\vec{x},t)$—of the social field model.

Before the specific modeling of the microscopic bicycle crossing behavior at the un-signalized intersections, a small-scale interview and questionnaire survey (see Section 8.1 in Chapter 8) was conducted for the Beijing cyclist to analyze the impact of previous various influencing factors on the cyclist's crossing behaviors at the un-signalized intersections. And we build the cyclist's microscopic behavior model based on interviews and survey results.

7.2.1 The cyclist's path planning behavior model (tactical level)

This section mainly models the cyclist's path planning behavior when crossing un-signalized intersections. Most of the variable definitions in the modeling process are the same as in the previous section unless otherwise specified.

In the behavioral model of a bicycle crossing the un-signalized intersection, we assume that the bicycle has no travel planning activity (PA) during the crossing of the un-signalized intersection. That is, the purpose of the cyclist entering the un-signalized intersection is to reach the destination D_f across the intersection. Therefore, the activity set A_Ω of the un-signalized intersection behavior model only contains expected temporary activity (ETA).

7.2.1.1 Inner state prediction model

In order to generate ETA at the tactical-level model and to make utility/cost estimates for the planned path, cyclist α needs an internal state prediction model to predict the motion state of itself and other related vehicles V_β in the un-signalized intersection.

Therefore, the cyclist first determines the positions $\vec{x}_\alpha(t)$ and $\vec{x}_\beta(t)$ of himself/herself and other vehicle V_β at current time t, records them as $\hat{x}_\alpha$ and $\hat{x}_\beta$, respectively, and predicts the future position (state) $\vec{x}_\alpha(\tau)$ and $\vec{x}_\beta(\tau)$, $\tau > t$.

In the microscopic behavior modeling of bicycle crossing un-signalized intersections, we assume that the bicycle considers itself and other vehicle movements to be random motion processes, but the vehicle running speed remains the same. Then the cyclist's subjective internal state prediction model is

$$\vec{v} = \vec{v}(t) \tag{7.42}$$

$$d\vec{x} = \vec{v}dt + \sigma d\vec{w} \quad s.t. \quad \vec{x}(t) = \hat{x} \tag{7.43}$$

where $\vec{v} = \vec{v}(\tau)$ denotes the speed vector of the vehicle at time $\tau > t$; $\vec{x} = \vec{x}(\tau)$ represents the position (state) of the vehicle at time $\tau > t$. The $\vec{x}(t) = \hat{x}$ in the formula (7.44) is the initial boundary condition.

The vector $\vec{w}$ represents a standard two-dimensional *Wiener* process. The so-called *Wiener* process is a mathematical model of *Brownian* motion. Under the microscope, the British botanist Brown observed the tiny particles floating on the calm liquid surface and found that they were constantly moving in disorder, which was later called Brownian motion.

The white noise vector $\vec{w}$ reflects the uncertainty of the traffic condition predicted by the cyclist and the resulting effect of this uncertainty on the trajectory $\vec{x}(\bullet)$ of the bicycle and other vehicles in the un-signalized intersection. The generation of uncertainty in the predictive model may be due to many uncontrollable factors such as the lack of experience of the cyclist on the traffic situation, the observation of the negligence of the intersection, and the randomness of the future situation.

The speed $\|\vec{v}\|$ of the bicycle i and other vehicles V_j depends on the physical speed limit of the vehicle, the control measures of the un-signalized intersection Ω, and the specific traffic flow conditions. Therefore, it is expressed as

$$\vec{v} \in \vec{V}_a(t,\vec{x},V) \quad \text{where} \quad \vec{V}_a(t,\vec{x},V) = \{\vec{v}, \|\vec{v}\| \le \vec{v}_{\max}(t,\vec{x},V)\} \subset \mathbb{R}^2 \tag{7.44}$$

where $\vec{V}_a$ is a set of possible velocity vectors. Due to the maximum speed of vehicle V travel and vehicle dynamic and static information (including vehicle type, driver characteristics such as gender, age and travel purpose), control measures for un-signalized intersection Ω, specific location (plane or slope, etc.) and traffic, the flow condition is closely related, so $\vec{V}_a$ is a function of t and $\vec{x}$.

7.2.1.2 Judgment of ETA

In the bicycle crossing un-signalized intersection behavior model, we assume that the ETA is mainly generated to avoid a potential conflict between the bicycle and the potential conflicting vehicle in the intersection. The cyclist judges whether he/she may have a potential conflict with other vehicles by predicting the state of the moving vehicles in the intersection.

We assume that if it is estimated in the cyclist's internal state prediction model that the closest distance of a certain vehicle V_j to bicycle α is less than a certain safety distance L_s, the cyclist believes that it might collide with V_j, thereby generating a corresponding ETA, then

$$i = \begin{cases} 1 & \text{if} \quad \min E\left(\|B(\vec{x}_\alpha(\tau)) - B(\vec{x}_\beta(\tau))\|\right) < L_s \\ 0 & \text{else} \end{cases} \tag{7.45}$$

where $i = 1$ represents the ETA and $i = 0$ means no ETA. $B(\vec{x}(\tau))$ represents the boundary of the vehicle at the $\vec{x}(\tau)$ position, and $E\left(\|B(\vec{x}_\alpha(\tau)) - B(\vec{x}_\beta(\tau))\|\right)$ represents the expected minimum distance between the bicycle α and other vehicle V_β boundaries.

Considering that different vehicle types (motor vehicles, bicycles, pedestrians) have different sizes, simply using a certain point position to represent the vehicle cannot fully reflect the real situation, so a boundary function is introduced. According to safety psychology data, L_s takes 1 m [9], i.e., $L_s = 1$ m.

7.2.1.3 Generalized general utility/cost function

The cyclist's planned path can be viewed as consisting of multiple activities (or TDs), so a utility cost function can be constructed for a situation where there is only one activity i corresponding to multiple feasible activity location D_{ij}.

Let $[t, t_1)$ be the planning period when the cyclist makes the path planning at the un-signalized intersection, that is, t and t_1, respectively represent the current time and the end of the planned path (also called terminal time); set T_a be the cyclist's initial arriving time at any of feasible location D_{ij} of activity i, and we define $T_i = \min(t_1, T_a)$.

Considering the speed $\vec{v}_{[t,T_i)}$ of the bicycle during the time period $[t, T_i)$, the bicycle position $\vec{x}(\tau)$ is regarded as the *system state* by the idea of the *stochastic process optimal control*, and the bicycle speed $\vec{v}_{[t,T_i)}$ can be regarded as the *control variable*.

Then, in the optimal control process, the cyclist needs to apply an internal state prediction model to predict his/her own position $\vec{x}(\tau)$ and other related vehicles' position $\vec{x}_j(\tau)$ (i.e., the system state), and $\vec{x}(\tau)$ is a random vector. It is assumed that given an initial state (the position state of each vehicle) $\vec{x}(t) = \hat{x}$, and a control process $\vec{v}_{[t,T_i)}$ (the bicycle speed), the cyclist can predict the utility/cost C_i

corresponding to the path $P(t)$ generated by the own motion speed $\vec{v}_{[t,T_i)}$. Here definition C_i is

$$C_i(T_i, \vec{v}_{[t,T_i)} \Big| T_{i-1}, \vec{x}(T_{i-1})) := E\left[\int_t^{T_i} L(\tau, \vec{x}(\tau), \vec{v}(\tau))d\tau + \phi(T_i, \vec{x}(T_i))\right] \tag{7.46}$$

The constraint is the formula (7.44), where L and ϕ represent the so-called *running cost* and *terminal cost*, respectively.

The running cost $L(\tau, \vec{x}(\tau), \vec{v}(\tau))$ represents the riding utility/cost of the bicycle being at $\vec{x}(\tau)$ and moving at a speed of $\vec{v}(\tau)$ during a very small time interval $[\tau, \tau + d\tau)$.

The terminal cost $\phi(T_i, \vec{x}(T_i))$ represents the utility/cost of the bicycle's arrival of $\vec{x}(T_i)$ at time T_i. This terminal cost usually reflects the magnitude of the subjective penalty (consequence) that the bicycle failed to reach the TD D_{ij} before the end of the scheduled time t_1.

Note the integrated cost formula for the path

$$C(A_\Omega, \{T_i\}_{i \in A_\Omega}, \vec{v}(\bullet) \Big| t_0, \vec{x}_0) = \sum_{i \in A_\Omega} C_i(T_i, \vec{v}(\bullet)|T_{i-1}, \vec{x}(T_{i-1})) + \psi(A_\Omega) \tag{7.47}$$

is a formula that can be added in the form of integrals.

This theoretically shows that the cost of the bicycle riding line has an additivity, that is, the total utility/cost of the path $P(t)$ can be obtained by summing the costs of the various small sections of the path $P(t)$.

This makes it easy to get the total utility/cost of the entire path across the intersection from the line utility/cost of a single activity.

7.2.1.4 Terminal utility/cost determination of the planned path

By definition, the terminal time T_i is either equal to the path planned terminal time t_1 or equal to the time T_a at which the bicycle initially arrives at any candidate area D_{ij} of the activity. Thus, the terminal cost ϕ has a spatial extent of $[t, t_1] \times \Omega$, and its formula is

$$\phi(T_i, x(T_i)) = \begin{cases} -U_{ij}(T_i) & \vec{x}(T_i) \in D_{ij} \\ \phi_i & \vec{x}(T_i) \notin D_{ij} \end{cases} \tag{7.48}$$

The terminal cost ϕ reflects the subjective penalty ϕ_i (consequence) that the bicycle did not reach the TD D_{ij} before the end of the planned time t_1, while U_{ij} reflects the subjective utility of the bicycle reaching the TD D_{ij} before the end of the planned time t_1. Note that the value of U_{ij} is related to the time T_i at which the bicycle reaches the TD D_{ij}, and the utility and cost are opposite definitions, so it is expressed as a negative value and written as $-U_{ij}(T_i)$.

For simplicity, ϕ_i can be considered a fixed value, but it is noted that ϕ_i is related to the cyclist's travel purpose (time pressure) and is related to the distance between the position $\vec{x}(T_i)$ and the target area D_{ij} of the bicycle at time T_i. Then there is

$$\phi_i = \beta_\phi(\|D_i - \vec{x}(T_i)\|) \tag{7.49}$$

where $\beta_\phi \geq 0$ is the parameter for calibration.

Here, by incorporating the utility and cost of different alternative activity areas (alternative TDs) D_{ij} into the integrated cost formula (7.47), a combination of cyclist's path planning and activity location selection issues is considered.

7.2.1.5 Running utility/cost determination of the planned path

According to small-scale interviews and questionnaire surveys, the main influencing factors affecting the running cost of the cyclist's planned path are *riding time*, *possible conflict* with other vehicles (safety factor), *riding distance* within a certain period of time (i.e., the riding speed) and *the congestion degree* of the path (i.e., the LOS of traffic flow).

The running costs are modeled next.

The running cost L reflects the cyclist α's subjective utility/cost of the different influence attributes on the planned path. For the sake of simplicity, it is considered that the utility/cost of each attribute of the planned path is linearly integrated, namely,

$$L(t,\vec{x},\vec{v}) = \sum_k c_k L_k(t,\vec{x},\vec{v}) \tag{7.50}$$

where L_k represents the utility/cost of the path affecting attribute k, and coefficient c_k represents the relative weight (i.e., the importance) of the attribute.

Note that although the utility/cost synthesis form used here is linear, not only the model uses the utility function of the linear form but also it is effective for the utility function integrated by other methods (such as multiplication and mixing integrated methods).

In addition, since the absolute value of each attribute weight cannot be obtained from the actual data, only the relative weight information can be obtained. What is more, corresponding to different groups of cyclists, due to the differences of various factors, such as age, gender, travel purposes and physiological conditions, usually different groups have different weight coefficients c_k.

Later we specify the utility/cost L_k of different attribute k for the planned path.

Expected riding time

According to formula (7.46), the running utility/cost of the planned path is the integral with respect to time. Therefore, the expected riding time is introduced as the attribute of the planned path utility/cost, and L_1 is defined as

$$L_1(t,x,v) = 1 \tag{7.51}$$

Substitute formula (7.51) into the path running cost general formula (7.47) to get the riding time cost of the planned path:

$$E\left[\int_t^{T_i} L_1(\tau, \vec{x}(\tau), \vec{v}(\tau))d\tau\right] = E\left[\int_t^{T_i} c_1 d\tau\right] = c_1 E[T_i - t] \tag{7.52}$$

Equation (7.52) shows that the planned path cost of the riding time is equal to the product of the expected time $(T_i - t)$ of the ride and the weight c_1, and the weight c_1 reflects the urgency of the cyclist α, which is related to factors such as the travel purpose.

Safety factors

The safety factor is actually that the distance between the bicycle and other vehicles (motor vehicles, bicycles, pedestrians) is too small.

We consider that in the decision-making process of the cyclist α's path planning behavior, the vehicle m that may be in conflict with himself/herself is regarded as an obstacle and is represented by the area $O_m \subset \Omega$, $m = 1, \ldots, M$.

Because in the tactical-level route planning behavior, the cyclist α will predict his/her own position and the conflicting vehicles m of each time τ based on the internal state prediction model, namely, the $\vec{x}_\alpha(\tau)$ and $\vec{x}_m(\tau)$.

In this way, from a rational analysis, as long as it is predicted that at any time τ, he/she will not collide with the conflicting vehicle m, the cyclist will regard the conflicting vehicle as a *stationary obstacle*. And the space that bicycle needs to circumvent may be related only to the type and size of the conflicting vehicle m and not directly related to its speed.

Therefore, we think that in the tactical-level path planning model of crossing the un-signalized intersection, the cyclist α does not consider the speed factor of the conflicting vehicle m.

Assume that the running cost item L_2 generated by any of the conflicting vehicles m is a monotonically decreasing function g with respect to the boundary distance $d(\vec{x}_\alpha, O_m)$ between the bicycle α and the conflicting vehicle m. Where $\vec{x}_\alpha$ represents the location of the bicycle α and O_m represents the location of the conflicting vehicle m. That is

$$L_2(t, \vec{x}, \vec{v}) = g_m(d(O_m, \vec{x}_\alpha)) = a_m e^{(-d(\vec{x}_\alpha, O_m)/b_m)} \tag{7.53}$$

where $a_m > 0$ and $b_m > 0$ are parameters to be determined, reflecting the range of influence of obstacles. The $d(\vec{x}_\alpha, O_m)$ is the minimum boundary distance between bicycle α and conflicting vehicle m:

$$d(\vec{x}_\alpha, O_m) = \min\|B(O_m) - B(\vec{x}_\alpha)\| \tag{7.54}$$

The functions $B(O_m)$ and $B(\vec{x}_\alpha)$ represent the boundary positions of the conflicting vehicle m and the bicycle α, respectively.

The exponential form of formula (7.53) is a functional form based on the repulsive influence of obstacles on pedestrian behavior as proposed by Helbing [7].

Expected riding speed

Since bicycles are human-powered vehicles, energy consumption is also a consideration in the cyclist's path planning.

Cyclists usually try to choose the shortest path to get to the destination with the lowest energy consumption in the same amount of time.

In the next section of the Beijing bicycle crossing un-signalized intersection behavior survey, we conducted this questionnaire to convert the attribute into the same distance travel distance.

Energy consumption also plays an important role in pedestrian travel speeds. For example, during pedestrian travel, there is a close relationship between energy, walking speed and pedestrian time pressure.

As Weidmann found, the speed of the "best energy consumption level" (i.e., the minimum energy consumed per mile by pedestrians) is 1.39 m/s, which is amazingly close to the average pedestrian walking speed of 1.34 m/s [10].

Therefore, we believe that the riding speed $\|\vec{v}\|$ of a bicycle is a balance between the cyclist's time pressure and energy level. From the physics knowledge, the energy consumption of the cyclist is proportional to the square of the riding speed, namely,

$$L_3(t,\vec{x},\vec{v}) = \frac{1}{2}\|\vec{v}\|^2 \tag{7.55}$$

Level of service of the planned path

The congestion and LOS of the planned path refer to the traffic congestion and corresponding LOS formed by the bicycle and pedestrian flow of other peers around the cyclist α when crossing the un-signalized intersection. The LOS also has an impact on the route selection behavior of cyclist α.

The LOS-related cost $\zeta = \zeta(t,\vec{x})$ is generated by the congestion that the cyclist α expects at the time t at the location $\vec{x}$. Therefore, we believe that ζ is closely related to the traffic flow density $k(t,\vec{x})$ at time t and location $\vec{x}$, namely,

$$L_4(t,\vec{x},v) = \zeta(k(t,\vec{x})) \tag{7.56}$$

Through field observations, the cyclist α may prefer to choose a route with a certain other bicycle or pedestrian choice rather than a route without other bicycles and pedestrians.

Therefore, in path planning a smaller nonmotorized traffic flow (bicycle and pedestrian) density value is attractive to the cyclist α, i.e., $\zeta(k) < 0$, when the nonmotorized traffic flow density is smaller than certain threshold ($k < k_{cr}$).

In this way, the factor LOS considers the impact of bicycles and pedestrian groups or other bicycles (pedestrians) on the cyclist's path planning and reflects the mimic behavior in psychology. In addition, too many similar behaviors also increase the density, which lower the LOS, thus reflecting another behavioral pattern of "avoidance behavior" in psychology.

Here, considering that pedestrians and bicycles often mix in the intersection of mixed traffic in Chinese cities. We converted the bicycle/pedestrian mixed group into crowd density by the following conversion formula:

$$K_p = \frac{(n_p + 1.657n_b)}{S} \tag{7.57}$$

where K_p is the pedestrian density converted from the mixed group of bicycles and pedestrians, the unit is p/m^2 (person per square meter); n_p is the number of pedestrians in the group, n_b is the number of bicycles, and S is area occupied by bicycles and pedestrians (m^2).

7.2.1.6 Normalization of the attributes

"Attribute" is the characteristic, quality or performance parameter of a choice scheme. In order to facilitate the evaluation and comparison of each choice scheme, the attributes are generally expressed by quantity. However, because different attributes represent different things, the unit of measurement and the size of the dimensions are different, which is not convenient for comprehensive comparison. Therefore, we need to normalize the attribute values so that all attribute values are expressed in a uniform dimension and unit and are generally standardized as dimensionless value (0-1), which is similar to the utility values.

The commonly used normalization methods mainly include the following: (1) linear transformation method; (2) interval transformation method and (3) vector transformation method.

To ensure that the standardized attribute values change in the same direction as the utility, that is, the larger the attribute value, the greater the utility, we adopt different transformation methods are for the utility and cost attributes. For the utility attributes, we use the linear transformation in the same direction, and for the cost attributes we use the inverse linear transformation.

In the following, we introduce the normalization of each attribute L_k of the planned path utility/cost C_i during the period $[t, T_i]$.

Terminal utility/cost of the planned path

As mentioned earlier, the terminal utility/cost of planned path includes the expected utility U_{ij} obtained by the activity i at the TD D_{ij} and the non-reaching TD penalty (consequence) ϕ_i. The former is a utility indicator and the latter is a cost one. The normalized values for U_{ij} and ϕ_i are calculated as

$$S_{U_{ij}} = \frac{U_{ij}}{U_{ij}^{max}} \tag{7.58}$$

$$S_{\phi_i} = \frac{\beta_\phi \|D_i - \vec{x}(T_i)\|}{\sum_i \beta_\phi \|D_i - \vec{x}}(T_i)\| = \frac{\|D_i - \vec{x}(T_i)\|}{L} \tag{7.59}$$

where L is the total distance of planned path $P(t)$, and other variables are defined as earlier.

Expected riding time

According to the definition $L_1(t, x, v) = 1$, it can be seen that the normalized value of the expected riding time is always equal to 1, that is

$$S_t \equiv 1 \tag{7.60}$$

Safety factors

The planned path's safety factor (the cost of being too close to other vehicles) is a cost indicator, and its normalized value calculation formula S_c^m is

$$S_c^m = \frac{a_m e^{-d(\vec{x}, O_m)/b_m}}{a_m e^{-d(\vec{x}, O_m)_{\min}/b_m}} = e^{[d(\vec{x}, O_m)_{\min} - d(\vec{x}, O_m)]/b_m} \tag{7.61}$$

where $d(\vec{x}, O_m)_{\min}$ represents the minimum distance between the boundary of bicycle α and the conflicting vehicle m. In this model, we think that $d(\vec{x}, O_m)_{\min}$ is equal to the minimum distance $L_{\min}$ of the bicycle bypassing the stationary motor vehicle. Since the $L_{\min}$ obtained from the video data is about 0.3 m, we take $d(\vec{x}, O_m)_{\min} = 0.3$. So we can set

$$e^{[d(\vec{x}, O_m)_{\min} - d(\vec{x}, O_m)_{\max}]/b_m} = \varepsilon \tag{7.62}$$

and

$$b_m = \frac{d(\vec{x}, O_m)_{\min} - d(\vec{x}, O_m)_{\max}}{\ln \varepsilon} \tag{7.63}$$

where ε is a very small number that can be ignored, taking $\varepsilon = 0.01$. $d(\vec{x}, O_m)_{\max}$ is the maximum distance of conflicting motor vehicles on bicycles. Considering that most un-signalized intersections do not exceed 25 m in width, we think the repulsive force is negligible when the vehicle is 25 m away in this situation, i.e., $d(\vec{x}, O_m)_{\max} = 25$ m. Calculated as $b_m \approx 5.4$, taking $b_m = 5$ in the model. Then

$$S_c^m = e^{[0.3 - d(\vec{x}, O_m)]/5} \tag{7.64}$$

Expected riding speed

The expected riding speed $\|\vec{v}\|$ is a cost indicator, normalized value S_v of which is calculated as

$$S_v = \left(\frac{\|\vec{v}(t)\|}{\|v^{\max}\|}\right)^2 \tag{7.65}$$

The maximum riding speed $v^{\max}$ is determined by the physical conditions of the cyclist α, it is usually taken as 7 m/s [11].

LOS of the planned path

From the earlier we can see that the cost ς generated by the congestion of non-motorized traffic (bicycles and pedestrians) is an interval cost indicator, so the normalized value of this attribute S_ς is calculated as

$$S_\varsigma = \frac{k(t,\vec{x}) - k_c}{k_{\max}} \tag{7.66}$$

where $k(t,\vec{x})$ represents the mixed bicycle–pedestrians traffic flow density of at time t at position $\vec{x}$; k_c represents the pedestrian traffic flow density threshold that attracts the cyclist when planning path; $k_{\max}$ represents the maximum bicycle–pedestrian mixed traffic density threshold at the un-signalized intersection.

In the model, the values of k_c and $k_{\max}$ are considered to refer to the density values of pedestrian traffic flow LOS A, B and LOS E as specified in the *Highway Capacity Manual* [11]. According to this manual, the bicycle (or pedestrian) LOS A, B has a density value of $k < 0.18$, and $k = 0.18 - 0.27\ \mathrm{p/m^2}$, and the density value k of LOS E is 1.1–1.33 p/m^2. Therefore, $k_c = 0.3\ \ \mathrm{p/m^2}$ and $k_{\max} = 1.5\ \mathrm{p/m^2}$ are taken in the model. Then

$$S_\varsigma = \frac{k(t,\vec{x}) - 0.3}{1.5} \tag{7.67}$$

- **Utility/cost function of the path planning model**

Substituting the calculation formula for the normalized value of each attribute of the path running cost (7.58)–(7.67) into (7.50), we get

$$L(t,x,v) = c_1 S_t + c_2 \sum_m S_c^m + \frac{1}{2} c_3 S_v + c_4 S_\varsigma \tag{7.68}$$

The normalized value S_ϕ of the terminal utility/cost ϕ is calculated as

$$S_{\phi_i} = \begin{cases} -\dfrac{U_{ij}}{U_{ij}^{\max}} & \vec{x}(T_i) \in D_{ij} \\ \dfrac{\|D_i - \vec{x}(T_i)\|}{L} & \vec{x}(T_i) \notin D_{ij} \end{cases} \tag{7.69}$$

Let the cyclist α path planning period be $(t_0, T_f]$ and $T_0 = t_0$, the subjective utility/cost function for the cyclist's planned path is

$$\begin{aligned} C &= \sum_{i\in A_\Omega} C_i\left(T_i, \vec{v}_{[T_{i-1},T_i)} | T_{i-1}, \vec{x}(T_{i-1})\right) = \sum_{i\in A_\Omega} E\left[\int_{T_{i-1}}^{T_i} L(\tau,\vec{x}(\tau),\vec{v}(\tau))d\tau + \phi(T_i,\vec{x}(T_i))\right] \\ &= \sum_{i\in A_\Omega} E\left[\int_{T_{i-1}}^{T_i} \left\{= c_1 + c_2 \sum_m e^{[0.3-d(\vec{x},O_m)]/5} + \frac{1}{2} c_3 \left(\frac{\|\vec{v}(t)\|}{7}\right)^2 + c_4 \frac{k(t,\vec{x}) - 0.3}{1.5}\right\} d\tau + S_{\phi_i}\right] \end{aligned} \tag{7.70}$$

It is noted that, in general, the cyclist's weighting coefficient c_k for each attribute of the line is similar to the "spiritual value" of the selection behavior, and its value has no practical significance.

For the cyclist α, the relative ratio between the relative weights c_k will have an effect on the path planning behavior, and only the relative ratio information of the undetermined coefficient c_k can be obtained from the actual data. Without loss of generality, we make $c_1 = 1$ in our model.

When there are different groups of the cyclists, the path planning behavior is also different due to differences in factors such as riding characteristics, physical conditions and travel destinations among different cyclist groups. In fact, it is precisely because different cyclists have differences in path planning behavior that these attributes are included and integrated in the path cost formula. Different groups of cyclists correspond to the relative weight coefficients c_k of different attributes, terminal utility functions ϕ and maximum speed limit v_{max}.

For example, in the group of commuters and riders, the riding time tends to dominate the path planning behavior. Therefore, the proportion of c_1 in the relative weight coefficient c_k is relatively high, while in the cyclist group for leisure and entertainment purposes, time is not an important factor, and the corresponding c_1 is relatively low in the relative weight coefficient c_k.

7.2.1.7 Optimal path planning for continuous time and space

In the NCB theory framework, the cyclist's path planning model when crossing the un-signalized intersection adopts a continuous space–time dynamic programming method.

This method can be used to model the selection/planning problem in continuous time and space. Since the alternate options are infinite, the choice of path is unlimited too. Therefore, it is more flexible than the discrete choice method and can more accurately reflect the cyclist's path planning behavior at the un-signalized intersections.

In addition, the multistage dynamic decision-making process using dynamic programming method better reflects the characteristics of people's decision-making behavior.

In addition, the continuous space–time model has great flexibility and universality. The planning start and end times can be set freely, such as the time interval (t_0, t_f)—from the cyclist entering the intersection t_0 to the crossing time of the intersection t_f, or at any time of the intersection (t_1, t_2), as long as $t_1 < t_2$ and $t_2 \leq t_f$ are satisfied, the time interval of the path planning can be freely set.

In addition, when performing model solving operations, the necessary discretization is performed.

In the discretization, the size of the time step h and the space step k can be determined freely and flexibly. The sizes of h and k can be adjusted on the basis of computational complexity, model accuracy requirements and behavioral assumptions.

By setting the discrete time step h and the dynamic planning period, the decision stage of dynamic programming can be controlled. In special cases, the multistage path planning decision process can be degraded into a single-stage choice decision process, so that our dynamic programming model degenerates into a discrete choice model.

The following is a specific formulation of the cyclist's continuous space–time optimal path planning problem at the un-signalized intersection.

It can be seen that the cyclist expects (predicts) the path and activity cost as a function of the initial time t_0, the cyclist's initial position $\vec{x}(t_0) = \hat{x}$ and the bicycle speed $\vec{v}_{[t,T_i)}$. Cyclists typically will choose the path (consisting of D_i and T_i) with the lowest expected path cost, maximum activity utility (or lowest terminal cost) and the corresponding speed $\vec{v}_{[t,T_i)}$.

The cyclist's path planning can be regarded as an optimal dynamic programming problem, so that the cyclist α's path planning period is $\left(t_0, T_f\right]$, and $T_0 = t_0$. Then the path planning model formula is as follows:

Objective function: $W(t_0, \hat{x})$:

$$= \min \sum_{i \in A_\Omega} E\left[\int_{T_{i-1}}^{T_i} L(\tau, \vec{x}(\tau), \vec{v}(\tau))d\tau + \phi(T_i, \vec{x}(T_i))\right]$$

$$= \min \sum_{i \in A_\Omega} E\left[\int_{T_{i-1}}^{T_i} \left\{c_1 + c_2 \sum_m e^{[0.3-d(\vec{x},O_m)]/5} + \frac{1}{2}c_3\left(\frac{\|\vec{v}(t)\|}{7}\right)^2 + c_4\frac{k(t,\vec{x}) - 0.3}{1.5}\right\}d\tau + S_{\phi_i}\right] \tag{7.71}$$

$$\begin{aligned}
\text{With constraints}: \; & d\vec{x} = \vec{v}dt + \sigma d\vec{w} \quad \text{(motion constraint)} \\
& \vec{x}(t_0) = \hat{x} \quad \text{(initial state constraints)} \\
& \vec{v} \in \vec{V}_a(t,\vec{x},V), \;\; \text{and} \;\; \vec{V}_a(t,\vec{x},V) \\
& = \{\vec{v}, \|\vec{v}\| \leq \vec{v}_{\max}(t,\vec{x},V)\} \subset \mathbb{R}^2 \quad \text{(speed range constraint)}
\end{aligned} \tag{7.72}$$

Since the time t and the space $\vec{x}(\tau)$ in the model are continuous variables, the path planning of continuous time and space is realized. As the solution of continuous space–time optimization problem is not the focus of this book, and the theoretical method is mature, interested readers can refer to the relevant reference materials of operations research and mathematics [12], and we will not repeat them here.

It is noted that the dynamic path planning model can be obtained by updating the model initial time t_0 and the initial position $\vec{x}(t_0) = \hat{x}$. In addition, by discretizing continuous time and space, that is, dividing the un-signalized intersection Ω into small squares of $\delta \times \delta$, the time is divided into $t_k = kh$. That is, δ and h, respectively, represent the minimum unit of space and time in the model. In this way, the optimal solution of the model can be implemented on a computer [6].

7.2.2 The cyclist's dynamic riding behavior model (operational level)

According to the specific situation of cyclist crossing the un-signalized intersection, the following four functional modules in the NCB operational level "social field" model are specifically formulated: goal tension, the cyclist's perception, effect tension and integrated social tension.

7.2.2.1 Goal tension along a given path

Because in the cyclist crossing the un-signalized intersection, the tactical level path planning model is discretized when solving, the input of the operational-level behavior model is actually a series of ordered space–time pairs (D_i, T_i). D_i and T_i, respectively, represent the location and expected arrival time of each TD of the path $P(t)$ planned in the tactical level:

$$Path(t) = \{(D_i, T_i)\}, \quad i = 1, 2, \ldots, k \tag{7.73}$$

where k is the number of TDs of the path $P(t)$ determined by the path planning model.

In the cyclist's dynamic riding behavior model crossing the un-signalized intersection, the cyclist α wants to cross the un-signalized intersection Ω, the current time is t, and the cyclist α is currently located at $\vec{x}(t)$. The nearest space–time pair (D_i, T_i) to the current time t in the planned path $P(t)$ is taken as the current destination D (TD) for cyclist α, that is, $Min\{(T_i - t) > 0\}$.

However, if the cyclist α's historic position $\vec{x}(\tau)(t_0 \leq \tau \leq t)$ has passed the TD D_i, that is, $D_i \in \vec{x}(\tau)$, $t_0 \leq \tau \leq t$, it is considered that the TD has been reached, and the next TD (D_{i+1}, T_{i+1}) is taken as the current destination for α. In summary, the current destination D is

$$D = D_j, \quad \text{if} \quad j = \arg\min\{(T_i - t) > 0\} \text{ and } D_j \notin \vec{x}(\tau),\ t_0 \leq \tau \leq t,\ j \in A_\Omega \tag{7.74}$$

According to the "shortcut" rule of people's behavior, the cyclist α is at time t and at the position $\vec{x}(t)$. If he/she wants to reach the TD D_i at the time T_i, his/her ideal speed $\vec{u}_\alpha^0(t)$ usually has the following properties:

- From the perspective of riding comfort, the cyclist α tends to ride at a constant speed to avoid unnecessary acceleration and deceleration, namely,

$$u_\alpha^0(t) \approx const. \tag{7.75}$$

where $u_\alpha^0(t)$ represents the magnitude of the ideal velocity $\vec{u}_\alpha^0(t)$, which is a scalar, and *const.* represents a constant variable. The following we use $\vec{e}_\alpha^0$ to indicate the direction of $\vec{u}_\alpha^0(t)$. The linear distance $s_\alpha(t)$ between the cyclist α and his/her current TD D_i is

$$s_\alpha(t) := \|D_i - \vec{x}(t)\| \tag{7.76}$$

If the cyclist wants to reach TD D_i at the expected time T_i, the riding time is $(T_i - t)$, so we have

$$u_\alpha^0(t) = \frac{s_\alpha(t)}{T_i - t} \tag{7.77}$$

- The cyclist α's current riding direction $\vec{e}_\alpha^0$ should be directed toward the current destination D_i. So we have

$$\vec{e}_{\alpha}^{0} := \frac{D_i - \vec{x}(t)}{\|D_i - \vec{x}}(t)\| \tag{7.78}$$

Combining the magnitude and direction of the ideal riding speed $\vec{u}_{\alpha}^{0}(t)$ earlier, we get the cyclist α's ideal speed $\vec{u}_{\alpha}^{0}(t)$:

$$\vec{u}_{\alpha}^{0}(t) := \frac{D_i - \vec{x}(t)}{T_i - t} = \frac{\|D_i - \vec{x}(t)\|}{T_i - t}\vec{e}_{\alpha}^{0} \tag{7.79}$$

We assumed that the cyclist α's current riding speed is $\vec{v}_{\alpha}(t) = d\vec{x}(t)/dt$, with the "mass" m_{α}. In order to achieve the ideal riding speed $\vec{u}_{\alpha}^{0}(t)$, the dynamic force $\vec{F}_{\alpha}^{0}$ is required to generate the acceleration $d\vec{v}_{\alpha}(t)/dt$. From the Newton dynamic formula we have

$$\vec{F}_{\alpha}^{0} \equiv m_{\alpha}\frac{d\vec{v}_{\alpha}(t)}{dt} := \gamma_{\alpha}\left[\left(\vec{v}_{\alpha}^{0}(t) - \vec{v}_{\alpha}(t)\right)\right] \tag{7.80}$$

Here we need to limit the ideal riding speed $\vec{v}_{\alpha}^{0}(t)$, such as the maximum riding speed $u_{\alpha}^{\max}$ specified by the cyclist's physical strength or intersection, and the minimum riding speed $u_{\alpha}^{\min}$ unique to the bicycle motion feature itself, the ideal speed after consideration is

$$\vec{v}_{\alpha}^{0}(t) := \vec{e}_{\alpha}^{0} \cdot \begin{cases} u_{\alpha}^{\min} & \text{for} \quad u_{\alpha}^{0}(t) < u_{\alpha}^{\min} \\ u_{\alpha}^{0}(t) & \text{for} \quad u_{\alpha}^{\min} \le u_{\alpha}^{0}(t) \le u_{\alpha}^{\max} \\ u_{\alpha}^{\max} & \text{for} \quad u_{\alpha}^{0}(t) > u_{\alpha}^{\max} \end{cases} \tag{7.81}$$

Note that the definition of cyclist α's "mass" m_{α} is different from that of physical Newtonian dynamics, because the dynamic force here is not the Newtonian force in the physical sense, but the cyclist α's Social Force form his/her "goal social field."

Therefore, in the social field model, similar to the influence of the mass on the motion of the object, parameter m_{α} is used to characterize the influence of the moving vehicle type (motor vehicle, bicycle or pedestrian) on the dynamics of the social field. There are corresponding "mass" parameters m_c and m_p for motor vehicles and pedestrians.

It can be seen from (7.80) that the magnitude of the goal force $\vec{F}_{\alpha}^{0}$ is proportional to the difference $\left(\vec{v}_{\alpha}^{0}(t) - \vec{v}_{\alpha}(t)\right)$ between the cyclist α's ideal speed $\vec{v}_{\alpha}^{0}(t)$ and the current riding speed $\vec{v}_{\alpha}(t)$.

The larger the difference between the ideal speed and the current riding speed (including the size and direction), the greater the driving force of the destination D_i to α. If the cyclist α's ideal speed $\vec{v}_{\alpha}^{0}(t)$ is the same as the current riding speed $\vec{v}_{\alpha}(t)$ in both the direction and magnitude, i.e., $\vec{v}_{\alpha}^{0}(t) - \vec{v}_{\alpha}(t) = \vec{0}$, then cyclist α does not need to change the current riding behavior, there is no behavior change force (without considering other influences and random interference). This is consistent with people's common sense and basic knowledge of psychology.

In (7.80), a motivation parameter γ_α is introduced, which is used to characterize the relationship between the behavioral dynamics $\vec{F}_\alpha^0$ produced by the target of the cyclist α and the current speed and the ideal speed difference $\left(\vec{v}_\alpha^0(t) - \vec{v}_\alpha(t)\right)$ of α. It can be seen that the motivation parameter γ_α characterizes the urgency of the cyclist α to achieve the desired speed $\vec{v}_\alpha^0(t)$, i.e., the degree of time pressure. That is, the $\gamma_\alpha \vec{v}_\alpha^o(t)$ characterizes the forward valence (i.e., the target tension) that the TD D_i produces for cyclist α, that is, the effect tension of the TD $\vec{f}_\alpha^0$:

$$\vec{f}_\alpha^0(t) = \gamma_\alpha \left(\vec{v}_\alpha^0(t) - \vec{v}_\alpha(t)\right) \tag{7.82}$$

In the latter integrate module, $\vec{f}_\alpha^0$ interacts with the effect tension of the objects to form an integrated behavioral force $\vec{F}_\alpha$.

For the sake of simplicity, in the following modeling process, the simple case $\vec{v}_\alpha^0(t) = \vec{u}_\alpha^0(t)$, i.e., the case of $u_\alpha^{\min} \le u_\alpha^0(t) \le u_\alpha^{\max}$, is considered, unless otherwise specified.

7.2.2.2 Cyclist perception function

The impact of the dynamic traffic environment at the intersection on the riding behavior of the cyclist α must be influenced by the cyclist's perception. According to psychological knowledge, 80% of the environmental information (stimulus) of ordinary people comes from the visual system, and other sensory systems only account for less than 20% [9].

In the behavioral model of the bicycle crossing the un-signalized intersection, it is considered that the cyclist's line of sight is toward the direction of the riding destination. That is, at the time t, α's line of sight coincides with that in formula (7.78). Let the effective angle of view of the cyclist α be 2φ and take $\varphi = 60°$ in the model (refer to the model of Helbing) [13], then $2\varphi = 120°$. The schematic diagram is shown in Figure 7.5.

It is assumed that cyclist α completely perceives the information of the effect object (vehicle) β_i within the effective angle of view 2φ of the α, and only the partial information of the object β_i other than the effective angle of view 2φ is

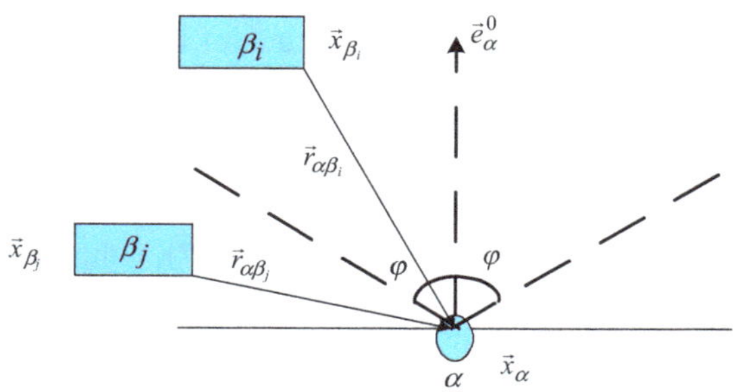

Figure 7.5 Schematic diagram of the cyclist's perspective

perceived. Then there is a discount corresponding to the effect tension of them, expressed as

$$P(\beta) = w\left(\vec{e}_\alpha^0(t), \vec{r}_{\alpha\beta}(t)\right) := \begin{cases} 1 & \text{if} \quad \vec{e}_\alpha^0(t) \cdot \vec{r}_{\alpha\beta}(t) \geq \|\vec{r}_{\alpha\beta}(t)\| \cos \varphi \\ c & \text{otherwise} \end{cases} \tag{7.83}$$

$$\vec{r}_{\alpha\beta}(t) = \vec{x}_\beta(t) - \vec{x}_\alpha(t) \tag{7.84}$$

where $\vec{r}_{\alpha\beta}(t)$ is the relative vector of the position of the object β and the bicycle α. c is the perceived discount coefficient, $0 < c < 1$, and $c = 0.5$ in the model. That is, the amount of information generated by objects that are not within the range of the cyclist's α angle of view will be reduced by half. And this is the perception function $P(\beta)$ of the object β to the cyclist α.

The perceptual function $P(\beta)$ represents the reduction of the perceptual information of the object β by the cyclist α at time t, acting on the "social field" $\vec{w}^{\beta}{}_{\alpha}(\vec{x}, t)$ and mental tension $\vec{f}_\beta(\vec{x}, t)$ produced by β, namely,

$$\vec{w}_\alpha^\beta(\vec{x}, t) = W_\alpha^\beta(P(\beta), \vec{x}, t) = P(\beta) \cdot W_\alpha^\beta(\vec{x}, t) \tag{7.85}$$

$$\vec{f}_\beta(\vec{x}, t) = P(\beta) \cdot \left[-\nabla W_\alpha^\beta(\vec{x}, t)\right] \tag{7.86}$$

7.2.2.3 Repulsive and attractive tension function

In this book, we study the model of conflict avoidance behavior between bicycles and other road users (motorcars, bicycles and pedestrians) as they cross un-signalized intersections. Compared with the Helbing's pedestrian "social field" model, the model is characterized by extending the concept of "social field" mental tension from pedestrian interactions to three different modes of transportation: motorcars, bicycles and pedestrians. These three types of road users not only have large differences in size but also have different speeds and behaviors. Therefore, the influence of the "social field" on the mental tension cannot apply the model of Heilbing [13] directly.

The model here only considers the repulsive effect social field and does not consider the attractive effect social field.

Considering that the middle of the un-signalized intersection is a plane, not lanes like a road section. The path choice of nonmotorized vehicles (bicycles and pedestrians) in the intersection is relatively free, and the way they avoid motor vehicles and other bicycle pedestrians is somewhat similar to the movement of charged particles in an electric field. In the model, it is considered that when a conflicting vehicle (motorcar, bicycle or pedestrian) or object β appears to form a certain repulsive "social field" $\vec{w}_\beta$, it has the following properties:

- The field strength (or potential energy) of "social field" $\vec{w}_\beta$ is related to the amount of information that the cyclist α perceives the object β, that is, $P(\beta)$.

The more complete the cyclist α perceives the amount of information on the object, the effect of the "social field" $\vec{w}_\beta$. The influence of the field strength is correspondingly larger. On the contrary, if the cyclist α does not fully understand the information amount of the object β, the influence field strength of $\vec{w}_\beta$ is correspondingly reduced.

- The field strength (or potential energy) of $\vec{w}_\beta$ is related to the size and type of object β (motorcar, bicycle, pedestrian) and the moving speed $\vec{v}_\beta$. The larger the object, the higher the speed of the object, the potential energy will be greater.
- The field strength equipotential line of $\vec{w}_\beta$ is elliptical, and its long axis direction is consistent with the velocity direction $\vec{e}_\beta$ of β; when β stands still (i.e., $\vec{v}_\beta = \vec{0}$), it is considered that the field strength equipotential line of $\vec{w}_\beta$ is circular at this time. The potential energy function is only related to the size and type of β.
- The field strength (or potential energy) of $\vec{w}_\beta$ is a decreasing nonnegative function of distance l. The further away from object β, the larger l is, the smaller the field strength is. At a great distance L, the field strength disappears to zero; and the rate of field strength reduction is also decreasing, that is, the field strength of $\vec{w}_\beta$ is a monotonically decreasing function and a concave function with respect to distance l.

Correspondingly, the effect of the object β's effect "social field" $\vec{w}_\beta$ on the cyclist α has the following properties:

- The size of $\vec{f}_\beta$ is related to the amount of information that the cyclist α perceives the object β, i.e., the result of $P(\beta)$. And the more complete the perception of the amount of information of the object β by the cyclist α, the greater the effect tension; if the cyclist's information about β is incomplete, the effect tension is relatively small.
- The size of $\vec{f}_\beta$ is related to the size and type of the affected object β (motorcar, bicycle, pedestrian), the magnitude and direction of β's moving speed $\vec{v}_\beta$, and the cyclist α's position $\vec{x}_\alpha$ and α's current riding speed $\vec{v}_\alpha$. The larger the size of the object, the higher the speed, the smaller the distance $\|\vec{r}_{\alpha\beta}\|$ between the cyclist α and the object β, and the greater the effect tension $\vec{f}_\beta$.
- The direction of $\vec{f}_\beta$ is along the line $\vec{r}_{\alpha\beta}$ between object β and cyclist α of $\vec{x}_\beta$ and $\vec{x}_\alpha$. The direction can be $\alpha \to \beta$ (attractive tension) or $\beta \to \alpha$ (repulsive tension).
- The size of $\vec{f}_\beta$ is a decreasing nonnegative function of the distance $\|\vec{r}_{\alpha\beta}\|$ between the cyclist α and the object β. The larger $\|\vec{r}_{\alpha\beta}\|$ is, the smaller the field strength is, the smaller $\|\vec{r}_{\alpha\beta}\|$ is, the larger the field strength is, but $\|\vec{r}_{\alpha\beta}\|$ cannot be 0 (we assume that no collision occurs), or $\vec{f}_\beta(\vec{r}_{\beta\alpha} = 0) = +\infty$. Moreover, the rate of decrease of the field strength is also decreasing, that is, the size of $\vec{f}_\beta$ is a monotonically decreasing function and a concave function with respect to the distance $\|\vec{r}_{\alpha\beta}\|$.

Based on the previous assumptions and properties of the effect "social field" $\vec{w}_\beta$ and the effect tension $\vec{f}_\beta$, we think that in the bicycle crossing the un-signalized intersection operation level model, the field strength equipotential line of the "social field" $\vec{w}_\beta$ generated by the conflicting object/obstacle β on cyclist α is shown in Figure 7.6.

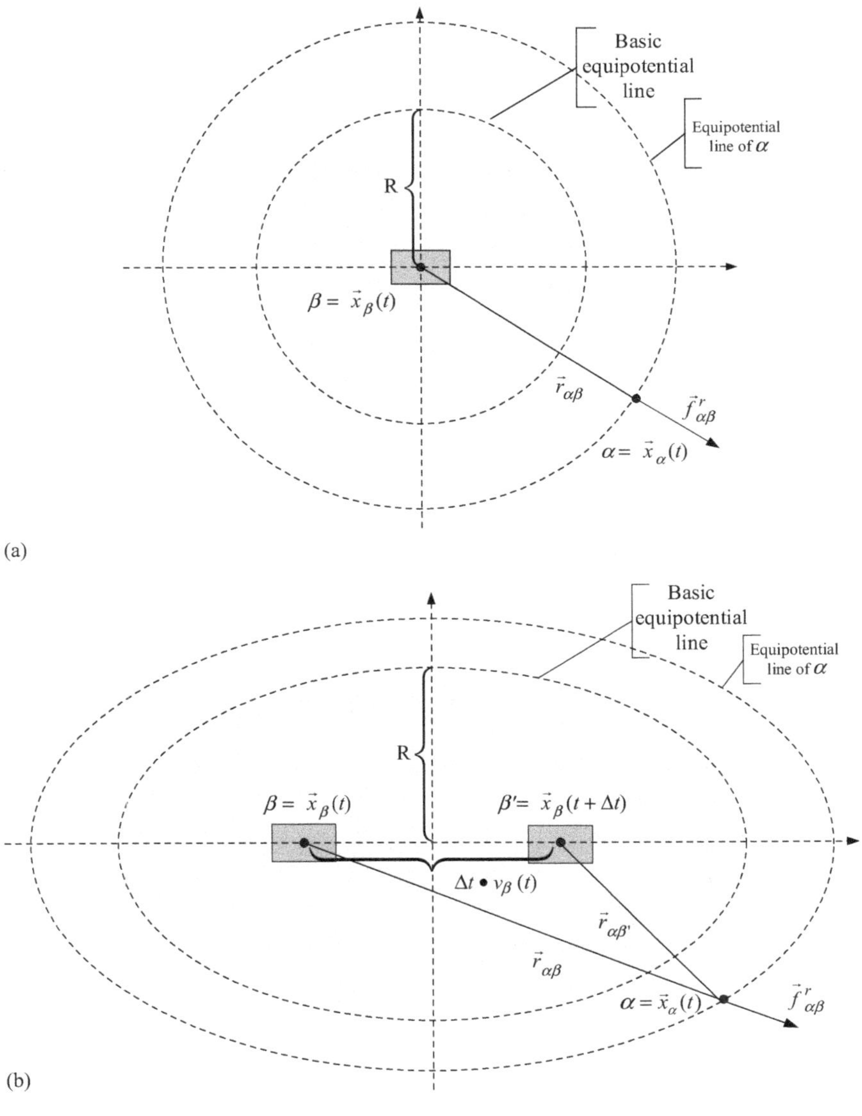

Figure 7.6 Effect "social field" strength diagram: (a) static object; (b) moving object

Wherein, the direction $\vec{e}_{\alpha\beta}$ of the effect tension $\vec{f}_\beta$ is from the effect object β to cyclist α at the current time t, the value is

$$\vec{f}_{\alpha\beta}(\vec{r}_{\alpha\beta}, \vec{v}_\beta) = -\nabla_{\vec{r}_{\alpha\beta}} w_\beta(R, a) \tag{7.87}$$

where $\vec{r}_{\alpha\beta}$ is the vector difference between position vectors $\vec{x}_\beta(t)$ and $\vec{x}_\alpha(t)$ of the effect object β and cyclist α at the current time t, with

$$\vec{r}_{\alpha\beta} = \vec{x}_\beta(t) - \vec{x}_\alpha(t) \tag{7.88}$$

where $\nabla_{\vec{r}_{\alpha\beta}}$ is the gradient of the field along the $\vec{r}_{\alpha\beta}$ direction.

$w_\beta(R, a)$ represents the social field potential energy generated by β relative to α. The parameter R is the short radius of the ellipse in the figure, which represents the field strength of the object β as static and is a parameter related to the type and size of the object β (motor vehicle, bicycle, pedestrian). The shape of the basic elliptical equipotential line is determined by the short radius R and the long radius a', with

$$2a' = \sqrt{4R^2 + (\Delta t \cdot v_\beta)^2} \tag{7.89}$$

And $a > 0$ represents the action parameter of the social field $w_\beta(R, a)$ to α, which is the ratio of the long radius L of the elliptical equipotential line where α is located and the long radius a' of the basic equipotential line. The calculation formula is

$$\begin{aligned} a = \frac{2L}{2a'} &= \frac{\|\vec{r}_{\alpha\beta}\| + \|\vec{r}_{\alpha\beta'}\|}{\sqrt{4R^2 + (\Delta t \cdot v_\beta)^2}} \\ &= \frac{\|\vec{x}_\beta(t) - \vec{x}_\alpha(t)\| + \|\vec{x}_\beta(t + \Delta t) - \vec{x}_\alpha(t)\|}{\sqrt{4R^2 + (\Delta t \cdot v_\beta)^2}} \end{aligned} \tag{7.90}$$

where Δt is the predicted time interval of cyclist α, and its size is related to the time pressure of α. It can be seen from (7.90) that the farther β is from α, the larger a is, so the magnitude of the effect "social field" strength (tension) can be regarded as a function of the variable a. The potential energy function $w_\beta(R, a)$ has the following mathematical properties:

1. $w_\beta(R, a)$ is a monotonically decreasing nonnegative function of the variable a ($a > 0$), namely,

$$w_\beta(R, a) \geq 0 \tag{7.91}$$

$$\frac{dw_\beta(R, a)}{da} < 0 \tag{7.92}$$

2. When a is quite large, that is, β is very far from α, it can be considered that the field strength is zero, namely,

$$\lim_{a->\infty} w_\beta(R, a) = 0 \tag{7.93}$$

3. The farther β is from α, the smaller the difference in field strength due to equal distance spacing, namely,

$$\frac{dw_\beta(R,a)^2}{d^2a} > 0 \tag{7.94}$$

After analysis, we found that there are two main functional forms that conform to the previous properties of $w_\beta(R,a)$:

1. Exponential function form:

$$w_\beta(R,a) = M_\beta Re^{-ba}, \quad a > 0 \tag{7.95}$$

2. Power function form:

$$w_\beta(R,a) = M_\beta Ra^{-b}, \quad a > 0 \tag{7.96}$$

If we let when $a = 1$, $w_\beta(R,a) = 1$, then substituting into formulas (7.95) and (7.96), respectively, we get

$$R = \frac{e^b}{M_\beta} \tag{7.97}$$

$$\text{and } R = \frac{1}{M_\beta} \tag{7.98}$$

Regardless of M_β (Suppose we take the default value $M_\beta = 1$, and R=1), when the value of b is not too large (such as $b < 3$), the strength values of the two function forms are quite close (see Figure 7.7).

As can be seen from the figure, the difference in value between the two functional forms is actually small, especially when a is about 1 (i.e., the distance between the cyclist α and the effect object β is not too close or too far) in the case of b> 1. Therefore, for the sake of simplicity, the power function form is chosen as

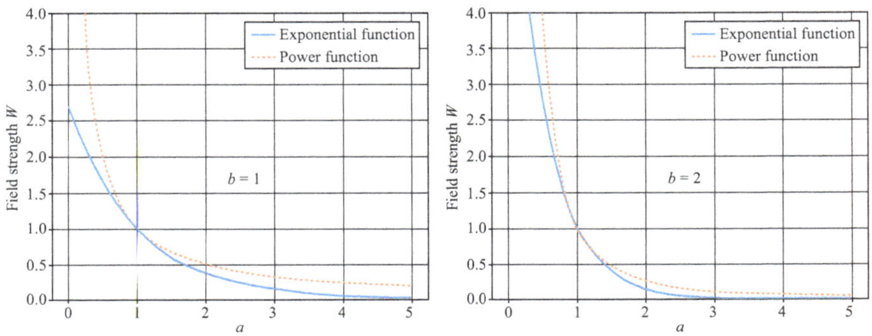

Figure 7.7 Comparisons of two field strength functions

the field strength form, and using $b = 1$, the corresponding effect of the tension amplitude function is

$$\frac{f_{a\beta} = \|dw_\beta(R,a)}{d\,a\| = b \cdot M_\beta \cdot Ra^{-(b+1)} = M_\beta \cdot Ra^{-2}, \quad a > 0} \tag{7.99}$$

7.2.2.4 Integration of different tensions

The effects of different mental tensions $\vec{f}^{a}_{\beta}$ and $\vec{f}^{r}_{\beta}$ (which may be conflicting) on the cyclist α's "social field," of course, cannot be simply superimposed as physics dynamics to obtain the cyclist α's final behavioral dynamics $\vec{F}_\alpha(\vec{x},t)$, though it is handled this way in the pedestrian "social field" model [13].

As mentioned earlier, the effect tension $\vec{f}_\beta$ (the lead distance value) of the object β on the cyclist α is similar to the goal tension $\gamma_\alpha v^o_\alpha(t)$ generated by the current destination D. Thus, by replacing the $\gamma_\alpha \vec{v}^o_\alpha(t)$ in formula (7.80), we have

$$m_\alpha \frac{d\vec{v}_\alpha(t)}{dt} := \vec{f}_\beta - \gamma_\alpha \vec{v}_\alpha(t) \tag{7.100}$$

Further generalizing to effect tensions generated by multiple objects β and destination D, we have

$$m_\alpha \frac{d\vec{v}_\alpha(t)}{dt} := \left[\sum_\beta \vec{f}_\beta + \gamma_\alpha \vec{v}^o_\alpha(t)\right] - \vec{v}_\alpha(t)\gamma_\alpha \tag{7.101}$$

However, the effect tensions generated by various objects (vehicles, obstacles, destinations, etc.) on the cyclist may be conflicting. According to Levin's "social field" theory, a new motivation $\vec{f}^{m}_{\alpha}$ is formed:

$$\vec{f}^{m}_{\alpha}(t) = \vec{f}^{m}_{\alpha}\left(\vec{f}_\beta, \gamma_\alpha \vec{v}^o_\alpha(t), t\right) \tag{7.102}$$

where $\vec{f}^{m}_{\alpha}(\bullet)$ is the integration function.

Then, at this time, the cyclist α's behavioral dynamics $\vec{F}_\alpha(t)$ would be

$$\vec{F}_\alpha(t) = \vec{f}^{m}_{\alpha}\left(\vec{f}_\beta, \gamma_\alpha \vec{v}^o_\alpha(t), t\right) - \gamma_\alpha \vec{v}_\alpha(t) \tag{7.103}$$

Then the riding behavior of the cyclist α is

$$m_\alpha \frac{d\vec{v}_\alpha(t)}{dt} = \vec{F}_\alpha(t) = \vec{f}^{m}_{\alpha}\left(\vec{f}_\beta, \gamma_\alpha \vec{v}^o_\alpha(t), t\right) - \gamma_\alpha \vec{v}_\alpha(t) \tag{7.104}$$

Later we determine the specific formula of the integration function $\vec{f}^{m}_{\alpha}(\bullet)$ based on the cyclist's microscopic behavior at the un-signalized intersection.

Consider the following situation, when cyclist α is expected to have a collision between motor vehicle β_i and bicycle β_j at the un-signalized intersection, if the repulsive tensions $\vec{f}^{r}_{\beta_i}$ and $\vec{f}^{r}_{\beta_j}$ of motor vehicle β_i and bicycle β_j are in the same direction. Then the total repulsive tension $\vec{f}^{r}_{\beta}$ felt by the cyclist α will be the sum of the repulsive tension $\vec{f}^{r}_{\beta_i}$ and $\vec{f}^{r}_{\beta_j}$, or the larger one of the two repulsive tensions.

According to common sense and experience, usually the cyclist's idea is to avoid the more dangerous object, and the less dangerous objects avoid it by the way. Therefore, the total repulsive tension $\vec{f}^r_\beta$ felt by the cyclist α would be the larger one of the tensions $\vec{f}^r_{\beta_i}$ and $\vec{f}^r_{\beta_j}$.

If the repulsive tensions $\vec{f}^r_{\beta_i}$ and $\vec{f}^r_{\beta_j}$ generated by the motor vehicle β_i and the bicycle β_j are opposite, the cyclist may balance the two conflicting repulsive tensions. What he/she feels in this case may be the offset tension between the two, that is, the superposition of $\vec{f}^r_{\beta_i}$ and $\vec{f}^r_{\beta_j}$.

Therefore, in this model, it is considered that the cyclist α has two different "social field" effect tension $\vec{f}_{\beta_i}$ and $\vec{f}_{\beta_j}$ generated by objects β_i and β_j, respectively. If $\vec{f}_{\beta_i}$ and $\vec{f}_{\beta_j}$ have the same direction (rather than the opposite direction) of the force components $\vec{f}^{x+}_{\beta_i}$ and $\vec{f}^{x+}_{\beta_j}$ (or $\vec{f}^{x-}_{\beta_i}$ and $\vec{f}^{x-}_{\beta_j}$), it is considered that the cyclist α will only consider the maximum tension (thumb principle). If the component tensions $\vec{f}^{x}_{\beta_i}$ and $\vec{f}^{x-}_{\beta_j}$ are in the opposite direction, it is considered that the two component tensions cancel each other out and perform a simple superposition (offset principle). This is formulated as

$$\vec{f}^x_\beta = \max\left(\vec{f}^{x+}_{\beta_i}, \vec{f}^{x+}_{\beta_j}\right) + \max\left(\vec{f}^{x-}_{\beta_i}, \vec{f}^{x-}_{\beta_j}\right) \tag{7.105}$$

where $\vec{f}^{x+}_\beta$ is the effect tension component in the positive direction of the X-axis and $\vec{f}^{x-}_\beta$ is the component of the negative direction of the X-axis. The Y-axis direction is similar to the X-axis formula:

$$\vec{f}^y_\beta = \max\left(\vec{f}^{y+}_{\beta_i}, \vec{f}^{y+}_{\beta_j}\right) + \max\left(\vec{f}^{y-}_{\beta_i}, \vec{f}^{y-}_{\beta_j}\right) \tag{7.106}$$

Generalize this to all affected objects $\beta_i \in \beta$ and the current destination D producing the goal tension $\gamma_\alpha \vec{v}^o_\alpha(t)$, we have

$$\vec{f}^{mx}_\alpha = \max_{\beta_i \in \beta}\left[\vec{f}^{x+}_{\beta_1}, \ldots, \vec{f}^{x+}_{\beta_m}, \gamma_\alpha \vec{v}^{ox+}_\alpha(t)\right] + \max_{\beta_i \in \beta}\left[\vec{f}^{x-}_{\beta_1}, \ldots, \vec{f}^{x-}_{\beta_m}, \gamma_\alpha \vec{v}^{ox-}_\alpha(t)\right] \tag{7.107}$$

and

$$\vec{f}^{my}_\alpha = \max_{\beta_i \in \beta}\left[\vec{f}^{y+}_{\beta_1}, \ldots, \vec{f}^{y+}_{\beta_m}, \gamma_\alpha \vec{v}^{oy+}_\alpha(t)\right] + \max_{\beta_i \in \beta}\left[\vec{f}^{y-}_{\beta_1}, \ldots, \vec{f}^{y-}_{\beta_m}, \gamma_\alpha \vec{v}^{oy-}_\alpha(t)\right] \tag{7.108}$$

where $\vec{f}^{mx}_\alpha$ and $\vec{f}^{my}_\alpha$ represent the X-axis and Y-axis component tensions of the integrated new goal tension $\vec{f}^m_\alpha$, respectively.

7.3 Summary

Based on the framework of microscopic behavior analysis of bicycles at un-signalized intersections in Chapter 6, this chapter introduces a concrete theoretical modeling of cyclists' behavior at un-signalized intersections.

In Section 7.1, we introduce the NCB theory and model, including the theoretical assumptions, the overall theoretical framework, the basic concepts and the features of the theory.

In Section 7.2, based on the NCB theoretical model framework, we analyze the specific traffic conditions of the un-signalized intersection and the microscopic behavior characteristics of bicycles are specifically modeled.

References

[1] E.C. Van Berkum and P.H.J. Van der Mede. The Impact of Traffic Information: Dynamics in Route and Departure Time Choice. Doctoral Thesis, TU Delft, 1993.

[2] M.R. Hill. Spatial Structure and Decision-Making of Pedestrian Route Selection Through an Urban Environment. University of Nebraska, 1982.

[3] P.H.L. Bovy and E. Stern. Route Choice: Wayfinding in Transport Networks. Dordrecht: Kluwer Academic Publishers, 1990.

[4] D. Helbing and M. Treiber. Gas-Kinetic-Based Traffic Model Explaining Observed Hysteretic Phase Transition. Physical Review Letters, 1998, 81:3042–3045.

[5] M. Ben-Akiva and S. Lerman. Discrete Choice Analysis: Theory and Application to Travel Demand. Cambridge, MA: The MIT Press, 1985.

[6] S. Hoogendoorn and P. Bovy. Pedestrian Route-Choice and Activity Scheduling Theory and Models. Transportation Research: Part B, 2004, 38:169–190.

[7] D. Helbing, J. Keltsch, and P. Molnar. Modelling the Evolution of Human Trail Systems. Nature, 1997, 388:47–50.

[8] V.J. Nordby and C.S. Hall. (Calvin Springer). A Guide to Psychologists and Their Concepts. W. H. Freeman; trade distributor: New York, Scribner, 1974.

[9] S. Hui and K. Wang. Safety Psychology. Beijing: Chemical Industry Press, 2004.

[10] U. Weidmann. Transportation technique for pedestrians, Schriftenreihe des Instituts fur Verkehrsplanung, Transporttechnik, StraÞen-und Eisenbahnbau number 90 (ETH Zurich, Switzerland), 1993.

[11] Federal Highway Administration (FHWA), US Department of Transportation. Highway Capacity Manual. Washington, DC, 2000.

[12] Editorial Board of Modern Applied Mathematics Handbook. Modern Applied Mathematics Handbook. Operations Research and Optimization Theory Volume. Beijing: Tsinghua University Press, 1998.

[13] D. Helbing and P. Molnar. Social Force Model of Pedestrian Dynamics. Physical Review E, 1995, 51:4282–4286.

Chapter 8
Empirical analysis of bicycle microscopic behavior model

In this chapter, we introduce the field data, data collection schemes and various field data collection and collation processes required for bicycle microscopic behavior modeling at the un-signalized intersection in Section 8.1. Accuracy analysis of acquired vehicle dynamic data from video files was performed in Section 8.2.

The data reliability analysis was performed on the actual collected stated preference (SP) and RP data in Section 8.3. Then, in Sections 8.4 and 8.5, the parameter identification of the tactical level model and operational level model of the bicycle crossing behavior model at the un-signalized intersection were introduced.

Finally, the model parameter identification results and related analysis conclusions are given.

8.1 Field data collection

The field data are very important for the behavior modeling process. In this chapter, we collect the data needed for the bicycle microscopic behavior model at un-signalized intersections built in the previous chapter.

First, we designed the data acquisition scheme; then we collected the field data and form a bicycle microscopic behavior database at the un-signalized intersections for bicycle microscopic behavior analysis. The database lays the foundation for the behavior analysis, model parameter identification and verification of our bicycle behavior model.

8.1.1 Purpose and significance of data collection

Accurate, reliable and large amounts of field data are of great importance to scientific research. True and reliable data do not necessarily lead to correct conclusions and models. However, without real and reliable data, scientific and rigorous conclusions must be obtained.

In this book, the purpose of our on-site data collection is to use a variety of advanced data collection methods to observe and obtain a large amount of field traffic data to establish a bicycles microscopic behavior database (BMBD) at un-signalized intersections. The BMBD provides a sound and solid data foundation for subsequent data analysis and modeling efforts.

The microscopic behavior of bicycle traffic through the un-signalized intersection studied in this book is still in its infancy and has not yet formed a mature, perfect theoretical system and corresponding behavior database.

The basic behavioral model of individual road users (motorcar drivers, cyclists, pedestrians) is based on a large number of on-site databases combined with some basic theories. Only a large collection of microscopic behavior data ensure that subsequent data analysis and modeling work is performed on a correct and stable basis. It can be said that the establishment of a traffic individual behavior database is the basis of theoretical research on traffic behavior.

In addition, from the perspective of the traffic simulation model, the actual data are the basis for the basic theoretical model construction, parameter calibration and verification of the microscopic traffic flow simulation system.

Therefore, it can be said that the traffic individual behavior database is also the key to the success of the entire simulation platform. The correctness, representativeness and reliability of the data in the traffic individual behavior database have a great impact on the basic behavior model of the traffic individual and the entire simulation platform.

8.1.2 Data/information required by the model

The NCB model at the un-signalized intersection includes two different levels of behaviors: the tactical level of path planning behavior and the operational level of dynamic riding behavior. These two different levels of behavioral models are based on different theories, and the types of data required are also very different.

At present, in the field of traffic modeling, data related to traffic behavior are mainly obtained by means of a probability sampling survey. This is a non-comprehensive survey method that randomly selects a part of the unit from the survey as a sample to investigate and based on the sample survey results to infer the overall quantitative characteristics.

There are generally two types of data obtained: first, revealed preference data or RP data. These are mainly related to the selection behavior data obtained by statistical real or observed traffic behaviors made by individual samples. Second, the SP data that are primarily obtained by asking the individual samples for the most likely choices that would be made in several hypothetical situations.

The RP data collect the actual selection behavior of the traveler and use the explicit and observable selection behavior to find the traveler's choice preference. The SP data acquisition method is to define the attributes of the alternatives in the choice set by the researcher and to find out the intrinsic and unobservable potential preferences of the traveler according to their ranking, scoring and selection of the choices.

The parameters required for the NCB tactical level behavior model mainly include the cyclist's subjective psychological process of crossing the un-signalized intersection; the main factors affecting the cyclist's path planning behavior and the relative weight or order of the main affecting factors.

The parameter information required for NCB operational level behavior model mainly includes dynamic and static information of vehicles around the cyclist at the intersection, that is, the motion information V_α^s, V_β^s, V_α^d and V_β^d of the cyclist α and its surrounding vehicles V_β in Ω. Among them, V_α^d and V_β^d are position $\vec{x}(t)$, velocity $\vec{v}(t)$ and acceleration $\vec{a}(t)$ of each related vehicle.

8.1.3 Data collection scheme

8.1.3.1 SP data collection scheme

We investigate and analyze the cyclist's psychology and behavior when crossing through the un-signalized intersections. The results of the survey and analysis can provide ideas for the cyclists' psychological and behavioral analysis. And further, it provide data foundation for the cyclist' path planning behavior modeling, parameter selection and validation.

Survey research is one of the important methods of data collection in psychology and behavior. There are usually two types of survey methods: one is the "questionnaire method" and the other is the "interview method" [1].

When modeling, we need to analyze the cyclists' subjective psychological process when crossing the un-signalized intersection and determine the factors affecting their subjective path planning. First, we use the "interview method" to conduct "unstructured interviews" on a small number of interviewees. The so-called unstructured interview is to set the theme "the psychological process when cycling through the un-signalized intersection" and "the main considerations in the path planning at the un-signalized intersections," so that the respondents can answer freely, and then we organized the useful information based on the interview record.

As the unstructured interview method is time-consuming, laborious and ambiguous, it is not suitable for a large-scale survey. This method is only used for a few typical objects. The number of subjects to be interviewed is 10–20.

The specific investigation plan is as follows:

1. **Interview on cyclists' crossing behaviors at un-signalized intersections**
   i. **Purpose**
   Through interviews and questionnaires, we understand and investigate the psychological activities, behavioral processes and characteristics of cyclists crossing non-signalized intersections. This is to understand the characteristics and influencing factors of the cyclist's path planning at un-signalized intersections and to provide ideas and clues for the subsequent microscopic behavior analysis.
   ii. **Method**
   We intend to start with interviews and questionnaires, which are more commonly used in behavior and psychology. Since the question of the survey is an open question, an open interview method is adopted, that is, the interviewees are free to elaborate on a theme, and then the investigators sort out and summarize the content of the respondents.

iii. **Interviewee**
The interviewees are very extensive and meet the conditions: adults who can ride a bicycle and have some experience riding bicycles on urban roads, who have once crossed the un-signalized intersections.

iv. **Topics**
The interview topics are as follows:
  a. The interviewees described the psychological and behavioral process of riding a bicycle through an un-signalized intersection; if the interviewee could not describe clearly, we could ask further questions: Does he/she generally plan the path when crossing an un-signalized intersection? If the answer is yes, then ask whether he/she makes a detailed path before crossing and then rides according to the established path, or has an approximate path in mind at the entrance of the intersection, then chooses the actual path while riding, or is there another way to plan a path to cross the un-signalized intersection?
  b. If there is a large traffic flow at the un-signalized intersection, and the situation is complicated, what factors are considered when planning a path?

v. **Questionnaire**
For the second topic, a questionnaire was designed for respondents to answer. The specific content is as follows:

Questionnaire for bicycles crossing un-signalized intersections (1)

When you are cycling through an un-signalized intersection, choose the considerations you consider important when planning the cross path:

□The weather
□Distance of the path
□Your own physical condition
□Potentially conflicting motor vehicle
□Potentially conflicting bicycles
□Potentially conflicting pedestrians
□Number of bicycles in the same direction
□Pedestrians around
□Mood at that time
□Riding time needed
□Congestion
□Pedestrians in the same direction
□Bicycle performance
□Number of turns of the path
□Whether it is rush hours
□Your own travel time
□Path of around other bicycles

Other important factors ______________________________

The results of this questionnaire will provide ideas for the design of a subjective questionnaire for cyclists to choose behavior preferences at un-signalized intersections.

2. **Data collection and analysis of the interview**

From March to May 2006, the authors interviewed 23 different cyclists. The interview process ranged from 10 min to half an hour. The age distribution of the interviewees ranges from the middle school students to retired elderly (as elementary school students rarely riding bicycle alone, it is not included in the study); commuter modes are from bicycles to commuting vehicles to occasional cycling, including various occupations (retirees, officers, blue-collar workers, shop assistants, college students, etc.). The interview data and collection results are mainly used in the bicycle microscopic behavior analysis framework and modeling.

Among the 23 interviewees (14 men and 9 women), 18 people (11 men and 7 women) described their psychological and behavioral process of riding a bicycle through un-signalized intersections.

When they are about to enter the intersection, they have a general understanding of the traffic flow conditions at the intersection and then quickly make a rough plan in their mind (crossing directly, or waiting, or crossing the intersection via the crosswalk), and then they will react and deal with it in detail.

The remaining five people (three men and two women) described that they usually did not make a predetermined path when crossing an un-signalized intersection and directly relied on their instinct to respond to riding or braking.

If the traffic conditions at un-signalized intersections are complicated, important factors when choosing a path (i.e., more than five people chose this factor) include

i. the distance of the path;
ii. riding time needed;
iii. potentially conflicting motor vehicle;
iv. potentially conflicting bicycles;
v. potentially conflicting pedestrians;
vi. your own travel time;
vii. number of bicycles in the same direction and
viii. path of around other bicycles.

3. **Questionnaire of cyclists' path planning behavior**

According to the main factors compiled by the "interview method," we use the "questionnaire method" to investigate the relative weight/sorting information of the influencing factors of the subjective utility function of the cyclist crossing the un-signalized intersection.

Interview analysis shows that the cyclists' most considered (most important) factors when they planning the path across un-signalized intersections include

i. the time T required to cross the intersection;
ii. conflicts with other vehicles (mainly motor vehicles), i.e., safety factor S;
iii. the bypass distance, this consideration can be understood as riding more distances at the same time, so it can be converted to a higher riding speed requirement, that is, the energy consumption factor V;
iv. interaction with other bicycles and pedestrians in the same riding direction, i.e., the *level of service (LOS)* of the bicycle.

Since the preliminary results of the "interview method" analysis are the four main influencing factors, the content of the questionnaire is small. Therefore, we intend to conduct roadside questioning surveys of different groups at un-signalized intersections, bus stations and metro stations. The advantage of this method is that the questionnaire has high quality and high recovery rate. The questionnaire consists mainly of the following parts:

i. the features of the respondents, including their gender, age, the main mode of transportation and travel purposes;
ii. the most important and least important factors in the planning of path crossing the un-signalized intersection.

The questionnaire is as follows:

Cyclists' crossing behavior at un-signalized intersection questionnaire

Gender Male☐ Female☐

1. Age:

Below 18☐ 18–35 years☐ 35–55 years☐ 55–70 years☐ Over 70 years☐

2. Trip purpose:

Work ☐ School ☐ Leisure, visiting friends☐ Shopping☐ Other☐

3. What is the most traffic common mode:

Bicycle☐ Bus☐ Private car☐ Metro☐ Taxi☐ Other☐

Suppose you are riding a bicycle crossing an un-signalized intersection:

4. You think the MOST *important* one of following four considerations is

Time☐ Distance☐ Safety☐ Comfort☐

5. And the LEAST *important* one of following four considerations is

Time☐ Distance☐ Safety☐ Comfort☐

8.1.3.2 RP data collection scheme

According to the NCB operational level modeling data requirements and the existing equipment in our laboratory, we intend to use video acquisition technology to collect data on the microscopic behavior of bicycles and other vehicles at the un-signalized intersections.

The basic requirements for video data collection are as follows:

1. **Data collection sites**: Multiple typical un-signalized intersections (including cross-shaped and T-shaped un-signalized intersections) should be selected, and there should be no special geographical and geometric factors, such as slope, sight obstacles. In order to ensure a sufficient sample size and representativeness, the intersection should have a sufficient mixed traffic flow, and there

should be conflicts between the motorcars, bicycles and pedestrians; what is more, the intersection should have a good shooting location, such as pedestrian overpass or high-rise buildings nearby.

2. **Weather**: The weather conditions on the investigation day shall be sunny (or cloudy as well) and windless, which will not have a significant impact on the traffic psychology and behavior of all road users (motorcars, bicycles and pedestrians); abnormal weather such as rain, fog, snow, hail and gale shall be avoided.
3. **Time period**: In order to ensure the representativeness of the sample data, the time period of video data collection should include the working day time and weekends, and the data of morning peak, evening peak and off-peak hours are collected for future analysis.
4. **Equipment**: Two HD cameras, eight spare batteries, two tripods, three walkie-talkies, and one tape measure (30 m).
5. **Personnel**: The number of field investigators should be at least three (one more than the number of field cameras). Make sure someone keeps an eye on the camera when shooting. The other person is responsible for communication, transmitting items and measuring the geometry of the intersection.

The basic process of live video data collection is as follows:

i. Intersection selection
First, we select the intersections that meet the previous basic conditions from the intersection database (digital map)—cross-shaped and T-shaped un-signalized intersections, no obvious slope, no sight obstruction, large mixed traffic flow, organic traffic flow and non-traffic conflict point, pedestrian bridge or high-rise buildings nearby.

ii. Preliminary site survey
Then, we selected two to three un-signalized intersections from the previous eligible intersections for data collection. And two investigators bring a camera and a tripod to these intersections for preliminary field investigations. The inspection contents include observing whether the intersection meets the data collection requirements, how the angle of view of the shooting location is effective, the mixed traffic flow operating conditions at the intersection, and determining the best shooting point of each camera.
At each shooting point, investigators collected more than 10 min of video data.

iii. Preliminary analysis
After a preliminary inspection of the intersection, we conduct a preliminary analysis of the collected video data to determine the reliable data that may be obtained and the corresponding data acquisition method. At the same time, determine the intersection geometry data needed to process the data and determine the markers at the intersection.

iv. Field data collection
Then we started formal video data collection. In the appropriate shooting weather, we placed two synchronized cameras at predetermined positions, adjusted the shooting point and lens angle and started shooting at the same time.

During the shooting process, we pay attention to continuous shooting without interruption and do not bump or move the cameras at will. At the same time, one investigator measured the relevant markers at the intersection to determine their size and location.

Finally, a panoramic photo of the intersection should be taken to record the basic geometric features of the intersection and other information. The point is to collect all the data needed for modeling.

8.1.4 Situations of field data collection

8.1.4.1 SP data collection

The cyclist interview

To analyze the cyclists' crossing psychological and behavioral processes at un-signalized intersections, we need to conduct interviews and surveys on cyclists.

The analysis results of the interviews are very important for the construction of the behavior analysis framework for bicycle crossing un-signalized intersections and the specific modeling of the NCB theoretical model that follows.

Therefore, from March to May 2006, the authors interviewed more than 20 different cyclists. The interview time ranged from 10 min to half an hour.

The age distribution of the interviewees ranges from middle school students to retired elderly (as elementary school students rarely travel by bicycle alone, it is not included in the study). Interviewees included those who used bicycles as commuting vehicles to occasionally travel by bicycle, including various occupations—retirees, white-collar workers, blue-collar workers, migrant workers, shop assistants and college students.

The interview data and collation results are mainly used in the microscopic behavior analysis framework and modeling of bicycles.

Questionnaire survey

The survey was conducted on May 19, 2006: 7.30–17.30 (Friday) and May 21, 2006: 7.30–17.00 (Sunday).

For the questionnaire, we selected the waiting zone of signalized intersections, overpasses, bicycle parking lots, motorcar parking lots and entrances to shopping centers. We also go to the nearby bus stations, subway stations near the transportation hubs in Beijing such as Fuchengmen, Xizhimen and Dongzhimen.

The method of the questionnaire survey is to ask questions on the spot by a trained investigator and then fill out the questionnaire based on the respondents' answers. This method can fully communicate with the interviewees, and the quality of the questionnaires recovered is high.

A total of 1,095 questionnaires were collected, of which 1,016 were valid questionnaires, with an effective rate of 92.8%.

8.1.4.2 Video data collection

We conducted a total of eight video data acquisitions at two different types of un-signalized intersections (cross-shaped and T-shaped un-signalized intersections). The basic situation of data collection at the two intersections is described next.

The basic situation of video data acquisition at non-signalized intersections (cross)

1. Time	May 16, 2006: 7.30–17.30 (Tuesday) May 17, 2006: 7.30–17.00 (Wednesday) May 18, 2006: 7.30–17.00 (Thursday) May 20, 2006: 7.30–17.00 (Saturday)
2. Location	Un-signalized intersection on the campus of Tsinghua University, Beijing
3. Weather	The weather is fine, the temperature is 15°C–18°C and the wind is 2–3
4. Traffic flow condition	This intersection is a typical cross-type no-signal intersection (see Figure 8.1). It is located on the campus of Tsinghua University. Traffic management methods used at intersections are motor vehicle yield to pedestrian and bicycle traffic. There are a lot of bicycles and pedestrians around the intersection, and there are many cars. Among them, during peak hours (7.30–9.00, 16.30–18.00), the vehicle traffic is about 220 vehicle/hour, the bicycle traffic is about 5,200 vehicle/hour and the pedestrian crossing is about 300 people/hour (regardless of direction). Off-peak motor vehicle traffic is about 300 vehicles/hour, bicycle traffic is about 2,100 vehicle/hour and pedestrian crossings are about 300 people/hour (regardless of direction). The distance between the two crosswalks at the intersection north and south is 20 m. The distance between the two crosswalks in the east–west direction is 20 m. There are no traffic management signs and objects at the intersection

Figure 8.1 Cross-shaped non-signalized intersection (actual data collection site 1)

(Continues)

(*Continued*)

5. Basic conflicts	Elaborated in Figures 8.2–8.5

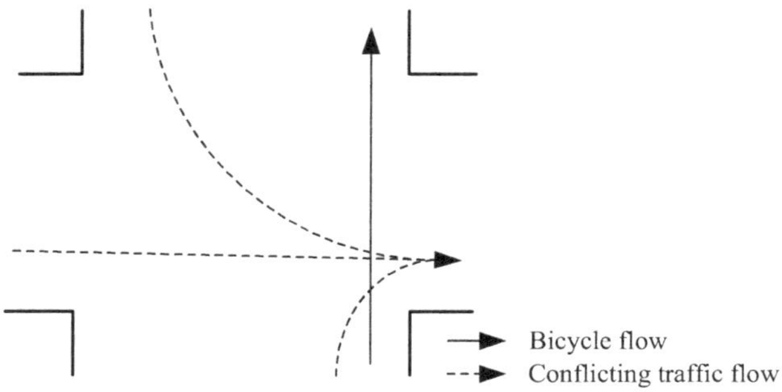

Figure 8.2 Conflicts between a straight bicycle flow and traffic flows from the left

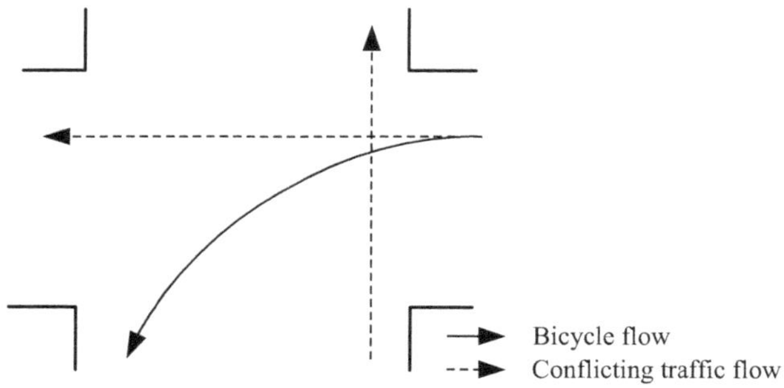

Figure 8.3 Conflicts between a straight bicycle flow and traffic flows from right

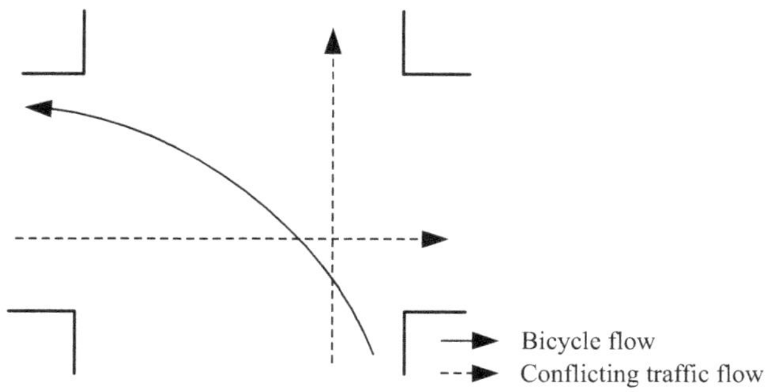

Figure 8.4 Conflict between a left-turn bicycle flow and traffic flows from left

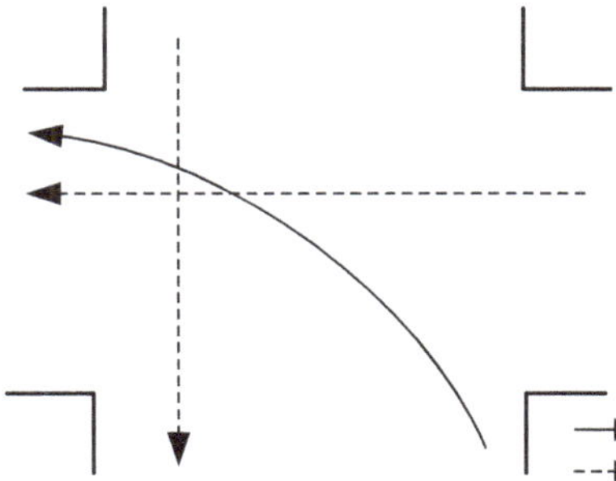

Figure 8.5 Conflict between a left-turn bicycle flow and traffic flows from right

The basic situation of video data acquisition at non-signalized intersections (T-shaped)

1. Time	May 10, 2006: 7.30–17.30 (Tuesday) May 11, 2006: 7.30–17.00 (Wednesday) May 12, 2006: 7.30–17.00 (Thursday) May 24, 2006: 7.30–17.00 (Sunday)
2. Location	Un-signalized intersection on the campus of Tsinghua University (East Gate), Beijing
3. Weather	The weather is fine, the temperature is 5°C–15°C and the wind is 2–3
4. Traffic flow condition	This intersection is a typical mixed traffic flow T-shaped signal intersection (see Figure 8.6). It is located at the East Gate of Tsinghua University.

Figure 8.6 T-shaped non-signalized intersection (actual data collection location 2)

(Continues)

(*Continued*)

	Due to the large number of high-tech companies near the site, it attracts a large number of motor vehicles, bicycles and people. At the peak of the intersection (7.30–9.00, 16.30–18.00), the vehicle traffic is about 800 vehicle/hour, the bicycle traffic is about 2,000 vehicle/hour and the pedestrian traffic crossing the intersection is about 1,000 people/hour (regardless of direction). Off-peak motor vehicle traffic is about 360 vehicle/hour, bicycle traffic is about 800 vehicle/hour and pedestrian traffic across intersections is about 400 people/hour (regardless of direction). The distance between the crosswalks in the east–west direction at the intersection is 32 m. The north–south road is 18 m wide
5. Basic conflicts	Elaborated in Figures 8.7 and 8.8

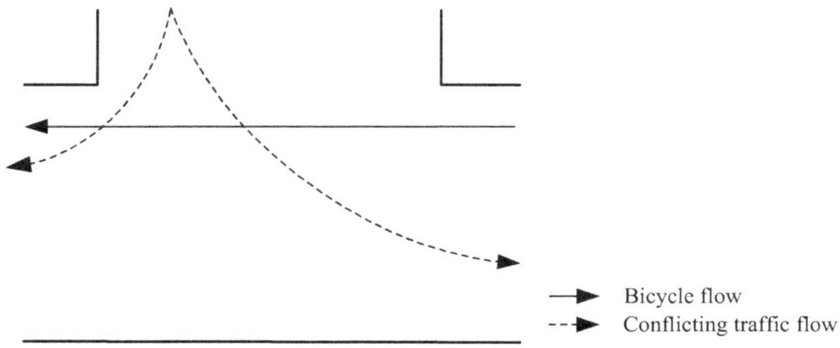

Figure 8.7 Conflicts of straight bicycle flow at T-shaped intersection

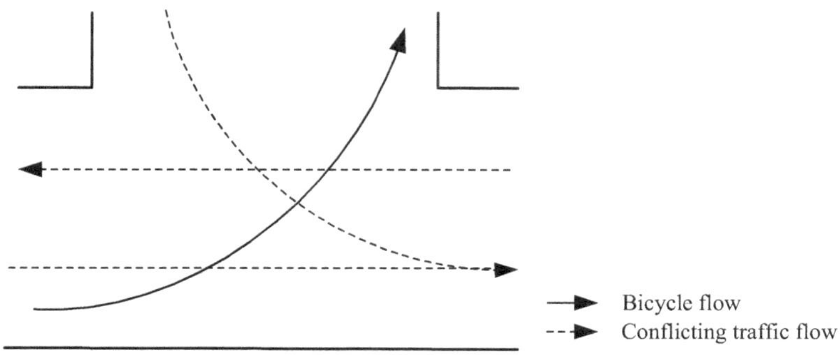

Figure 8.8 Conflicts of left-turn bicycle flow at T-shaped intersection

8.2 Data collection and preprocessing

8.2.1 The SP data and RP data

8.2.1.1 The preprocessing of SP data

The preprocessing of SP data is relatively simple, and it mainly depends on manual work. The information on interviews and surveys can only be analyzed by manual analysis. The process of sorting out the data and information of the questionnaire mainly includes (1) first, select valid questionnaires, (2) encode the questionnaire options and (3) then enter the encoded information to form a database of preferences for bicycle path planning at un-signalized intersections.

8.2.1.2 The preprocessing of SP data

Video data collected at non-signalized intersections is first converted to AVI format video files. The converted AVI video file will be sampled and segmented first. Typical samples of bicycles crossing non-signalized intersections (including conflict-free crossing, motor vehicle–bicycle conflict, bicycle–bicycle conflict, pedestrian–bicycle conflict and obstacles around).

The AVI files cut into sample cases can be processed by the laboratory's special video data capture software VSpeed system (Figure 8.9). The VSpeed system was developed in cooperation with Beijing Hanwang Technology Co., Ltd. and is used to study the automatic acquisition of traffic parameters in traffic scenarios.

VSpeed software can easily output the required traffic flow data, and the data are given in the form of files.

Figure 8.9 VSpeed interface

The output data include the position and speed (size, direction), acceleration, headway, headway and traffic of the specified moving vehicle relative to the world coordinate system (position of the differential GPS base station) at each moment.

In the study, we only need to use the first three motion parameters provided by the VSpeed system: the time t, the position vector $\vec{x}(t)$, speed vector $\vec{v}(t)$ and acceleration vector $\vec{a}(t)$ of the target vehicle.

Most of the code of the VSpeed is programmed in C++. Its main functions include:

1. read video file as input (AVI format);
2. camera calibration method for traffic scenes (Figure 8.10);
3. according to different traffic conditions, users can use manual or automatic two different positioning methods; manual method is the user to select the target vehicle tracking point with the mouse, the system automatically tracks based on the tracking point; the automatic method is the system to automatically determine the center for tracking points, automatic tracking is performed (Figure 8.11);
4. robust tracking algorithm based on the fusion of Meanshift and Kalman filters;
5. the output method is a data file (Table 8.1);
6. analysis of traffic flow parameters.

Readers who are interested in the technical details of the VSpeed system can refer to Dr. Bai's doctoral dissertation "Research on Video-Based Traffic Flow Parameter Measurement in Intelligent Transportation" [2].

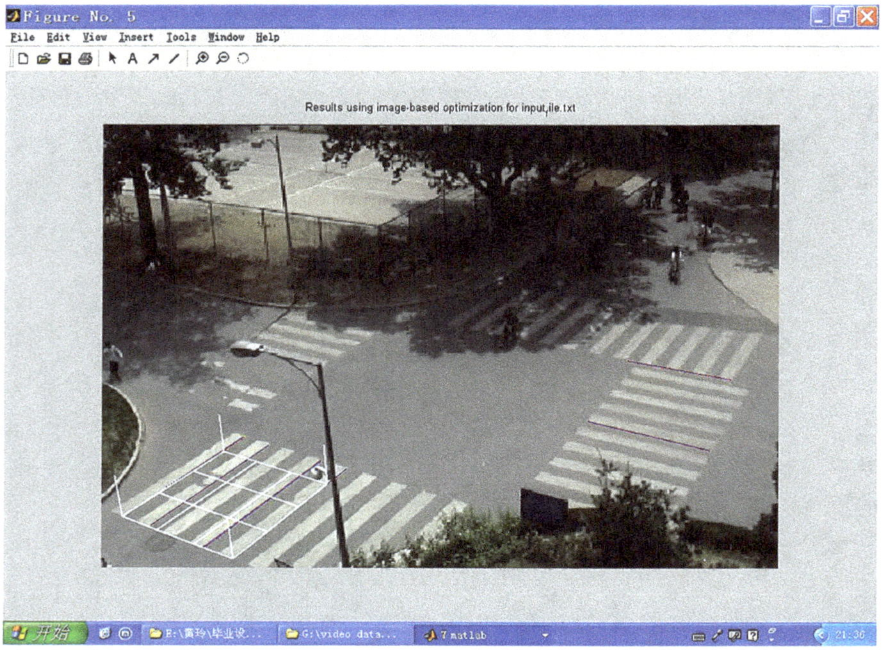

Figure 8.10 VSpeed traffic scene coordinate system calibration results (white grid lines are the camera coordinate system)

Figure 8.11 VSpeed tracks and locates bicycles (green points are track points)

Table 8.1 is an example of VSpeed's output raw positioning data file (txt file). The first term is time, the second term is frame order, the third term is the *X*-axis coordinate of the tracking point and the fourth term is the *Y*-axis coordinate.

8.2.2 Accuracy analysis of the RP data

There are two main sources of error in vehicle dynamic data obtained from video files: one is the error between the coordinate system obtained when the camera is calibrated in the VSpeed, and the real-world coordinate system (position of the differential GPS base station) when the target vehicle is tracking and acquiring data, the positioning point error due to the tracking point drift.

For the first error, Dr. Bai discussed in detail the accuracy of VSpeed's positioning of camera coordinates in his doctoral dissertation [2]. According to his data analysis results, when the camera is shooting at a height of 70 m, the error between the camera coordinate system obtained by the VSpeed from the video file, and the world coordinate system (position of the differential GPS base station) is about 8%. When shooting at a height of 25 m, the error between the obtained coordinate system and the world coordinate system is about 3%.

In the video acquisition of traffic flow at two un-signalized intersections, the height of the camera is about 25 and 30 m (on five and six floors, respectively), so the error of the estimated coordinate system maybe 3%–8%, about 5%.

It is noted that in the NCB behavior model at un-signalized intersection, mostly data applied are the relative position, speed and acceleration of dynamic vehicles,

Table 8.1 Examples of VSpeed output raw data

Time	Frame	*X*	*Y*
00::00::04::440	111	−14.6183	12.28541
00::00::04::480	112	−14.6251	12.47712
00::00::04::520	113	−14.6251	12.47712
00::00::04::560	114	−14.649	12.65571
00::00::04::600	115	−14.6733	12.83659
00::00::04::640	116	−14.6874	12.97773
00::00::04::680	117	−14.7017	13.12025
00::00::04::720	118	−14.7442	13.29104
00::00::04::760	119	−14.7442	13.29104
00::00::04::800	120	−14.7764	13.42056
00::00::04::840	121	−14.8199	13.59519
00::00::04::880	122	−14.8529	13.72764
00::00::04::920	123	−14.8683	13.87788
00::00::04::960	124	−14.9019	14.01302
00::00::05::000	125	−14.9019	14.01302
00::00::05::040	126	−14.9177	14.1662
00::00::05::080	127	−14.9518	14.30413
00::00::05::120	128	−14.9863	14.44342
00::00::05::160	129	−14.9912	14.55404
00::00::05::200	130	−15.0078	14.71288
00::00::05::240	131	−15.0078	14.71288
00::00::05::280	132	−14.9945	14.84262
00::00::05::320	133	−15.0113	15.00437
00::00::05::360	134	−14.9978	15.13635
00::00::05::400	135	−15.0334	15.28358
00::00::05::440	136	−15.0199	15.41777
00::00::05::480	137	−15.0199	15.41777
00::00::05::520	138	−15.0251	15.53528
00::00::05::560	139	−15.0115	15.67141
00::00::05::600	140	−15.0288	15.84172
00::00::05::640	141	−15.0151	15.98027
00::00::05::680	142	−15.0203	16.10179
00::00::05::720	143	−15.0203	16.10179
00::00::05::760	144	−15.0065	16.24237
00::00::05::800	145	−14.9925	16.38404
00::00::05::840	146	−14.9176	16.26267
00::00::05::880	147	−14.9226	16.38601
00::00::05::920	148	−14.8961	16.47547
00::00::05::960	149	−14.8961	16.47547
00::00::06::000	150	−15.0348	16.66902
00::00::06::040	151	−15.0667	16.7044
00::00::06::080	152	−15.0667	16.7044
00::00::06::120	153	−15.0597	16.77703
00::00::06::160	154	−15.0597	16.77703
00::00::06::200	155	−15.0597	16.77703

so the system error generated by the camera coordinate system has been offset to a certain extent.

Another error is caused by tracking point drift. In the process of using VSpeed to obtain the target vehicle motion data, there are people to monitor the entire process. Once the tracking point drift is found, it will be manually corrected, and the mark in the video file (crosswalk mark line width) is compared, and the error will not exceed ±0.3 m.

8.3 Data reliability analysis

In order to ensure the correctness and rigor of the study of the behavior model of bicycles crossing un-signalized intersections, the reliability of the collected data needs to be sampled and analyzed before analyzing the data and identifying the model parameters. The following is a reliability analysis of the questionnaire survey (SP data) and video data (RP data) collected by the author in Beijing.

8.3.1 SP data reliability analysis

Because the questionnaire data are the responsibility of the investigator to ask the respondent, explain and fill in the answer, in the data collection process, there may be reasons for the surveyor (e.g., the unconscious influence of the sample selection, interpretation or attitude of the interviewee) to affect the surveyor's attitude and the survey results, thereby affecting the reliability of the survey data.

Therefore, it is necessary to make statistical comparisons of sample data of different researchers. If there is a significant difference in sample data between the data of one investigator and the data of most other investigators, and there is no reasonable reason and explanation, it may indicate that the sample collected by the investigator may have some deviation. At this time, the investigator's survey data may not be included in the data analysis.

There are five investigators in the questionnaire survey, who are represented by investigators 1, 2, 3, 4, 5, respectively. An investigator is responsible for different areas. For example, investigator 1 is mainly responsible for the area around the Fuchengmen intersection, including roadsides, intersections, bicycle parking areas at the mall entrance and bus stops. Investigator 3 is responsible for the people around the Xizhimen intersection, bus stops and subway stations.

In the reliability analysis of questionnaire data, first, the sample data collected by different investigators are grouped and analyzed, and the overall distribution between the sample data of each investigator is compared; and then the overall information of a typical group of the sample, such as the youth group, is compared.

8.3.1.1 Comparison of investigator sample groups

We analyzed and compared survey sample data from investigators 1, 2, 3, 4 and 5. The specific statistics are shown in Appendix A.

The comparison results of different comparison items of the sample data of each investigator are shown in Figure 8.12.

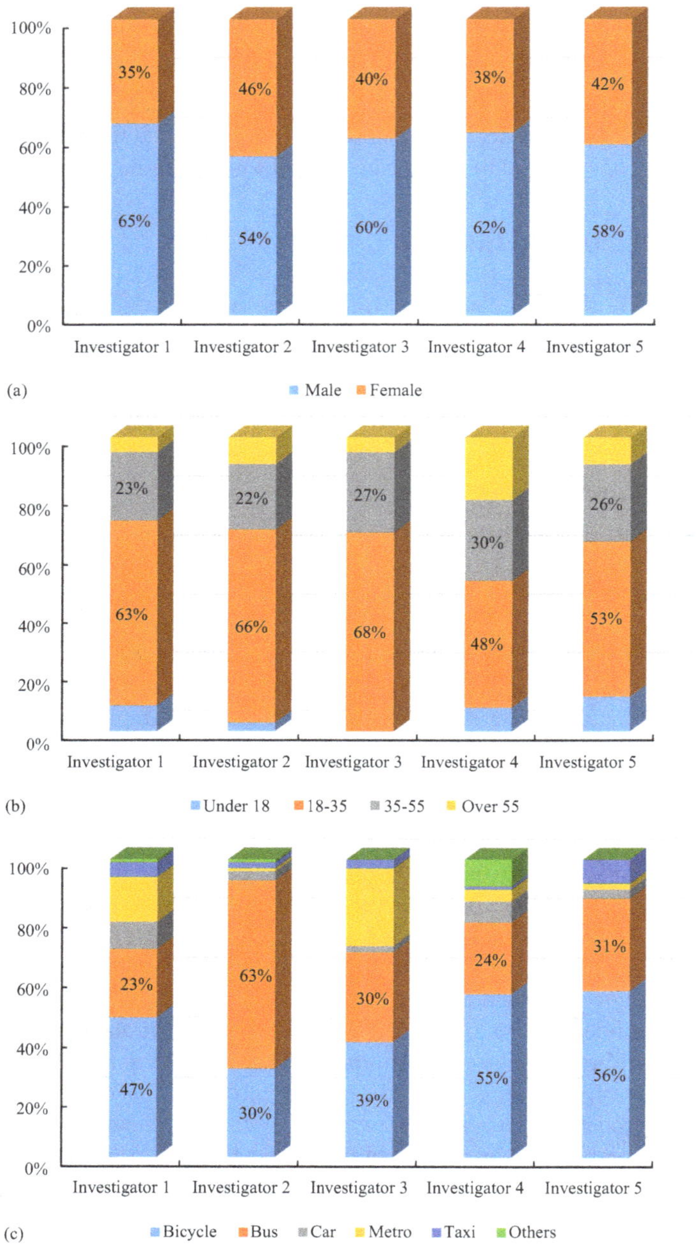

Figure 8.12 Comparisons of the sample data grouped by each investigator: (a) gender group comparisons; (b) age group comparisons; (c) commuter traffic mode group comparisons; (d) the most important factors of the path planning group comparisons; (e) the least important factors of the path planning group comparisons

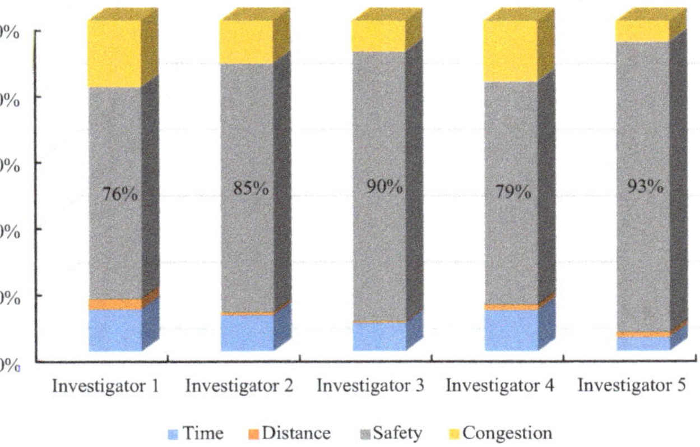

(d)

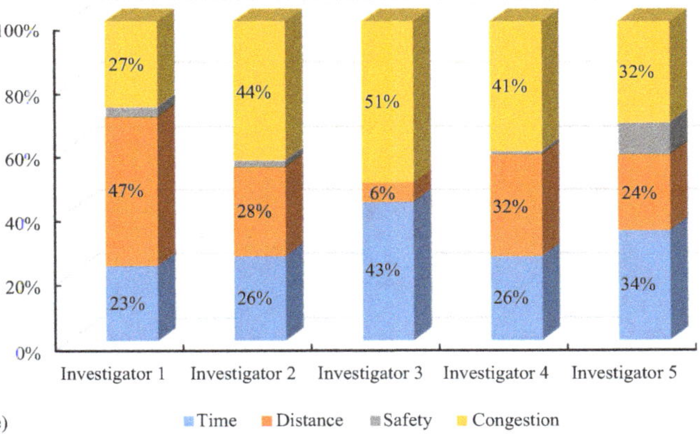

(e)

Figure 8.12 (Continued)

From the previous group comparison analysis, it can be seen that the sample groups of investigators 1–5 are quite consistent, but the travel mode of investigator 3 is mainly a bus, which is consistent with his survey location. Most of them are related to bus stations, and in the 18–35 age group, the researcher's sample age group accounted for a large proportion. Therefore, there are some differences in the order of important factors that influence the selection of paths crossing un-signalized intersections, which can also be explained.

8.3.1.2 Group comparison of youth groups of various investigators

In order to further analyze whether different investigators have introduced personal subjective influence into the questionnaire survey, the youth group (age 18–35, regardless of gender) of the sample questionnaire was analyzed.

To investigate whether there is a significant difference in the ranking of the factors affecting the cyclists' path planning behaviors at un-signalized intersections

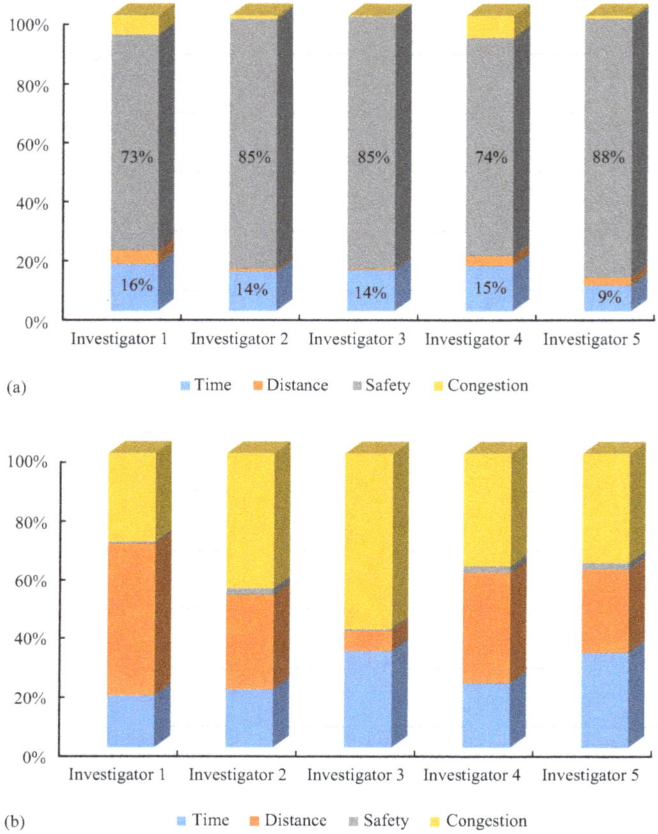

Figure 8.13 Comparison of influence factors of different investigators (youth group): (a) comparison of the most important influencing factors; (b) comparison of the least important influencing factors

by the youth sample of each investigator, the comparison results are shown in Figure 8.13.

As can be seen from Figure 8.13, the distribution of factors affecting the path planning behavior of bicycle crossing un-signalized intersections among the youth samples of each investigator is generally consistent, and there are no significant differences.

To sum up, the author can think that among the collected survey data, the sample of each investigator is reasonably distributed, and no subjective influence of the investigator has been found to have a significant impact on the quality of the data.

Therefore, the author believes that the survey questionnaire data of all investigators are reliable and can all be used as a basic information source for behavior analysis and model parameter identification.

8.3.2 RP data reliability analysis

The positioning accuracy of the vehicle dynamic motion data obtained from the video files captured by the VSpeed from the scene has been completed during the design process of the VSpeed. Here, we only sample the vehicle motion trajectory, speed and acceleration trajectory from the perspective of traffic engineering and analyze to see if they meet the reasonable value range of traffic engineering.

Here, we examine the data of motor vehicles, bicycles and pedestrians at un-signalized intersections.

We analyze the motion data of several motor vehicles, bicycles and pedestrians from the data samples and conduct a macroscopic examination of their trajectories, speeds and accelerations.

According to the *Traffic Engineering Manual* [3], the design speed of motor vehicles passing through un-signalized intersections (on-road sections) usually does not exceed 60 km/h (about 17 m/s), and the acceleration range of vehicles is generally less than ± 8 m/s^2; and the design speed of bicycles at un-signalized intersections (on-road sections) usually does not exceed 30 km/h (about 8 m/s), and the acceleration range is generally less than ± 5 m/s^2; the design speeds of pedestrians at un-signalized intersections (on-road sections) usually do not exceed 12 km/h (about 3 m/s), and the acceleration range is generally less than ± 2 m/s^2.

We found that whether it is a motor vehicle, bicycle or pedestrian, its movement displacement, speed and acceleration are within normal ranges.

The samples of motor vehicles, bicycles and pedestrians that we spot checked for motion displacement travel speed, and acceleration at un-signalized intersections are shown in Figures 8.14–8.16, respectively.

8.4 Path planning model parameter identification

8.4.1 SP data analysis

8.4.1.1 Overall sample data analysis

A total of 1,016 valid questionnaires were collected in this questionnaire. Among the respondents, 612 were males, accounting for 60.2%, and females were 404, accounting for 39.8%. Among the age groups, the youth group (age <18 years) had the smallest proportion, with 74 people, accounting for 7.3%; the youth group (18–35 years old) had the most, with 606 people, accounting for 59.6%, the middle-aged group (35–55 years) had 258, accounting for 25.4%, and the elderly group (> 55 years) had 78 people, accounting for 7.7%. The sample's travel purpose distribution is as follows: respondents have the most travel purpose for commuting, with 521 people, accounting for 51.3%, 129 people for leisure and entertainment, accounting for 12.7%, 150 people for shopping, accounting for 14.8%; other travel purposes 216 people, accounting for 21.3%. The distribution of the common travel modes of the sample is as follows: the most traveled by bicycle, with 465 people, accounting for 45.8%, followed by the bus, with 334 people, accounting for 32.9%, taking subway/light rail, self-driving cars, taxis and others are 99, 52, 42 and 24, respectively.

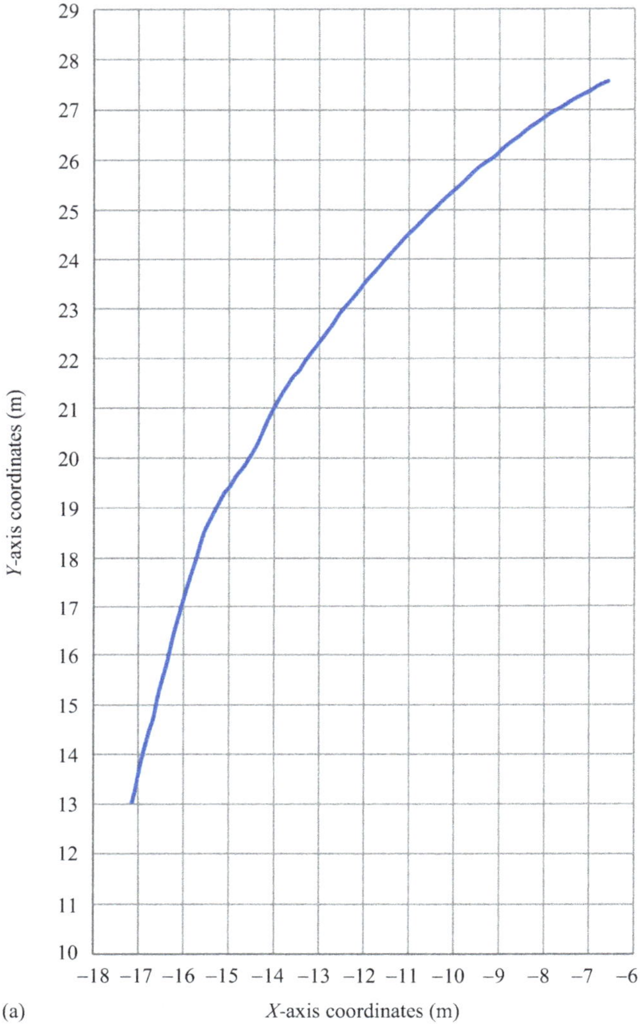

(a)

Figure 8.14 Motor vehicle sample dynamic motion data: (a) motor vehicle's trajectory; (b) speed in X direction; (c) speed in Y direction; (d) motor vehicle's speed; (e) acceleration in X direction; (f) acceleration in Y direction; (g) motor vehicle's acceleration

In general, most people (83.86%, 852 people) consider the safety factor (i.e., the conflict with motor vehicles when passing through an intersection, noted as S_{afe}) is the most important consideration; 11.6% (118 people) consider the time factor (T) to be the most important consideration; very few people (2.17% and 2.36%) think the distance (L) and bicycle and pedestrian congestion (LOS) factors

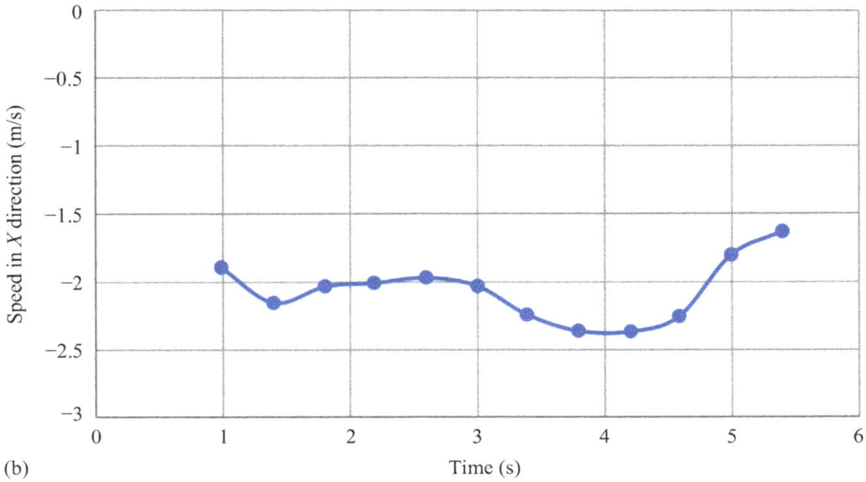

(b)

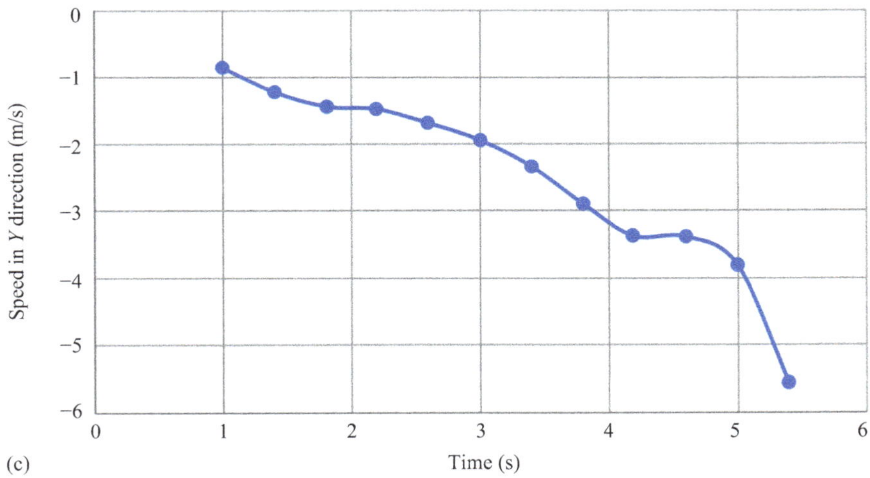

(c)

Figure 8.14 (*Continued*)

are the most important considerations. In terms of the least important (indifferent) influencing factors for bicycle path planning when passing through un-signalized intersections, 37.89% (385 people) think that bicycle and pedestrian congestion (*LOS*) are the least important considerations; 29.63% (301 people) believe that the time factor (T) to cross the intersection is the least important consideration; 29.33% (298 people) think that the riding distance (L) at the intersection is the least important consideration, only a few people (0.98%, 10 people) consider the safety factor (S_{afe}) to be the least important consideration. See Figure 8.17 for the distribution of the overall influencing factors of the youth group.

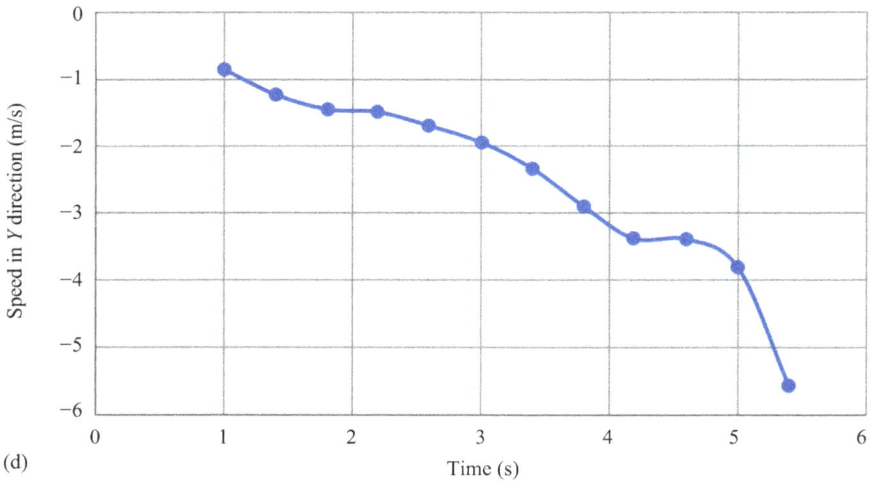

(d)

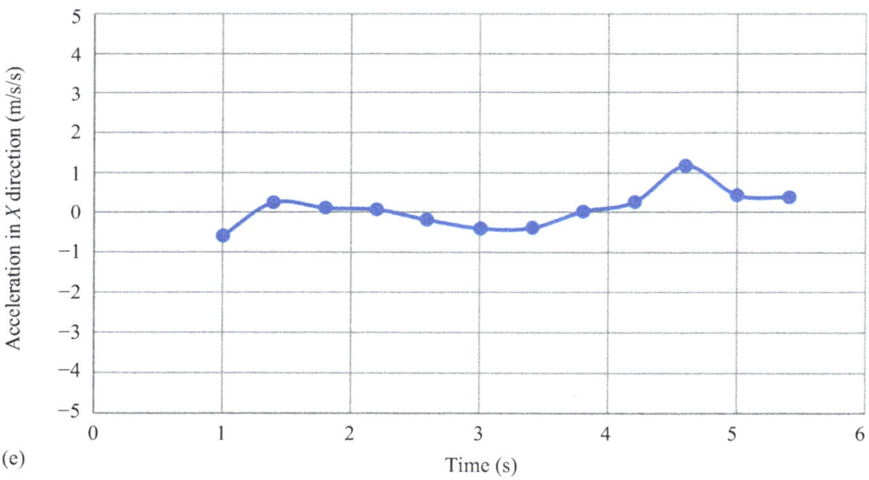

(e)

Figure 8.14 (Continued)

From the overall analysis earlier, it can be seen that when a cyclist plans a path through an un-signalized intersection, the order of importance of the previous four considerations (S_{afe}, T, L and LOS) is rough as follows:

$$S_{afe} \gg T > L > LOS \tag{8.1}$$

where the symbol ">" means greater than, and "≫" means much more important.

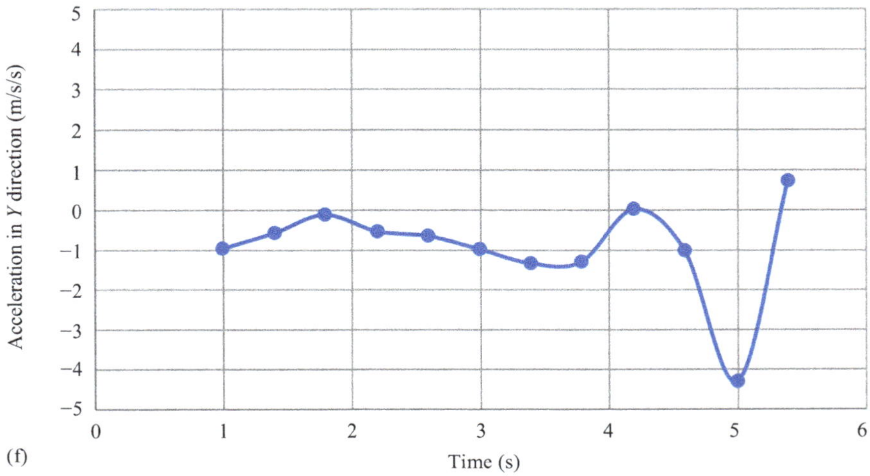

(f)

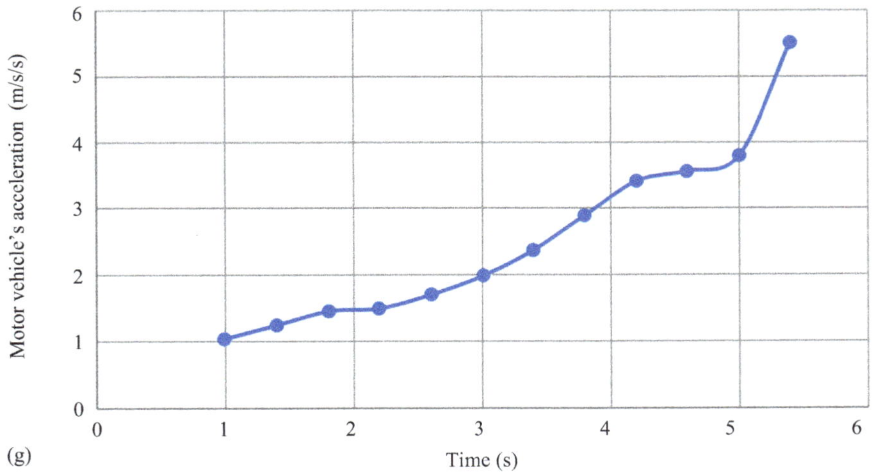

(g)

Figure 8.14 *(Continued)*

8.4.1.2 Influencing factor analysis

In order to further analyze the characteristics of cyclists, such as gender, age, trip purpose and the usual ways of travel affect the path planning behavior of cycling across un-signalized intersections, the samples were analyzed in groups. We grouped the sample by gender, age group, trip purpose, and commute traffic mode analyzed the ranking of important and nonimportant factors in the path planning behavior, and then analyzed the influencing factors. Detailed grouping analysis is presented in Tables A.1–A.5.

Here are the results of the analysis.

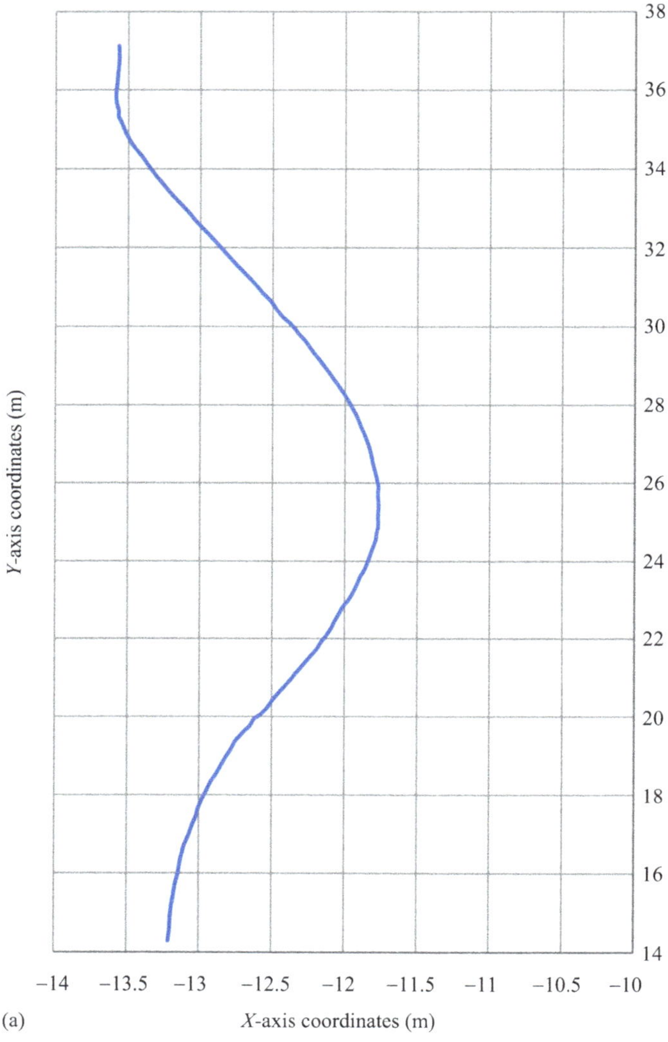

Figure 8.15 Bicycle sample dynamic motion data: (a) bicycle's trajectory; (b) speed in X direction; (c) speed in Y direction; (d) bicycle's riding speed; (e) acceleration in X direction; (f) acceleration in Y direction; (g) bicycle's acceleration

Gender

We found that gender factors have some influence on important considerations for cyclists' path planning behavior at un-signalized intersections, but they have not affected the importance order of the four factors. Latter we discuss the differences between various considerations for male and female cyclists:

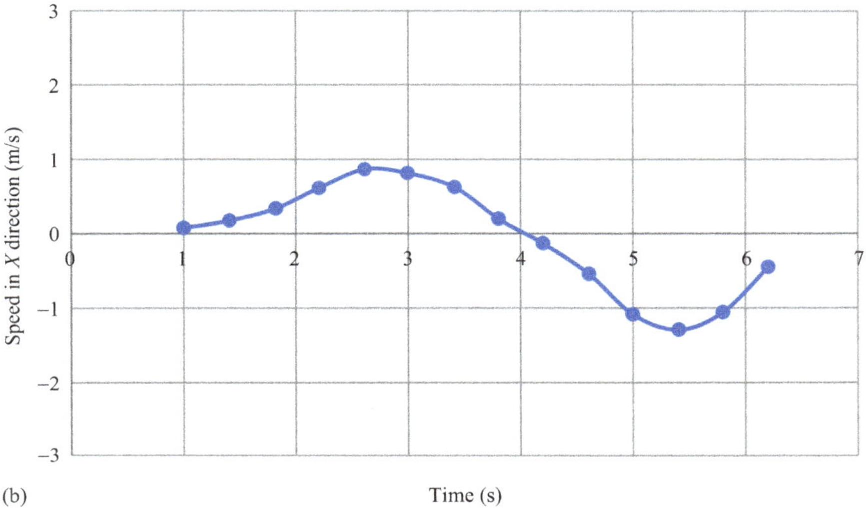

(b)

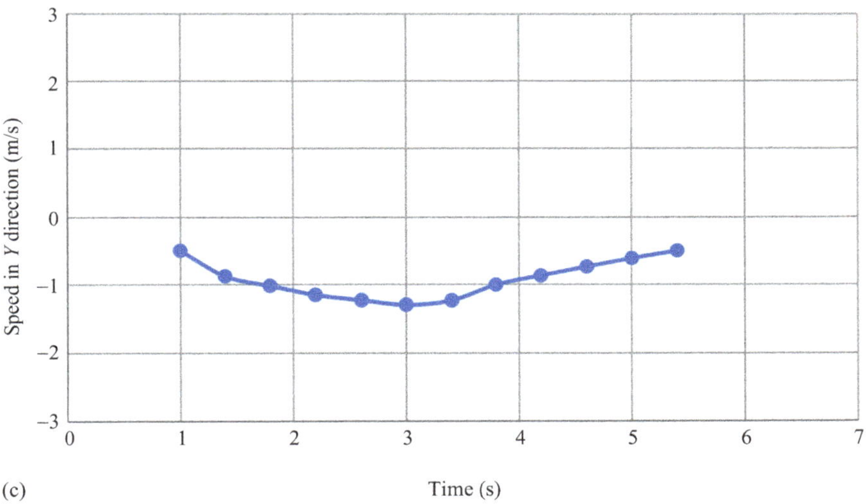

(c)

Figure 8.15 (*Continued*)

1. From the percentage of the most important and least important options, it can be seen that male riders value time more than female ones. Among the most important factors, men are 7% higher than women and in the least important 2% lower.
2. Male cyclists seem less intrusive than female cyclists to the congestion factor (*LOS*). In the most important factors, the attitudes of men and women are similar (a slight difference of 1.5%) and in the least important factors are higher than women 7% point.

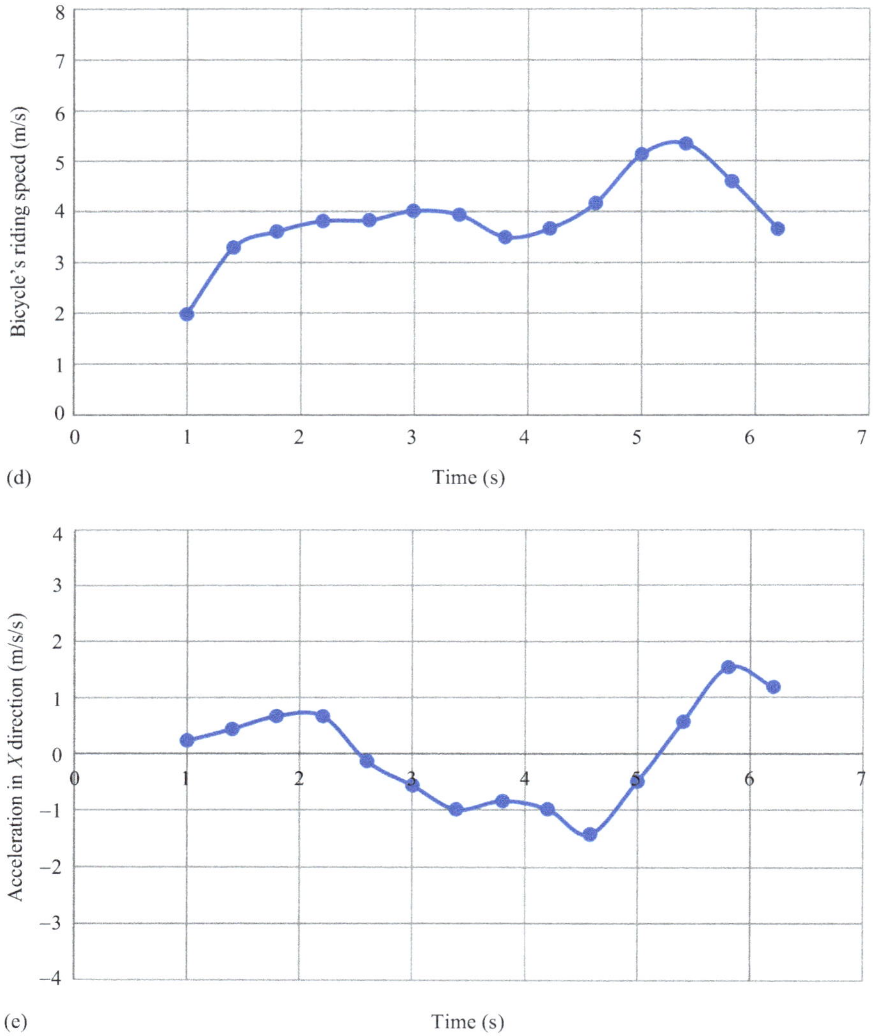

Figure 8.15 (Continued)

3. Female cyclists consider the distance factor (i.e., riding speed) more than male cyclists—1.5% higher than males in the most important factor, and male and female attitudes in the least important factor are almost the same (with a difference of 0.2%. The likely cause is that women have less physical strength than men. Another thing that illustrates this is that in the mode of travel, most of the men interviewed chose bicycles as the daily transportation mode, while most women chose buses as the daily transportation mode.

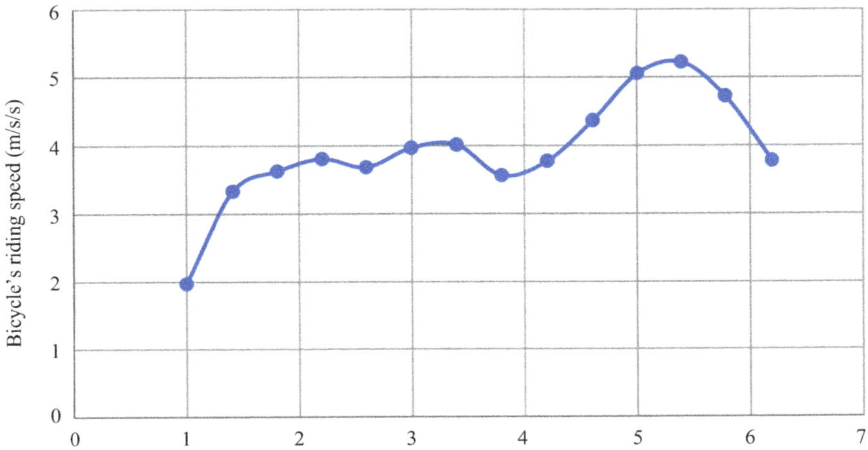

(f)

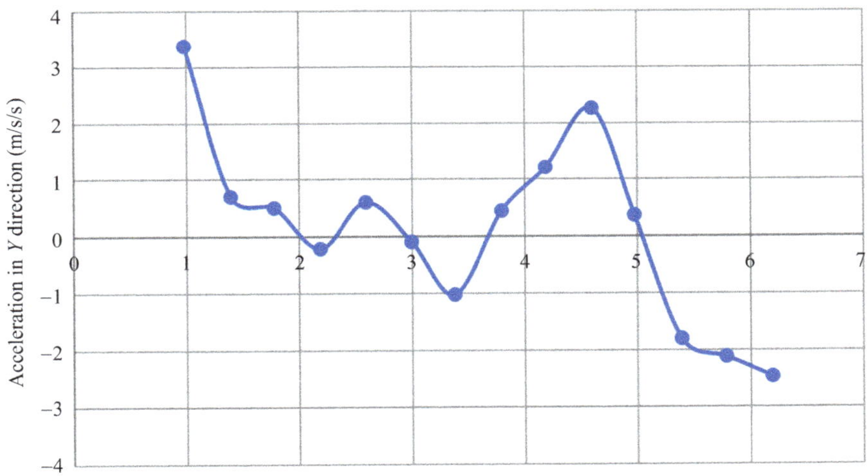

(g)

Figure 8.15 (Continued)

Age

Age has a significant effect on important considerations in cyclists' route selection behavior at un-signalized intersections.

For example, the importance ranking of the factors affecting the elderly group is different from the preference ranking of the other three age groups, which is "$S_{afe} \gg L > LOS > T$," while the ranking of other age groups is "$S_{afe} \gg T > L > LOS$."

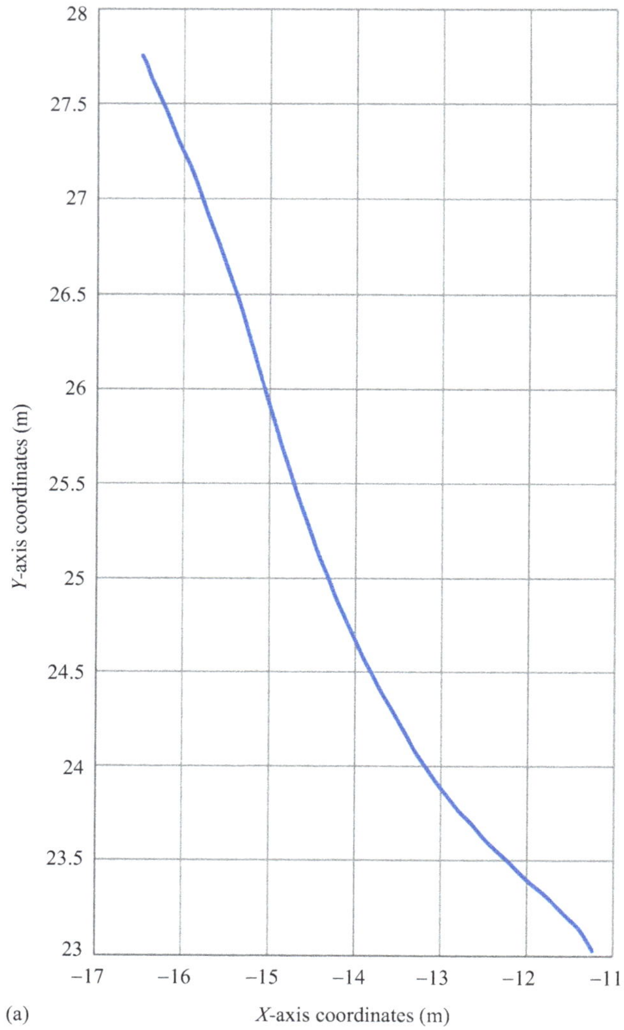

Figure 8.16 Pedestrian sample dynamic motion data: (a) pedestrian's trajectory; (b) speed in X direction; (c) speed in Y direction; (d) pedestrian's walking speed; (e) acceleration in X direction; (f) acceleration in Y direction; (g) pedestrian's acceleration

The different considerations for cyclists of different ages in path planning are as follows:

1. Respondents of different age groups have an awareness of the importance of safety. As age increases, the four age groups (<18 years, 18–35 years, 35–55

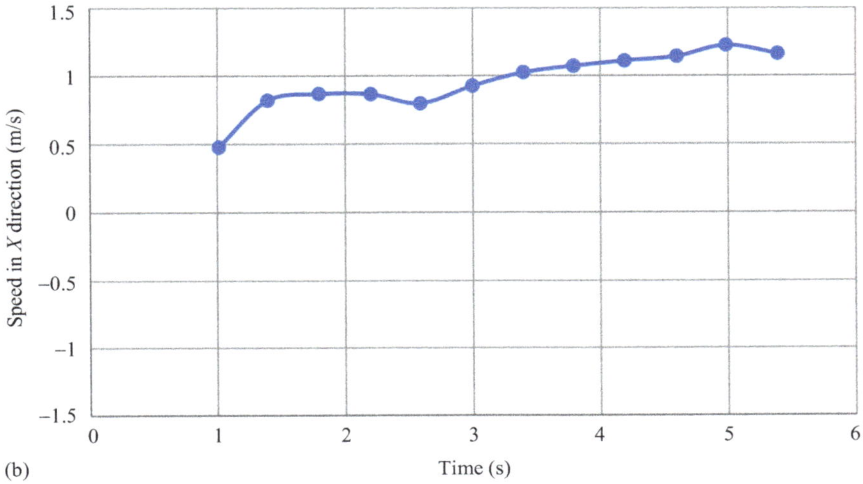

(b)

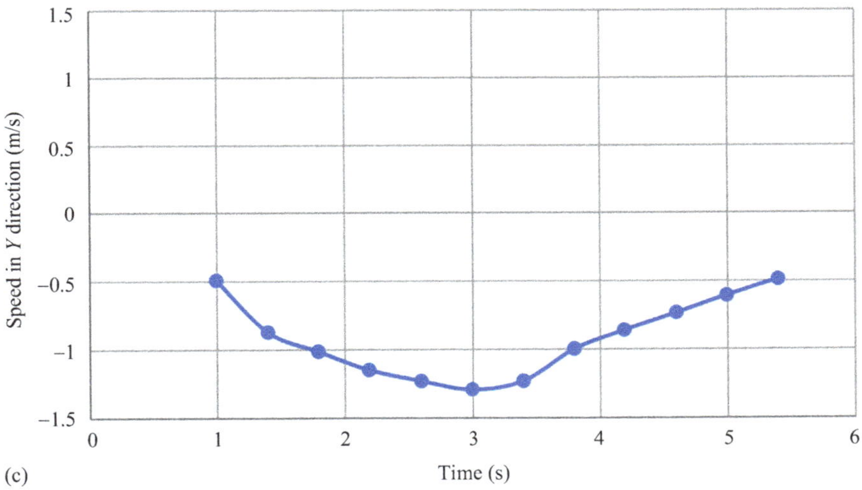

(c)

Figure 8.16 *(Continued)*

years, >55 years) regard safety as the most important. The proportion of important considerations is rising, 78%, 80%, 90% and 92%, respectively, while the proportion of choices as the least important consideration in safety is decreasing, respectively, 1.4%, 1.3%, 0.4% and 0%.

2. The importance of the time factor of respondents at different ages has declined significantly with age. The proportion of four age groups (<18 years, 18–35 years, 35–55 years, >55 years) that time is the most important consideration is decreasing, respectively, 21%, 14%, 8% and 4%. And the proportion of choices that are the least important considerations of time is rising, which are 19%,

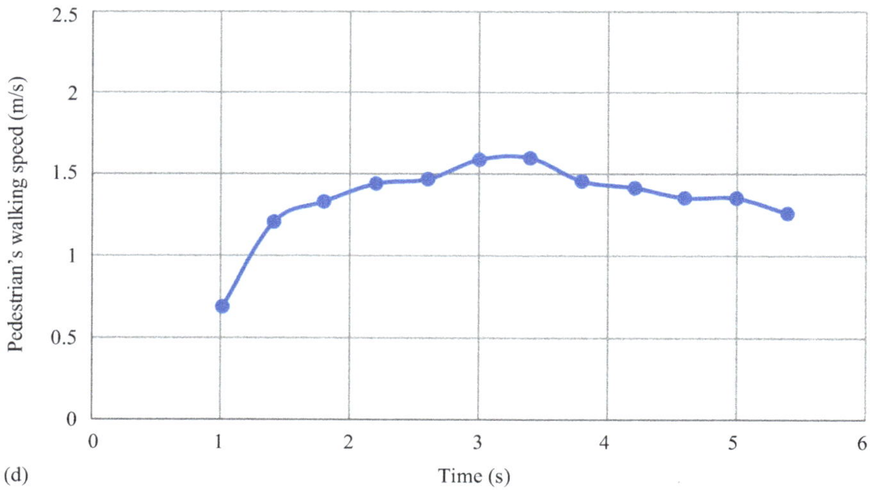

(d)

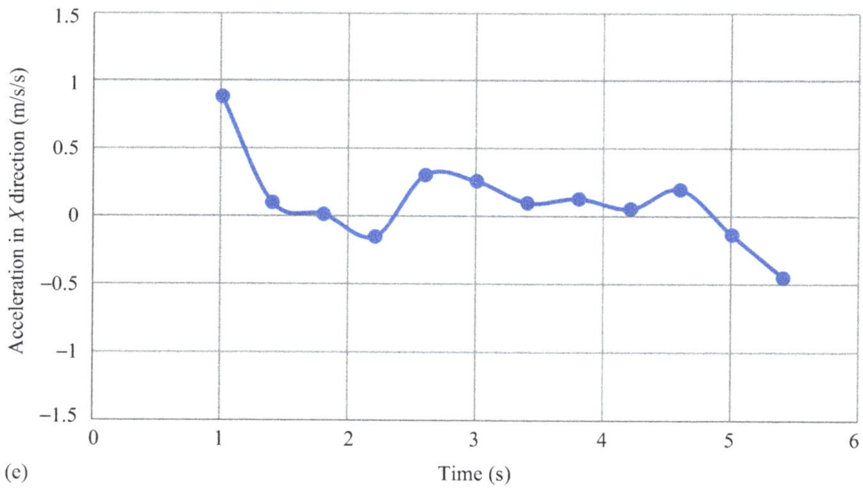

(e)

Figure 8.16 (Continued)

24%, 38% and 63%, respectively. The largest differences reached 17% and 23%, respectively.

3. The riding distance will increase a little with age, and the upward trend of the impact is small, which cannot be seen in the important factor options (because they are all very small). The proportions of the least important factors are declining, which are 40%, 33%, 22% and 17%, respectively. It can be seen that as age increases, the weight of physical factors also increases when planning a path.

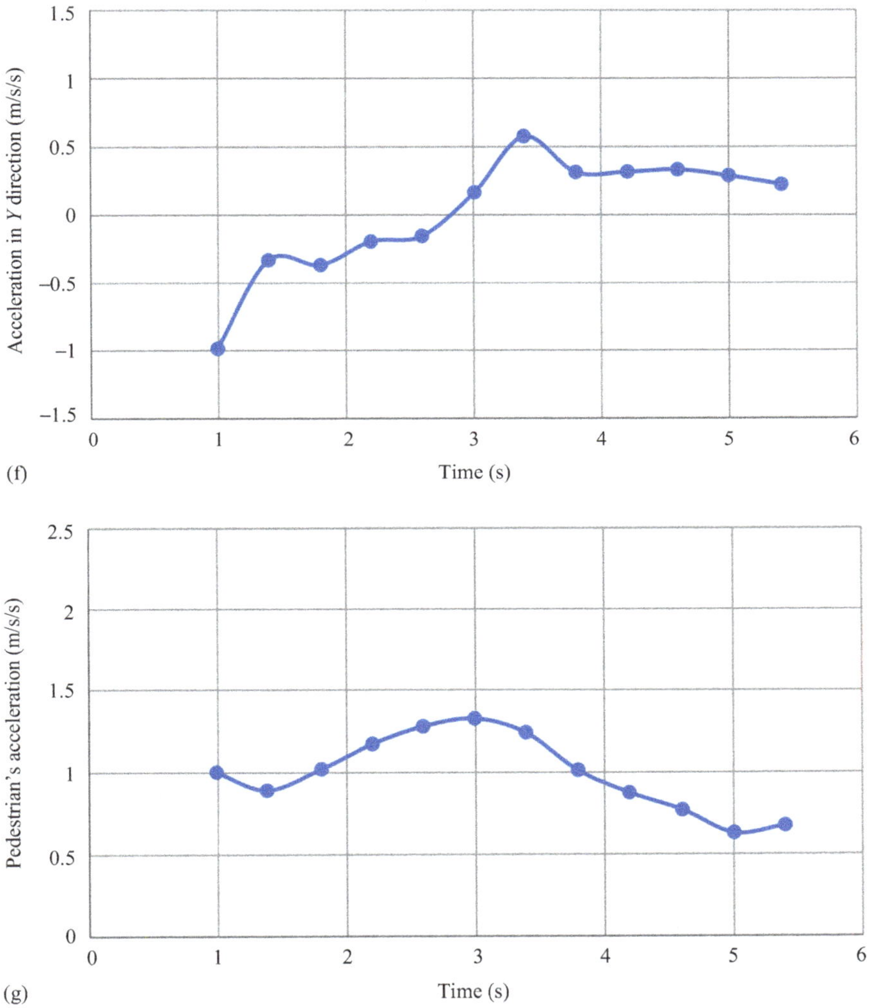

Figure 8.16 (*Continued*)

4. Elder cyclists pay more attention to the congestion (*LOS*) of crossing unsignalized intersections than other age groups. Among the least important factors, only 1.2% of older cyclists chose the *LOS*, while other age groups chose it at 41%, 41% and 35%, respectively.

Trip purpose

On trip purpose, we divided the questionnaire into two categories: commuting to work/school and non-commuting.

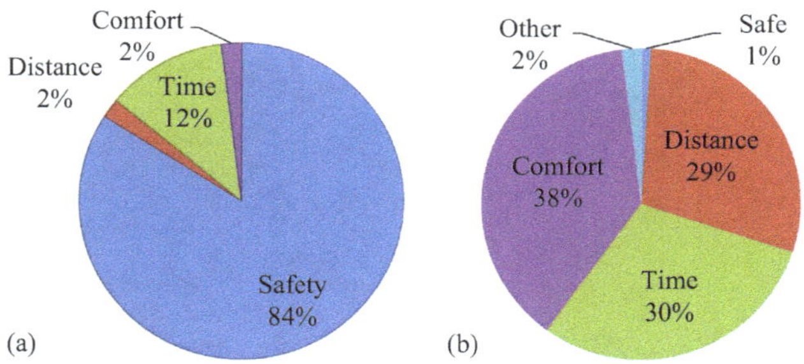

Figure 8.17 Distribution of influencing factors (youth group): (a) distribution of the most important factors; (b) distribution of the least important factors

1. For safety factors, commuter cyclists weigh less than non-commuter cyclists. They are 5% less in choosing it as the most important factor, while in the least important factor, commuter and non-commuter cyclists have a similar attitude.
2. For time factor, commuter cyclists have greater considerations than non-commuter ones—they are 5% more in choosing it as the most important factors and 12% less in choosing it as the least important factors.
3. For distance factor, commuting cyclists have a slightly lower consideration than non-commuting ones—they have a similar proportion in choosing it as the most important factors (0.7% less) and 7% more in the least important factors.
4. For the congestion (LOS) factor, commuting cyclists also have a slightly lower consideration than non-commuting cyclists—they have a similar proportion in choosing it as the most important factors (0.1% less) and 5% more in the least important factors.

Usual traffic mode

1. The statistical test results show that respondents who use different usual traffic modes have no significant differences in attitudes on most and least important factors in the path planning of cycling through un-signalized intersections, which indicates that different traffic modes have no effect on attitudes.

In summary, the results of SP data analysis are in line with common sense in traffic engineering.

8.4.2 Parameter identification

8.4.2.1 Parameter identification from SP data

The SP data obtained from the questionnaire data mainly examines the importance and relative weight of the four main considerations when cyclists planning their path through un-signalized intersections. And the latter defines the coefficients of the operational utility/cost attribute of each path in the path utility/cost function (Equation (7.71)) of the tactical-level path planning behavior model of NCB bicycle at un-signalized intersections.

Generally, the method of determining the attribute weight coefficient of the utility/cost function in economics is relatively simple, qualitative and fuzzy. The specific methods and processes are shown in Section 6.4 and [4].

From the overall analysis earlier, we already know the importance ranking of the four attributes. The SP data are analyzed next to further discuss the relative weighting of the four main operating utility/cost attributes of the bicycle path.

Let P_t, P_L, P_S and P_{LOS} represent the percentage of samples in the questionnaire that consider the *time* factor, *distance* factor, *safety* factor and *LOS* factor to be the most important considerations in the path planning of cycling through un-signalized intersections; similarly, let NP_t, NP_L, P_S and P_{LOS} represent the percentage of samples in the questionnaire that consider the *time* factor, *distance* factor, *safety* factor and *LOS* factor as the least important considerations in the previous path planning.

When we perform weight analysis on the important influencing factors of path planning behavior, the most important factor percentage P_i is the main information, and the least important factor percentage NP_i is the auxiliary information. Due to the sample size, model validation only identifies and discusses the path planning behavior of young cyclists.

From the analysis of the questionnaire data, we can know the factor weight w when the cyclists plan a path to cross the un-signalized intersection:

- From P_i, we get $P_{LOS} \approx P_L$; and from NP_i, we get $NP_{LOS} = 37.9 > NP_L = 29.3$. We can infer that the s*afety* factor is the most important consideration, and the *LOS* factor is the least important consideration, that is, $w_S >> w_t > w_L > w_{LOS}$.
- Because $P_S > P_t + P_L + P_{LOS}$ and $NP_S \approx 0.01$, the *safety* factor is far more important than the other three factors. We infer that its weight may be greater than the sum of weights of the other three factors, which is $w_S > w_t + w_L + w_{LOS}$.
- From $P_t < P_{LOS} + P_L$ and $NP_t \approx NP_L$, we can infer that although the importance of the *time* factor is greater than the *distance* and *LOS* factors, it is still less than the sum of the weights of the latter two factors, that is, $w_t < w_L + w_{LOS}$.

Inspired by the information entropy theory of Shannon, the author uses the most important factor percentages P_i and the least important factor percentages NP_i in the questionnaire survey data to quantify the relative weight w_i of factor i as

$$w_i = -\frac{1}{\left[\ln(P_i) + \frac{1}{\ln(NP_i)}\right]} \tag{8.2}$$

In this way, we use the SP data of the questionnaire to calculate the weight values w_t, w_L, w_S and w_{LOS} which are 0.33, 0.22, 2.54 and 0.21, respectively. The calculation results are consistent with the previous qualitative analysis properties.

And the corresponding model path attribute coefficients are

- for **young cyclist**: $c_t = 1$, $c_L = 0.9$, $c_S = 6.2$ and $c_{LOS} = 0.7$;
- for **young male cyclist**: $c_t = 1$, $c_L = 0.8$, $c_S = 5.3$ and $c_{LOS} = 0.7$;
- for **young female cyclist**: $c_t = 1$, $c_L = 1$, $c_S = 7.8$ and $c_{LOS} = 0.8$.

8.4.2.2 Parameter adjustment by RP data

It is noted that the SP data reflect the subjective preference tendency of the respondent, but the actual behavior of the person often differs from the subjective willingness or understanding. Therefore, we need to use the actual trajectory data (RP data) of bicycles crossing un-signalized intersections to adjust the value of the path selection/planning behavior attribute coefficients obtained from the SP data.

In the theory of discrete choice behavior, there are mature technical methods for identifying the attribute coefficient of the utility/cost function using RP data, and the main theoretical basis is based on the statistical *maximum likelihood method*. We first assume that most of the sample selection behaviors reflect the optimal choice, and then we can estimate the parameter values by examining the selective distribution of the samples in the same selection scenario set.

In the route selection behavior model of bicycle crossing un-signalized intersections, although the intersection plane can be discretized by dividing the intersection plane into small squares, there are a large number of influential factors in the route selection behavior of bicycle crossing non-signalized intersections. The dynamic changes, such as the type, size, speed direction and size of the conflicting vehicle, and the relative distance and position to the bicycle are not the same and cannot be summarized into simple cases. Therefore, the parameters of discrete selection behavior cannot be directly used. The identification method can only be designed according to the characteristics of the actual situation.

We adopt the idea of the *maximum likelihood statistics* in the parameter identification of the discrete selection model. Assuming that the final selected path of the collected bicycle is the optimal line P^*, each sub-path $\{D_{i-1}, D_i\}$ is also the optimal one. The cost of this sub-path is U_i^*. Then the cost U_j of choosing another TD D_j for the bike at this time T_i of the path is not less than U_i^*, that is, $U_j > U_i^*$, given $U = LC$, there are

$$L_jC - L_iC \geq 0 \tag{8.3}$$

Among them, L is the normalized running cost attribute vector of the path, and C is the weight coefficient vector of each attribute vector.

At time T_i, the cyclist has n (n can be infinite) TD alternatives, forming a system of equations:

$$\begin{cases} L_1C - L_iC \geq 0 \\ L_2C - L_iC \geq 0 \\ \vdots \\ L_nC - L_iC \geq 0 \end{cases} \tag{8.4}$$

where the value of L_i can be calculated. It is known that $c_t = 1$ in C. By solving the previous inequality equations, C can be obtained.

In the following, a certain component c_S in the vector C is taken as an example to discuss the parameter identification results under different solutions of the equations. Let us solve the system of equations and get the following results:

$$\begin{aligned}&c_S \le b_1 c_t, c_S \le b_2 c_t, \ldots, c_S \le b_m c_t \\ &\text{and } c_S \ge a_{m+1} c_t, \ldots, c_S \ge a_n c_t, a > 0, b > 0\end{aligned} \tag{8.5}$$

1. If $a_{\max} \le b_{\min}$, then (8.5) has a solution $a_{\max} c_t \le c_S \le b_{\min} c_t$, let c_S take the median value, that is, $c_S = (a_{\max} + b_{\min}/2)c_t$.
2. If $a_{\max} > b_{\min}$, then (8.5) has no solution. At this time, we draw the cumulative distribution curve of a_i and b_i, and let the value k corresponding to the intersection of the two curves be the equivalent value of the parameter c_S, that is, $c_S = kc_t$ (Figure 8.18).

This method of using curve intersections as equilibrium points is widely used in economics. It can be said that the two curves a and b reflect the fuzzy equivalent relationship between the two attributes safety factor S_{afe} and time factor T in the minds of the cyclist. However, the inequality method is used here, and the two curves represent different trends (one is greater than, the other is less than). The equilibrium point of the two trends is the intersection of the curves, which represents the best reflection of undifferentiated equivalence points of this preference.

It should be noted that the use of RP data to adjust the parameters of the bicycle tactical-level path planning behavior model is mainly based on the results of SP data analysis. That is, adjustment values derived from RP data are acceptable only if they do not violate the qualitative information obtained from the SP data.

8.4.3 Parameter identification results

Calculated by the SP data obtained from the questionnaire, the attribute coefficient c_k of the NCB path planning model are as follows:

- for **young cyclist**: $c_t = 1$, $c_L = 0.9$, $c_S = 6.2$ and $c_{LOS} = 0.7$;
- for **young male cyclist**: $c_t = 1$, $c_L = 0.8$, $c_S = 5.3$ and $c_{LOS} = 0.7$;
- for **young female cyclist**: $c_t = 1$, $c_L = 1$, $c_S = 7.8$ and $c_{LOS} = 0.8$.

The adjusted attribute coefficient c_k by the RP data are as follows:

- for **young cyclist**: $c_t = 1$, $c_L = 0.8$, $c_S = 3.2$ and $c_{LOS} = 0.7$;
- for **young male cyclist**: $c_t = 1$, $c_L = 0.8$, $c_S = 2.7$ and $c_{LOS} = 0.7$;
- for **young female cyclist**: $c_t = 1$, $c_L = 0.9$, $c_S = 4.5$ and $c_{LOS} = 0.8$.

See Table 8.2 and Figure 8.19 for details.

The order of the important attributes of the path selection and most of the weighting coefficients have not changed significantly, except for the weighting coefficient c_S of the safety factor, which reflects that people's awareness of safety is higher than that in behavior. This is in line with previous research results on traffic safety behaviors [5].

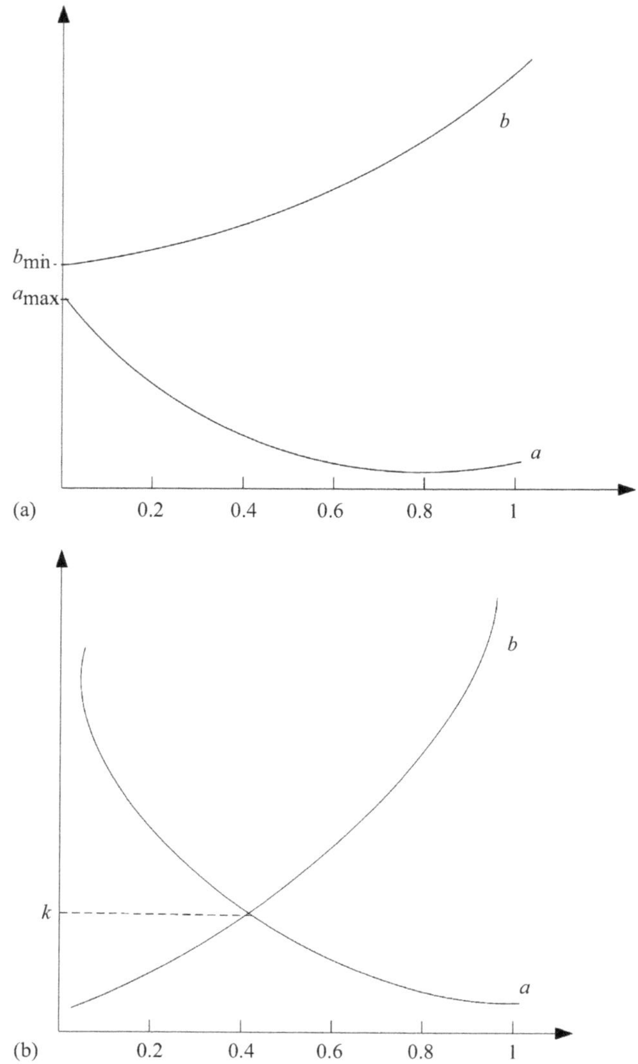

Figure 8.18 The adjustment of parameter c_S in two situations: (a) $a_{max} \leq b_{min}$; (b) $a_{max} > b_{min}$

The RP-data-adjusted parameters of the path selection model also show that the urgency and purpose of the riding behavior of female cyclists are not as obvious as that of males, that is, the proportion of T weights is relatively low. For male cyclist $c_t/\sum_k c_k = 0.19$, while for female cyclist $c_t/\sum_k c_k = 0.14$.

Table 8.2 Comparison of parameter identification results of SP and RP data

Groups	SP data analysis				RP data adjustment				Sample size (questionnaire number/case)
	c_t	c_L	c_S	c_{LOS}	c_t	c_L	c_S	c_{LOS}	
Young male cyclist	1	0.81	5.31	0.67	1	0.8	2.7	0.7	612/22
Weight percentage	13%	10%	68%	9%	19%	15%	52%	13%	
Young female cyclist	1	1	7.77	0.84	1	0.9	4.5	0.8	404/15
Weight percentage	9%	9%	73%	8%	14%	13%	63%	11%	
Young cyclist	1	0.88	6.20	0.73	1	0.8	3.2	0.7	1,016/37
Weight percentage	11%	70%	10%	8%	18%	14%	56%	12%	

At the same time, we found that male cyclists have a smaller safety coefficient proportion than female cyclists: the $c_S/\sum_k c_k$ for male and female cyclists are 0.52 and 0.63, respectively.

8.5 Dynamic riding model parameter identification

8.5.1 RP data analysis

8.5.1.1 Qualitative analysis of RP data

Before carrying out specific model parameter identification on the data of bicycle crossing un-signalized intersections, that is, quantitative analysis, we first make a qualitative analysis of dynamic motion data of various collision avoidance behaviors of bicycle crossing un-signalized intersections.

We randomly select a sample case to examine the bicycle's motion characteristics (trajectory, speed and acceleration) curve when it avoids a conflicting vehicle (motor vehicle, bicycle or pedestrian) and obstacle.

Figure 8.20 is the motion trajectory of a bicycle sample at an un-signalized intersection (Case ID: BC44). In this case, the bicycle bypassed the lateral potentially conflicting motor vehicle in its front.

The bicycle's trajectory clearly shows the process of the bicycle avoiding the approaching motor vehicle. Here, the bicycle advances in the *Y* direction in the intersection coordinate system.

We examine the relationship between the acceleration $\vec{a}$ of the bicycle during the avoidance of the lateral conflicting motor vehicle. Furthermore, we analyze the destination driving force $\vec{f}_0$ and the psychological effect tension $\vec{f}_1$ of the conflicting vehicle in the NCB operation-level model (Figure 8.21).

In Figure 8.21, Ax, f_{0x} and f_{1x}, respectively, represent the *X*-direction components of the bicycle acceleration $\vec{a}$, the goal-driven force $\vec{f}_0$ and the psychological effect force $\vec{f}_1$ of the conflicting vehicle of the intersection coordinate system. And

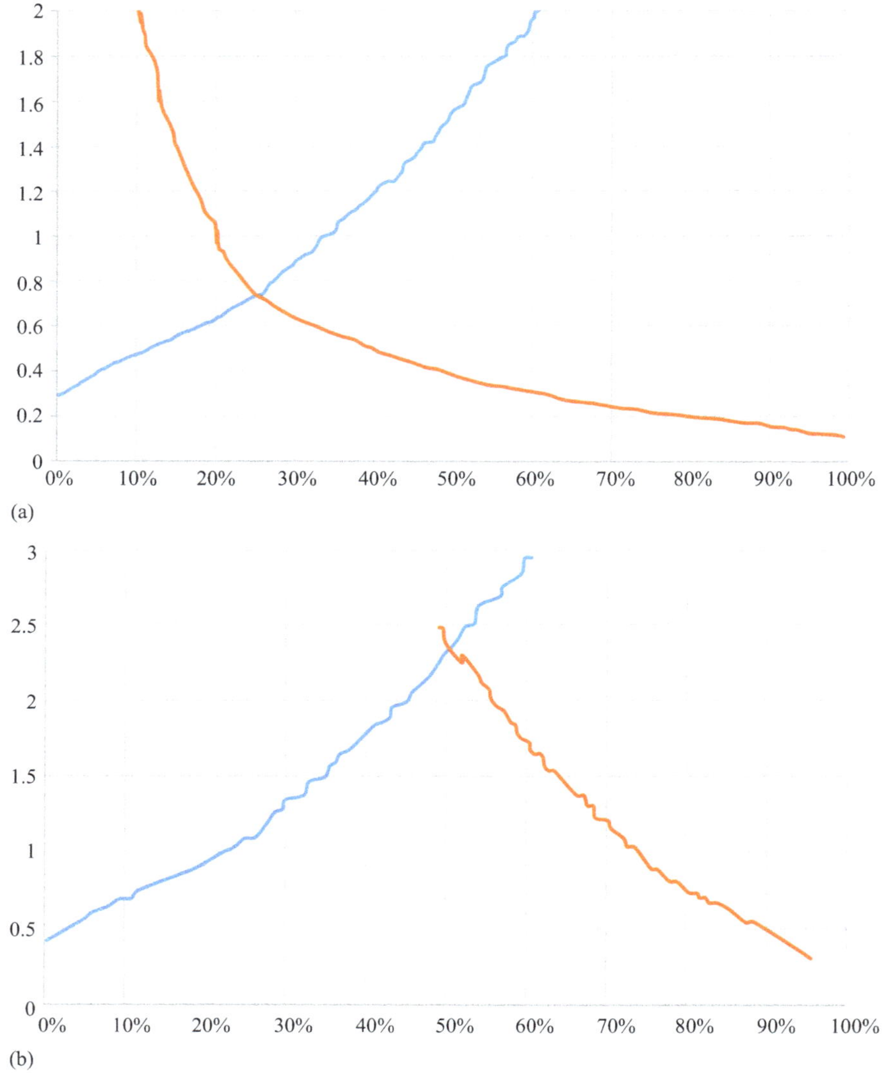

Figure 8.19 The equivalent conversion result of each attribute coefficient of path riding utility/cost function: (a) equivalent point of c_L to c_t; (b) equivalent point of c_S to c_t; (c) equivalent point of c_{LOS} to c_t

Ay, f_{0y} and f_{1y} represent *Y*-direction components in the intersection coordinate system.

It can be seen from the figure that the acceleration of the bicycle riding has a fluctuating characteristic, and the fluctuations of the *X* and *Y* direction components

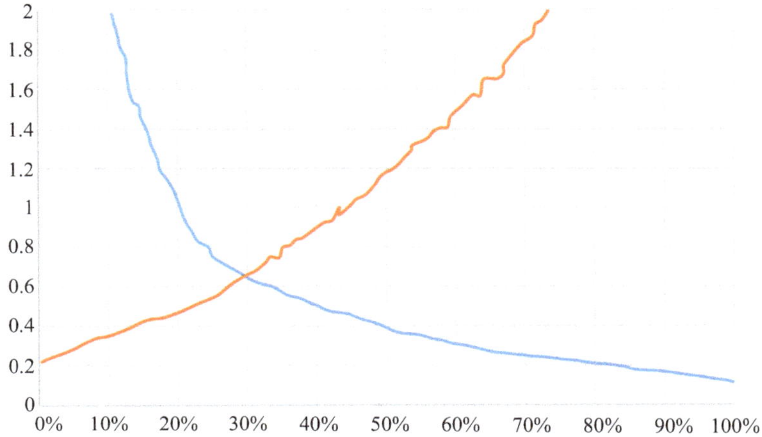

(c)

Figure 8.19 (Continued)

have time consistency (see time points 7, 12, 17 and so on). This is due to the riding characteristics of the bicycle: power is generated by humans periodically stepping on the bicycle, thereby generating corresponding wave-shaped acceleration.

In addition, the bicycle's lateral direction (X direction) is a *negative→positive→negative* acceleration change process in the course of avoiding conflicting motor vehicles. The first change is that the bicycle accelerates to the right to avoid the conflicting vehicle, and the second change is that the bicycle accelerates back to the left after avoiding the conflicting vehicle and then continues to ride to the expected destination.

The longitudinal acceleration of the bicycle has always been positive, but it has gradually flattened later, which indicates that the cyclist also accompanies the longitudinal acceleration during the process of avoiding the conflicting vehicle. After bypassing the conflicting vehicle, he/she will no longer accelerate.

The red line in Figure 8.21 indicates the driving force component of the destination on the bicycle. When the bicycle's current speed differs from the cyclist's expected speed, the amplitude of the curve increases. We can see that the fluctuations of the acceleration curve of the bicycle and the driving force curve of the destination are basically the same, though there is a certain delay in time between them. This phenomenon is more obvious in multiple cases, with a time interval of 0.5 s in a multi-sample data set (Figure 8.23).

Because the effect force components f_{1x} and f_{1y} of the conflicting motor vehicle are too small, they are enlarged by 50 times and compared with the bicycle acceleration Ax, Ay and the destination driving force f_{0x}, and f_{0y}, as shown in Figure 8.22.

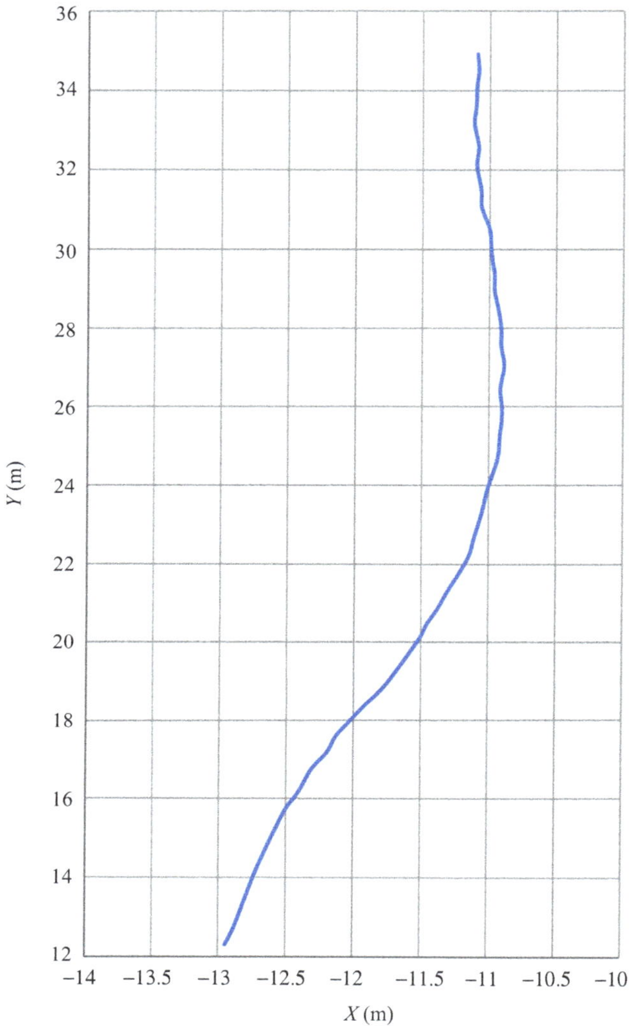

Figure 8.20 The bicycle's motion trajectory of avoiding a conflicting vehicle (BC44)

We can also see from the figure that the acceleration of the bicycle is also affected by the effective force $\vec{f}_1$. As shown in Figure 8.22(a), in the Y direction component of acceleration around the time point 13 is more obvious. A similar situation occurs in a large number of other examples (see Figure 8.23). This point shows that the influence design in the operation hierarchy model is quite in line with the actual situation.

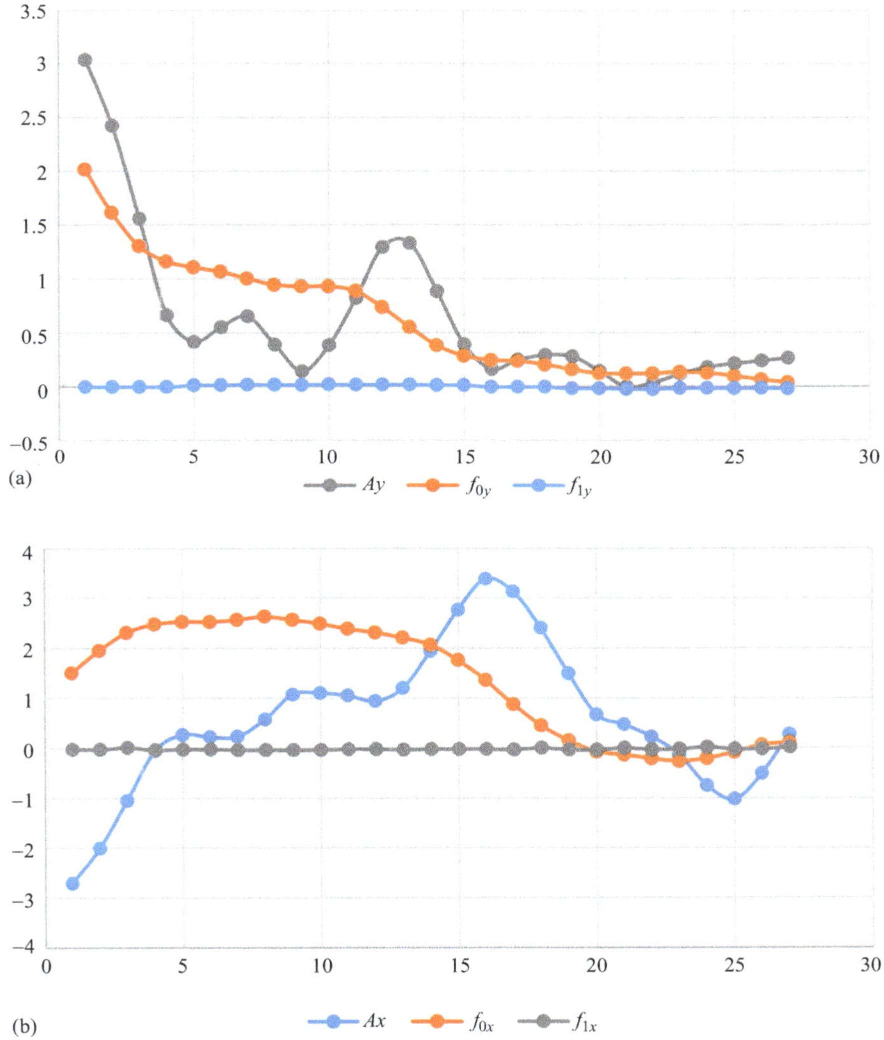

Figure 8.21 Bicycle's acceleration and SF in confliction avoidance (time interval 0.2 s, BC44): (a) bicycle's acceleration and SF components in the Y direction; (b) bicycle's acceleration and SF components in the X direction

8.5.1.2 Summary

Through the previous investigation and analysis of a large number of examples, the author draws the following main qualitative preliminary analysis results on the dynamic characteristics of cycling:

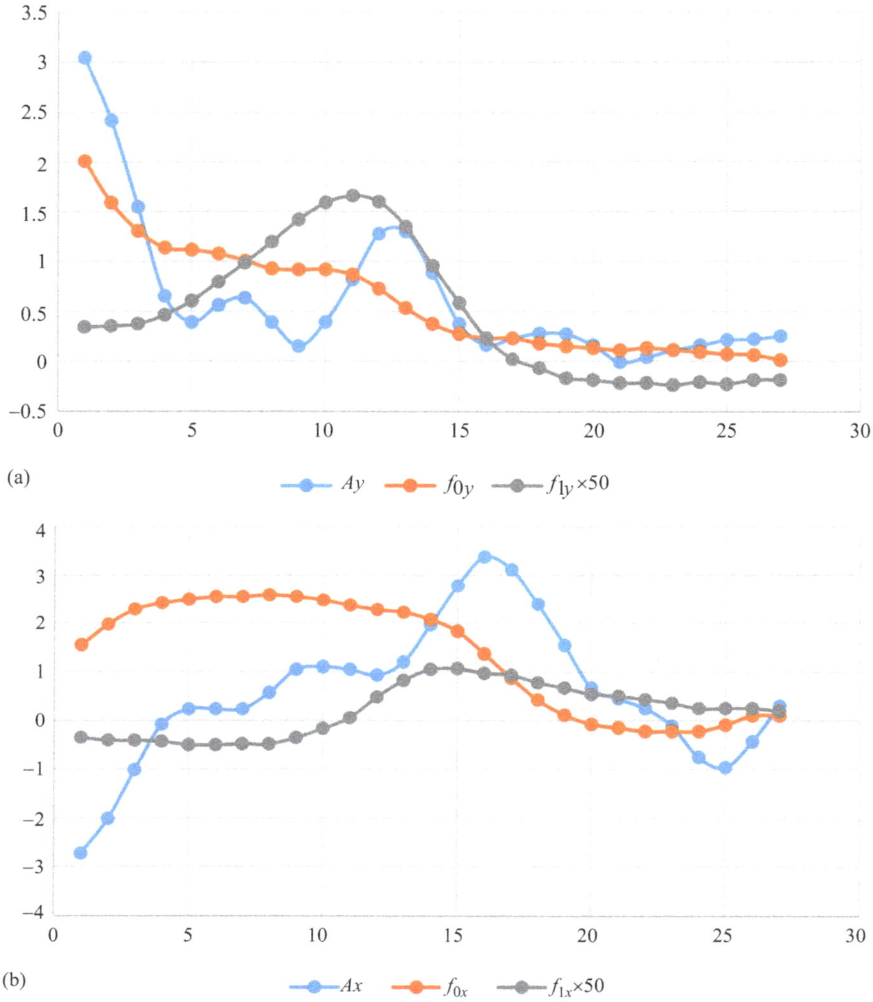

Figure 8.22 Bicycle's acceleration and SF in confliction avoidance ($\vec{f_1}$ enlarged by 50 times, BC44): (a) bicycle's acceleration and SF components in the Y direction (f_{1y} enlarged by 50 times); (b) bicycle's acceleration and SF components in the X direction (f_{1x} enlarged by 50 times)

1. The acceleration of cycling is characterized by periodic fluctuations, and the fluctuations in the *X*- and *Y*-direction components are consistent in time. This is due to the cycling characteristics of the bicycle itself: the power is periodically generated by the human pedaling, which generates the corresponding wave-shaped acceleration. This is fundamentally different from the dynamic movement of motor vehicles and pedestrians.

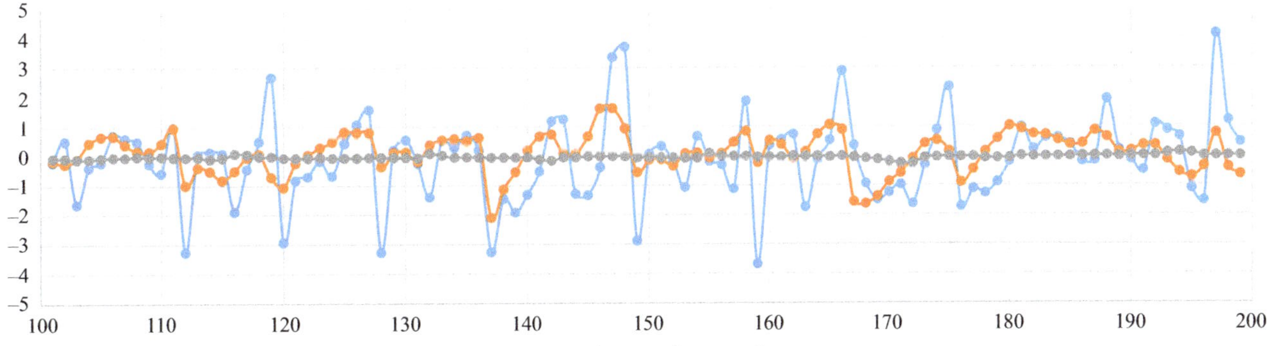

(a)

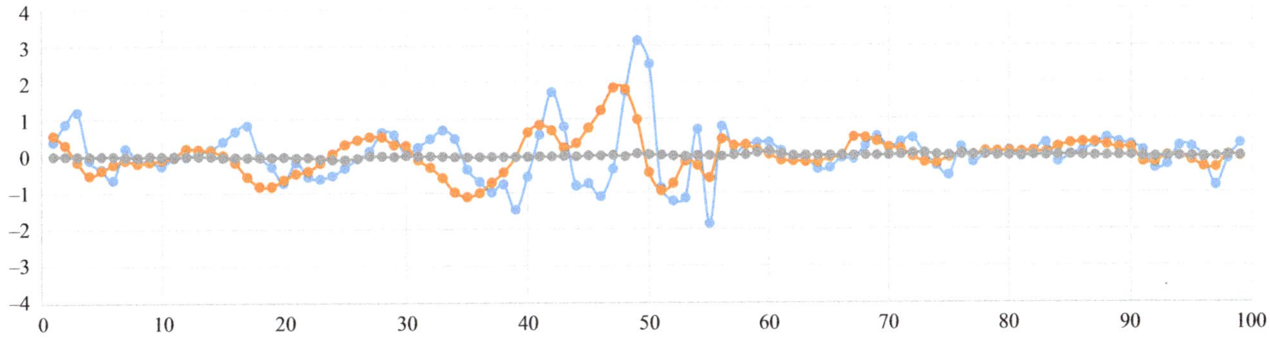

(b)

Figure 8.23 *Bicycle's acceleration and SF force curves in various conflict avoidance situations (time interval 0.5 s): (a) B–C cases in X direction (PART-1); (b) B–C cases in X direction (PART-2); (c) B–C cases in Y direction; (d) B–B cases in X direction; (e) B–B cases in Y direction; (f) B–P cases in X direction; (g) B–P cases in Y direction; (h) B–O cases in X direction; (i) B–O cases in Y direction*

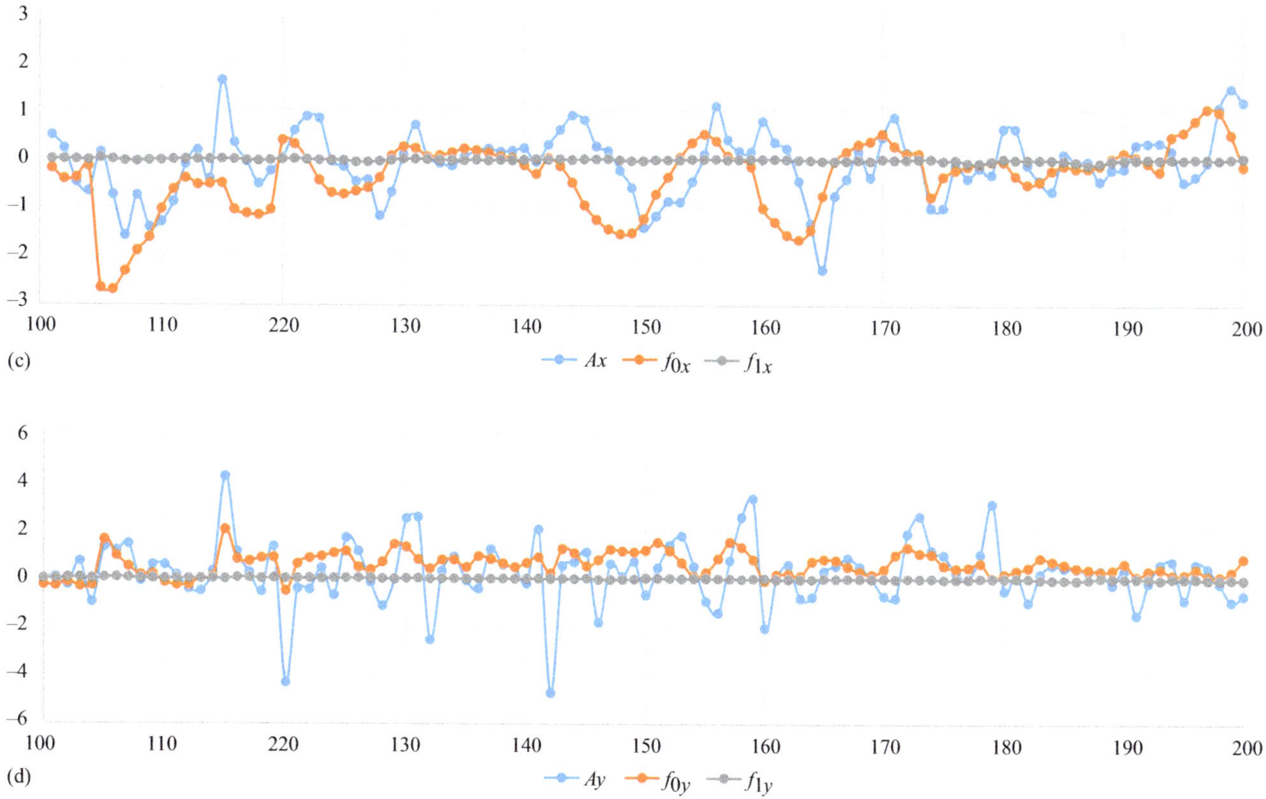

Figure 8.23 (Continued)

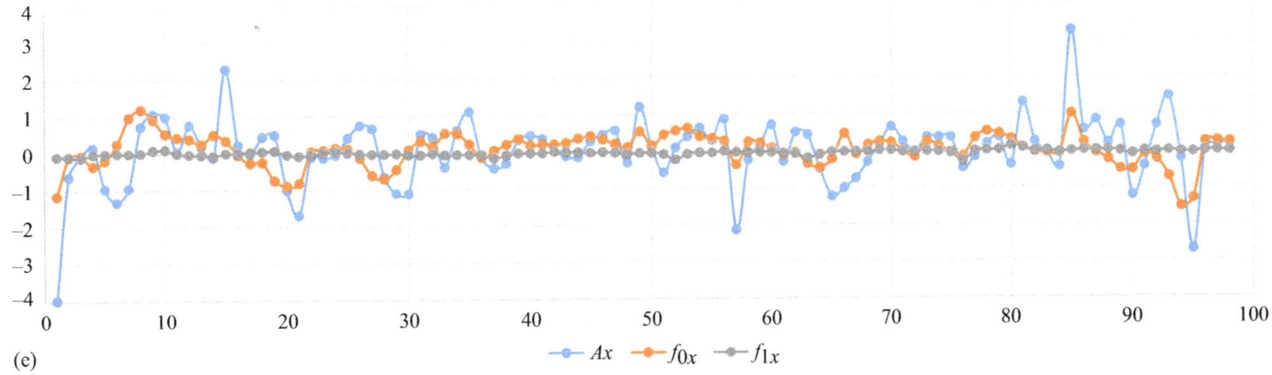

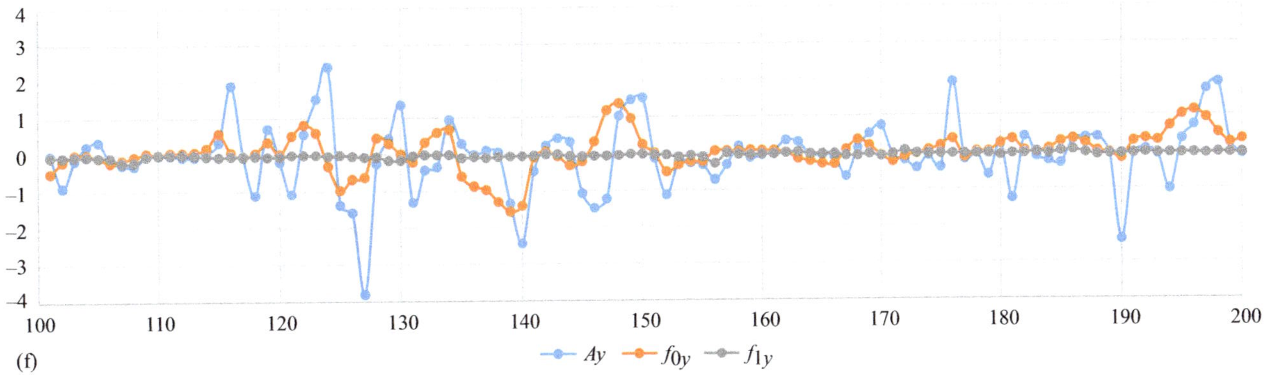

Figure 8.23 (Continued)

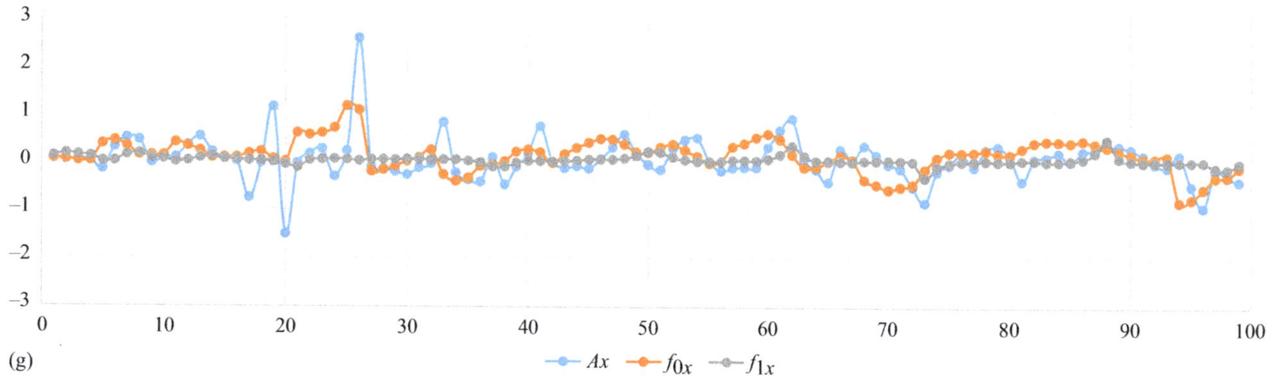

(g)

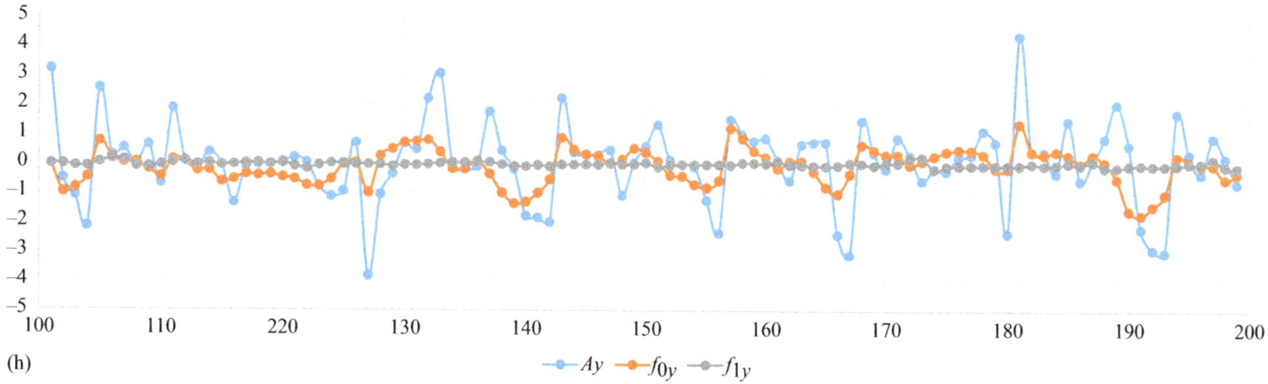

(h)

Figure 8.23 (Continued)

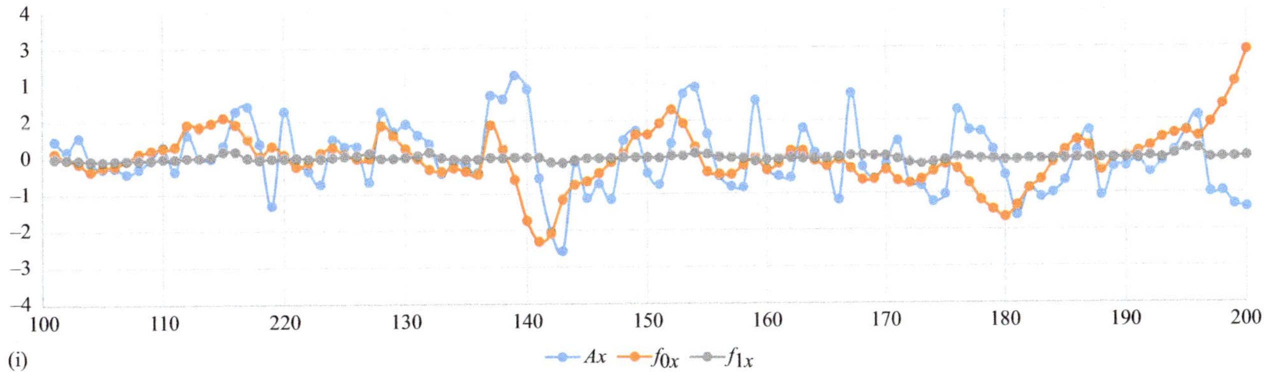

Figure 8.23 (Continued)

Because the bicycle is landing on two wheels, the movement is unstable, and often the acceleration actually generated by the bicycle may also have undulating fluctuations, which is different from the acceleration expected by the rider. This should be paid attention to when performing quantitative data analysis.

2. The fluctuation of the bicycle's acceleration curve and the cyclist's goal driving force $\vec{f}_0$ are basically the same, except there is a certain time delay.

   This phenomenon occurs in many different examples, including bicycle–motorcar conflict cases (B–C conflict), bicycle–bicycle conflict cases (B–B conflict), avoid bicycle–pedestrian conflict cases (B–P conflict) and bicycle–obstacle cases (see as B–O conflict), as shown in Figure 8.23.

   This shows that after the cyclist perceives the difference between his own speed and the ideal expected speed and the impact of possible conflicting objects (motor vehicle, bicycle, pedestrian and obstacle), it will take some time for him/her to implement the acceleration results of the brain/psychological judgment. That is, the various parts of the body (hands, legs, feet, etc.) coordinate movement so that the bicycle generates corresponding acceleration.

   We think that this delay phenomenon justifies the rationality of applying the psychological field theory model in the operation-level model.

3. The acceleration of the bicycle is also affected by the effective force $\vec{f}_1$ of the conflicting vehicle, and from the final three variables all tend to be *zero*, we can infer that the accelerations of the bicycle always tend to eliminate (or relax) the effect force $\vec{f}_1$ and the destination driving force $\vec{f}_0$. Therefore, the conflict avoidance behavior process of the bicycle can also be regarded as a dynamic control process.

4. From the phenomenon of the delay of bicycle acceleration change and the driving force and effect force, we can observe the cyclist's *perception–reaction time (PRT)* at the un-signalized intersections. After large numbers of observations, we initially estimated the cyclist's PRT time to be 0.4–1.0 s.

The AASHTO's suggested response time for cyclists is 2.5 s [6], while Taylor's recommended response time for cyclists at signalized intersections is 1 s [7].

8.5.2 Model parameter identification

We use dynamic data to identify the parameters of the operation-level model, using a simple regression idea. In the micro-behavior database of bicycle crossing un-signalized intersections, there is information about the spatial movement status of bicycles and the conflicting vehicles (motor vehicles, bicycles and pedestrians), and the speed and acceleration information of bicycles and the conflicting vehicles can be further obtained.

By substituting the previous dynamic information data into the model formula in Chapter 7, various social field dynamics of the cyclist at each time point t can be calculated: the goal driving force $\vec{f}_0$, and the effective force $\vec{f}_1$. By analyzing the effects of these psychological fields and the bicycle's behavior change at time t (i.e., acceleration $\vec{a}$), the parameters of the NCB operation level model can be obtained: the conflicting vehicle-type parameter M (M_C, M_B, M_P and M_O), and the cyclist's psychological field radius R and goal driving coefficient γ_0.

From the earlier qualitative analysis, it can be seen that bicycle movements have periodic fluctuations. In the quantitative analysis of the data, the acceleration of the bicycle caused by this fluctuation must be removed to meet the effective acceleration of the model parameter identification. Examining the straight motion characteristics of the bicycle (Figure 8.24), we found that the bicycle's lateral acceleration (Figure 8.24(b)) still fluctuates periodically even in the absence of interference.

With reference to relevant foreign studies [3] and statistical analysis of bicycle movement data, it is considered that the acceleration wave has a sinusoidal curve with an amplitude of about 0.05–0.2 m/s^2 and a period of about 0.5–2 s.

We think of the bicycle acceleration information $\vec{a}$ as consisting of the bicycle's effective acceleration $\vec{a}_{valid}$ and the wave acceleration $\vec{a}_s$. Because the wave acceleration has a sinusoidal curve, the wavelet information denoising

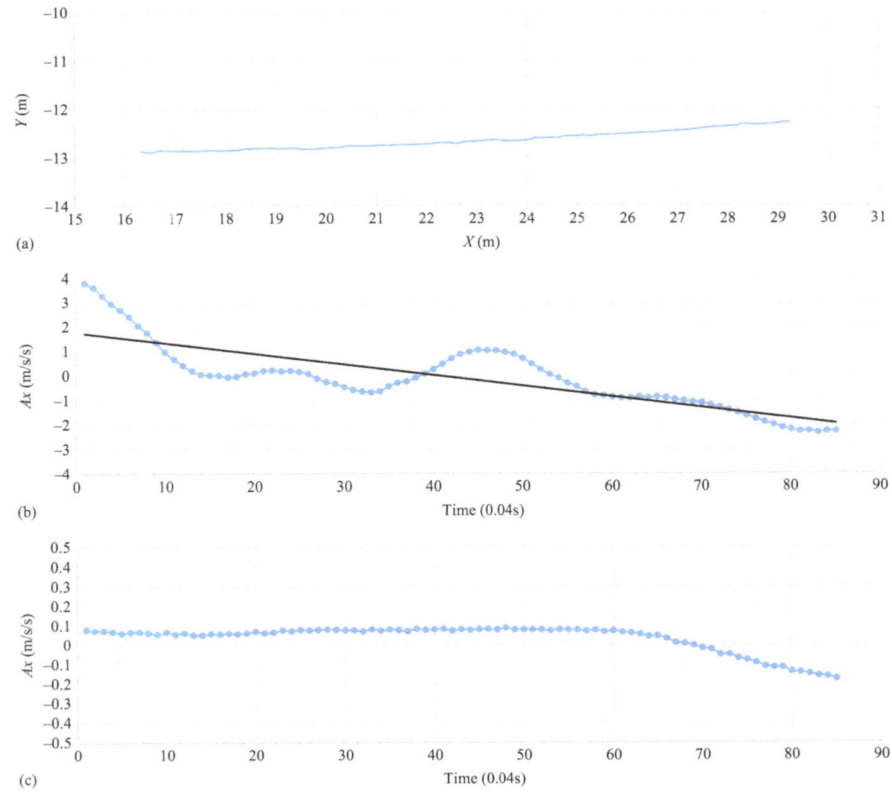

Figure 8.24 The original motion characteristics of interference-free bicycles: (a) the bicycle's original motion trajectory; (b) the bicycle's original acceleration in X-direction; (c) the bicycle's original acceleration in Y-direction

method is used. We first construct a wavelet $\vec{W}_s$, amplitude h and period c of which have certain randomness, and both are assumed to conform to a normal distribution.

Then $h = 0.1\ I_2\ I$ and $c = 1\ I_2$, where I is a two-dimensional random variable that conforms to the standard normal distribution. After using this wavelet denoising method, the bicycle acceleration value is shown in Figure 8.25.

After eliminating periodic fluctuations, there are two ways for regression analysis of the effective acceleration of the bicycles:

The first is to identify the parameters of each bicycle's conflict avoidance situation. Then analyze the behavior parameters of each bicycle individual, such as the conflicting vehicle-type parameter M, distribution and influencing factors.

The other method is to aggregate the four main types of behavioral data of conflict avoidance situations (B–C, B–B, B–P and B–O situations), and then perform a regression analysis to obtain the behavioral parameter identification results for each type of conflict situations, and then analyze the parameter identification results.

The results of the two analysis methods are discussed and compared next.

When the behavior model parameter identification is performed for a single specific conflict avoidance situation, the correlation between the behavior model parameters and the dynamic behavior of the bicycle (i.e., effective acceleration) is quite high in most cases, and the correlation coefficients are almost all above 0.8, generally around 0.9. The situation also reached a level of 0.98.

This shows that the NCB behavior model is quite effective in describing the collision avoidance behavior of bicycles crossing un-signalized intersections.

However, it is also found that there are significant differences in the behavior parameters of different cyclists: the identification result of the objective driving

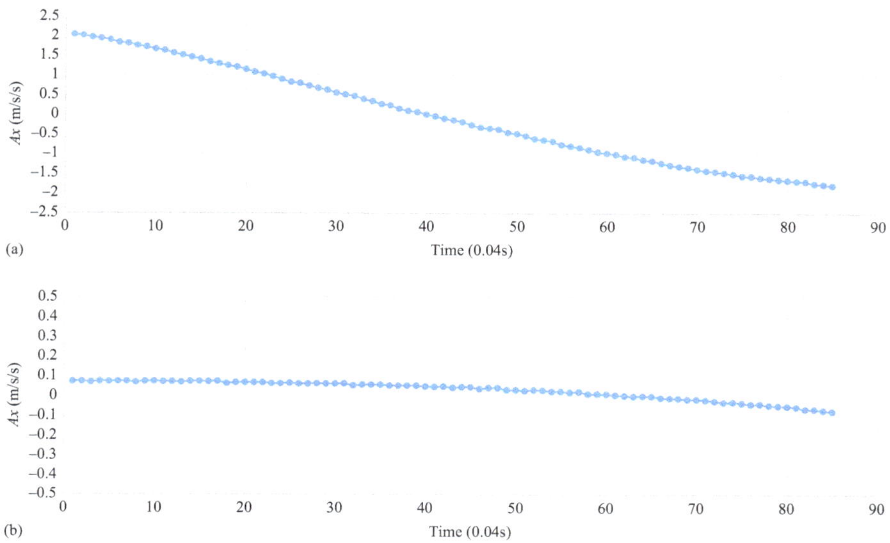

Figure 8.25 The effective accelerations of the interference-free bicycle: (a) the bicycle's effective acceleration in X-direction; (b) the bicycle's effective acceleration in Y-direction

coefficient V varies between 0.3 and 2; the type parameter M: M_C ranges from 1 to 10 (here may conflict with the size of the motor vehicle). Type-dependent, the largest parameter is a large bus vehicle, M_B changes from 0.5 to 5, M_P changes from 0.5 to 2 and M_O changes from 0.2 to 2; the cyclist's psychological radius R changes from 1 to 3.

In addition, the collision avoidance behavior when a bicycle crosses an un-signalized intersection has a short duration, usually 2–5 s. During data analysis, to eliminate data fluctuations, and ensure accuracy and reliability, the time interval Δt for calculating the acceleration data of the bicycle must not be too small.

Here, we used two calculation methods, $\Delta t = 0.2$ and $\Delta t = 0.5$. Even with the $\Delta t = 0.2$, the sample size for each individual bicycle's specific situation is not too large (usually 10–30 samples).

Regression analysis without a large number of samples will cause the problem of regression instability, that is, the existence of some data in the regression data has a decisive influence on the results of the regression. In this case, the parameter results obtained by regression calculation are also unstable, and the reliability of the results must be compromised, which cannot be used as the data basis for behavior analysis.

Therefore, we adopt the second method to identify and analyze the behavior parameters of the cyclist.

When using the second analysis method, due to the influence of individual differences, the correlation coefficient R of the model parameter regression results is obviously not as ideal as the individual bicycle situation. The regression correlation coefficient R is between 0.7 and 0.8 and R^2 is about 0.6.

It can be seen from the analysis results of the abovementioned single situation that this is due to the significant differences in behavior between different riders, for example, different motivations lead to different driving forces, which is different from the psychology of avoiding conflicting vehicles. If we classify different conflicting vehicle types and cyclists by gender, the regression effect will be better.

For specific parameter identification results, see the next section.

8.5.3 Parameter identification and analysis results

8.5.3.1 Riding behavior parameter identification results

The specific parameter identification results of bicycle operational level model are shown in Table 8.3 on the next page.

8.5.3.2 Riding behavior parameter analysis

In the operational-level behavior model of a bicycle crossing an un-signalized intersection, the parameter γ_0 represents the motivation of the cyclist and the degree of time urgency—the larger γ_0, the stronger the motivation of the cyclist and the greater the time pressure (Table 8.3).

The social field parameter R^2 shows the intensity of the impact of the stationary conflict vehicle on cyclist's social field. The larger R^2 is, the greater the field strength of the object to the cyclist's social field; correspondingly, at the same distance, the influence tension of the social field is greater.

In all four cases, male cyclist's social field parameters R^2 are smaller than their female counterparts, and in the tactical-level path planning behavior model, male

Table 8.3 Parameter regression results of bicycle operational level model

Conflict situations	Gender	Correlation coefficient R_2	Sample size	Parameter	Value	Standard deviation	*t*-Statistics	*P*-Value	95% Lower limit	95% Upper limit
Bicycle–motor car	–	0.543	1,420	γ_0	0.532	0.112	4.763	0.000	0.311	0.753
				$M \cdot R^2C$	5.228	0.974	5.365	0.000	3.301	7.154
	Male cyclist	0.650	1,132	γ_0	0.671	0.085	1.962	0.054	0.433	0.898
				$M \cdot R^2C$	5.186	0.690	7.520	0.000	3.811	6.560
	Female cyclist	0.505	268	γ_0	0.461	0.120	4.371	0.000	0.022	0.899
				$M \cdot R^2C$	5.821	2.143	2.716	0.008	1.541	10.102
Bicycle–Bicycle	–	0.764	1,270	γ_0	0.658	0.117	5.637	0.000	0.427	0.889
				$M \cdot R^2B$	0.904	0.150	6.031	0.000	0.608	1.201
	Male cyclist	0.749	936	γ_0	0.645	0.218	2.961	0.005	0.205	1.085
				$M \cdot R^2*B$	0.853	0.194	4.390	0.000	0.460	1.245
	Female cyclist	0.751	334	γ_0	0.591	0.162	3.651	0.000	0.270	0.913
				$M \cdot R^2B$	1.073	0.265	4.045	0.000	0.546	1.600
Bicycle–pedestrian	–	0.713	834	γ_0	0.651	0.106	8.026	0.000	0.242	1.060
				$M \cdot R^2P$	0.521	0.197	2.646	0.010	0.129	0.912
	Male cyclist	0.754	708	γ_0	0.806	0.172	4.860	0.000	0.061	1.550
				$M \cdot R^2P*$	0.432	0.269	1.704	0.115	−0.108	0.972
	Female cyclist	0.765	126	γ_0	0.562	0.153	6.167	0.000	0.041	1.083
				$M \cdot R^2P*$	0.613	0.355	1.728	0.096	−0.116	1.342
Bicycle bypass the obstacle	–	0.633	752	γ_0	0.824	0.082	10.011	0.000	0.662	0.987
				$M\ R^2O$	0.913	0.200	4.568	0.000	0.518	1.307
	Male cyclist	0.637	632	γ_0	0.942	0.118	7.984	0.000	0.708	1.177
				$M\ R\ ^2O$	0.669	0.255	2.622	0.010	0.163	1.175
	Female	0.640	120	γ_0	0.640	0.101	6.344	0.000	0.437	0.843

* Indicates that the confidence of this parameter item is 90%, and the confidence of the remaining parameter items is 95%.

cyclists' safety factors are also smaller than their female counterparts (the $c_S/\sum_k c_k$ ratios of male and female young cyclists are 52% and 63%, respectively, see Table 8.2).

Therefore, we can consider that there is good consistency in the tactical and operational levels of behavior models of the NCB behavior theory. Traffic psychologists have also pointed out that in people's traffic behavior, there is a link between complex behavior and simple automatic behaviors [5].

In the model, the parameter M_C, M_P, M_B and M_O are set to be similar to the mass coefficients of motorcars, pedestrians, bicycles and obstacles in physical dynamics, respectively. Without loss of generality, let $M_B = 1$. For all cyclists, the value of M_C/M_B is approximately 5.5.

If divided by gender, the M_C/M_B ratios for male and female cyclists are 6 and 4.8, respectively.

The value of this parameter indicates that the influence range or intensity of a motorcar on the cyclist's behavior is about 5–6 times that of a bicycle.

For all cyclists, the value of M_P/M_B is approximately 0.6, and the M_P/M_B ratios of male and female cyclists are 0.5 and 0.57, respectively.

For all cyclists, the value of M_O/M_B is approximately equal to 1, and the difference of this parameter between male and female cyclists is significant. The values M_O/M_B for male and female cyclists are 0.8 and 1.4, respectively (see Table 8.4).

From the estimated value of the type parameter, it can be considered that when a bicycle crossing an un-signalized intersection, if the conflicting of speed is not considered, the impact of a conflicting motor vehicle on the cyclist's riding behavior is equivalent to that of five to six conflicting bicycles; the impact of a conflicting pedestrian is equivalent to the impact of half a conflicting bicycle; a stationary obstacle is equivalent to the impact of a conflicting bicycle.

Further investigation revealed that in addition to obstacles, the type parameter value of the conflicting vehicle is very close to the proportion of the vehicle's static occupied area (Table 8.5).

This verifies the intuitive thinking of people's daily behavior, that is, regardless of the speed factor, the impact of a moving object on an individual is directly related to the size of the object and has a proportional relationship.

Therefore, we can exclude the impact of different conflicting vehicle types M_C, M_B, M_P, M_O and the moving speed, by taking $M_C = 6$, $M_B = 1$, $M_P = 0.5$, $M_O = 1$. Table 8.6 lists the values of parameters γ_0, R^2 and R of the bicycle operational-level behavior model in different conflict situations after excluding the influence of conflicting vehicle type and speed.

Table 8.4 Conflict vehicle and obstacle type parameter estimates

Parameter	M_C	M_B	M_P	M_O
Young cyclist	5.783	1	0.576	1.010
Young male cyclist	6.080	1	0.506	0.784
Young female cyclist	4.872	1	0.571	1.363
Parameter estimate	5.5	1.0	0.5	1.0

Table 8.5 The relative proportion of land occupied by various types of vehicles and the calibrated type parameter M_j

Item	Motor vehicles	Bicycles	Pedestrians	Obstacles
Length (m)	3.5	2	0.8	2
Width (m)	2	0.6	0.8	1.5
Occupied area (m^2)	7	1.2	0.64	3
The ratio of occupied area to bicycle	5.8	1	0.53	2.5
Type parameter M_j	5.78	1	0.58	1.01

Table 8.6 Parameters γ_0, R^2 and R of the bicycle operational-level behavior model in different conflict situations

	Gender	γ_0	R^2	R
B–C conflict $M_C = 6$	–	0.532	0.951	0.98
	M	0.671	0.943	0.97
	F	0.461	1.058	1.03
B–B conflict $M_B = 6$	–	0.658	0.904	0.95
	M	0.645	0.853	0.92
	F	0.591	1.073	1.04
B–P conflict $M_P = 6$	–	1.651	1.042	1.02
	M	1.806	0.864	0.93
	F	1.562	1.226	1.11
Bicycle bypass the obstacle $M_O = 1$	–	0.824	0.913	0.96
	M	0.942	0.669	0.82
	F	0.64	1.463	1.21

As can be seen from Table 8.6, the values of the social field model parameter R in the four conflict avoidance situations are quite stable, about 1.

In different conflict situations, the R parameter values for male cyclists are between 0.8 and 1, while the values for female cyclists are between 1 and 1.2.

This is a good confirmation of our assumption about the field strength parameter R when modeling a social field. That is, there might be a relatively stable field strength coefficient R for the influence of an object on the social field strength of the cyclist. The expression of this coefficient is the radius of the field strength of the psychological influence of a stationary object, and the magnitude of the field strength of a conflicting object is also related to its speed, moving direction and size of the influencing object.

In addition, the model parameters γ_0 and R seem to have an opposite relationship.

That is, in the same type of conflict avoidance situation, a higher goal-driven parameter γ_0 is usually accompanied by a lower social field strength coefficient R (Table 8.6).

It is also found that the goal-driven parameter γ_0 of female cyclists is usually smaller than that of male cyclists, and the social field strength coefficient R is relatively high.

This "opposite" relationship of the parameters γ_0 and R can be explained from the perspective of traffic psychology and behavior.

When a cyclist is driven more by the destination in the planned path due to time pressure, his/her riding behavior will be more "aggressive." At this time, the conflicting vehicle also has a smaller psychological impact on him/her.

Usually in the "aggressive" situation, the driver (cyclist) tends to subjectively think that his reaction is faster than usual, that is, the reaction time Δt is shorter than usual estimates [5]. Therefore, according to this model, the influence radius a' of the social field is also reduced (see the formula of the social field influence field strength model in Chapter 7).

In the process of model parameter identification, we assume the reaction time of ALL the cyclists as 0.5 s. Therefore, this may cause the recognition result of R to be different. On the other hand, female cyclists usually ride less aggressively than their male counterparts, and their subjective estimated response time is also larger, which also explains why female cyclists have a smaller goal-driven parameter γ_0 and a larger social field strength coefficient R.

In summary, after discussing the influence on the vehicle type, size, speed and direction of the conflicting vehicles, as well as the individual characteristics of the cyclists such as gender, goal-driven and aggressiveness, we can conclude that the social field strength parameter R of the NCB theoretical model is a fairly stable social field behavior parameter.

The social field strength parameter R can be regarded as a quantitative expression of the influence of a static physical "point" on the cyclist's psychology or behavior.

This stable parameter seems to provide a theoretical basis for the unified modeling of the effects of conflicting objects of different types, sizes, movement directions and speeds on the cycling behavior of various types of cyclists (such as gender, travel purpose, age and reaction time).

8.6 Summary

In this chapter, we use interviews, questionnaire surveys and video data to analyze the cyclists' crossing behavior at the un-signalized intersections and determine the model parameters.

Data analysis shows that both the SP and RP data have good reliability.

Then, we determined and discussed the parameters of the tactical-level model and the operational-level model of the bicycle crossing behavior model at the un-signalized intersection.

We also analyzed the psychological, behavioral and riding characteristics of bicycles crossing un-signalized intersections.

Appendix A: Questionnaire group analysis statistics

A.1 Comparison of investigator sample groups

Table A.1 Comparisons of different investigator sample groups

Item	Groups	Investigator 1 (%)	Investigator 2 (%)	Investigator 3 (%)	Investigator 4 (%)	Investigator 5 (%)
Gender	Female	34.74	46.02	40.33	38.42	42.13
	Male	65.26	53.98	59.67	61.58	57.87
Age	<18 years	9.12	2.84	0.55	9.60	12.69
	18–35 years	63.16	66.48	67.40	47.46	52.79
	35–55 years	23.51	21.59	27.07	29.94	25.89
	>55 years	4.21	9.09	4.97	12.99	8.63
Trip purpose	Commute	51.93	60.23	52.49	40.68	50.76
	Leisure and visiting friends	17.19	12.50	10.50	8.47	12.18
	Shopping	15.44	13.64	20.44	15.82	8.63
	Other	15.44	13.64	16.57	35.03	28.43
Commuting traffic mode	Bicycles	47.37	29.55	38.67	55.37	55.84
	Bus	22.81	62.50	29.83	24.29	31.47
	Private car	8.42	3.98	2.76	6.21	2.54
	Metro	15.09	0.57	25.41	3.39	1.52
	Other	1.40	1.14	0.55	9.60	0.00
MOST important consideration	Time	15.09	12.50	9.94	14.69	4.57
	Distance	4.21	1.14	0.55	2.26	1.52
	Safety	76.14	85.23	89.50	78.53	93.40
	Comfort	4.56	1.14	0.00	4.52	0.51
LEAST important consideration	Time	22.81	26.14	42.54	25.99	34.01
	Distance	47.37	27.84	5.52	31.64	24.37
	Safety	0.70	1.70	0.55	1.13	1.02
	Comfort	27.37	44.32	51.38	41.24	31.98

A.2 Group analysis of the questionnaire influencing factors

In order to further analyze the impacts of the characteristics of cyclists, such as gender, age, trip purpose and the usual commuter travel mode, on their path planning behavior at the un-signalized intersections, we analyzed the samples in groups.

In the following, we analyze the ranking of the MOST important factor and the LEAST important factor of path planning of the sample by gender, age group, trip purpose and commuter traffic mode.

A.2.1 Group analysis by gender

Table A.2 Analysis of influencing factors of the questionnaire by gender

Gender group—female, sample size 404			
MOST important consideration		**LEAST important consideration**	
Time	9.90%	Time	33.91%
Distance	2.97%	Distance	29.21%
Safety	85.64%	Safety	0.50%
Comfort	1.49%	Comfort	35.40%
		Other	0.99%
Age		**Commute traffic mode**	
<18	6.44%	Bicycle	35.40%
18–35	56.68%	Bus	42.08%
35–55	28.96%	Private car	5.69%
>55	7.67%	Metro	8.42%
>70	0.25%	Taxi	5.94%
		Other	2.48%
Gender group—male, sample size 612			
MOST important consideration		**LEAST important consideration**	
Time	12.75%	Time	26.80%
Distance	1.63%	Distance	29.41%
Safety	82.68%	Safety	1.31%
Comfort	2.94%	Comfort	39.54%
Other	0%	Other	2.94%
Age		**Commute traffic mode**	
<18	7.84%	Bicycle	52.61%
18–35	61.44%	Bus	26.80%
35–55	23.20%	Private car	4.74%
>55	6.37%	Metro	10.62%
>70	1.14%	Taxi	2.94%
		Other	2.29%

A.2.2 Group analysis by ages

Table A.3 Analysis of influencing factors of the questionnaire by age

Age group—<18 years, sample size 74			
MOST important consideration		**LEAST important consideration**	
Time	20.55%	Time	19.18%
Distance	1.37%	Distance	39.73%
Safety	78.08%	Safety	1.37%
Comfort	1.37%	Comfort	41.10%
Trip purpose		**Commute traffic mode**	
Work	9.46%	Bicycle	60.87%
School	79.73%	Bus	31.13%
Leisure and visiting friends	1.35%	Private car	2.67%
Shopping	4.05%	Metro	2.67%
Other	5.41%	Taxi	2.67%
		Other	0%
Age group—18–35 years, sample size 606			
MOST important consideration		**LEAST important consideration**	
Time	13.86%	Time	23.76%
Distance	2.81%	Distance	32.51%
Safety	80.20%	Safety	1.32%
Comfort	3.14%	Comfort	40.92%
		Other	1.49%
Trip purpose		**Commute traffic mode**	
Work	46.20%	Bicycle	43.40%
School	8.75%	Bus	38.45%
Leisure and visiting friends	14.03%	Private car	3.96%
Shopping	13.04%	Metro	10.89%
Other	17.99%	Taxi	2.48%
		Other	0.83%

Age group—35–55 years, sample size 258			
MOST important consideration		**LEAST important consideration**	
Time	7.75%	Time	37.60%
Distance	1.16%	Distance	22.09%
Safety	90.70%	Safety	0.39%
Comfort	0.39%	Comfort	35.27%
		Other	4.65%
Trip purpose		**Commute traffic mode**	
Work	43.41%	Bicycle	48.45%
School	2.33%	Bus	23.26%
Leisure and visiting friends	11.63%	Private car	8.91%
Shopping	15.89%	Metro	9.69%
Other	26.74%	Taxi	6.98%
		Other	2.71%

Age group—>55 years, sample size 78			
MOST important consideration		**LEAST important consideration**	
Time	3.85%	Time	62.82%
Distance	0.00%	Distance	16.67%
Safety	92.31%	Safety	0.00%
Comfort	3.85%	Comfort	19.23%
		Others	1.28%
Trip purpose		**Commute traffic mode**	
Work	6.41%	Bicycle	42.31%
School	0.00%	Bus	23.08%
Leisure and visiting friends	16.67%	Private car	3.85%
Shopping	33.33%	Metro	7.69%
Other	43.59%	Taxi	8.97%
		Other	14.10%

A.2.3 Group analysis by trip purpose

Table A.4 Analysis of influencing factors of questionnaire group by trip purpose

Trip purpose—commute, sample size 521			
MOST important consideration		**LEAST important consideration**	
Time	14.01%	Time	23.99%
Distance	2.50%	Distance	32.63%
Safety	81.19%	Safety	0.77%
Comfort	2.30%	Comfort	40.12%
		Others	2.50%
Commute traffic mode			
Bicycle	49.90%		
Bus	31.67%		
Private car	4.03%		
Metro	11.90%		
Taxi	1.92%		
Other	0.58%		
Trip purpose—non-commute, sample size 495			
MOST important consideration		**LEAST important consideration**	
Time	9.29%	Time	35.96%
Distance	1.82%	Distance	25.45%
Safety	86.46%	Safety	1.21%
Comfort	2.42%	Comfort	35.56%
		Others	1.82%
Commute traffic mode			
Bicycle	41.41%		
Bus	34.14%		
Private car	6.26%		
Metro	7.47%		
Taxi	6.46%		
Other	4.24%		

A.2.4 Group analysis by commute traffic mode

Table A.5 Analysis of influencing factors of questionnaire group by commuter traffic modes

Commute traffic mode—bicycle, sample size 465			
MOST important consideration		**LEAST important consideration**	
Time	11.40%	Time	29.03%
Distance	2.37%	Distance	29.89%
Safety	84.52%	Safety	1.08%
Comfort	1.72%	Comfort	36.56%
Others	0.00%	Others	3.44%
Trip purpose			
Work	44.52%		
School	11.40%		
Leisure and visiting friends	10.97%		
Shopping	11.40%		
Other	21.72%		
Commute traffic mode—bicycle, sample size 334			
MOST important consideration		**LEAST important consideration**	
Time	12.87%	Time	28.74%
Distance	2.10%	Distance	29.04%
Safety	82.04%	Safety	1.20%
Comfort	2.99%	Comfort	39.82%
		Other	1.2%
Trip purpose			
Work	33.83%		
School	15.57%		
Leisure and visiting friends	23.17%		
Shopping	17.66%		
Other	9.77%		
Commute traffic mode—others, sample size 217			
MOST important consideration		**LEAST important consideration**	
Time	10.14%	Time	32.26%
Distance	1.84%	Distance	28.57%
Safety	85.25%	Safety	0.46%
Comfort	2.76%	Comfort	37.79%
		Time	0.92%
Trip purpose			
Work	37.79%		
School	6.45%		
Leisure and visiting friends	15.67%		
Shopping	17.51%		
Other	22.58%		

References

[1] Y. Huang and T. Li. Introduction to Behavioral Science. Guangzhou, South China University of Technology Press, 2000.
[2] H. Bai. Video-Based Traffic Flow Parameter Measurement in Intelligent Transportation. PhD Dissertation, Beijing, Institute of Automation, Chinese Academy of Sciences, 2006.
[3] J. Xu and X. Chen. Chapter 10 Pedestrian and Bicycle Traffic. In: Traffic Engineering Manual, Beijing, China Communications Press, 1997.
[4] J. Hu. Economic Analysis of Transport Choice Behavior. PhD Dissertation, Beijing, School of Economics and Management, Beijing Jiaotong University, 2004.
[5] R. Näätänen and H. Summala. Road-User Behaviour and Traffic Accidents. Amsterdam, Publication of North-Holland Publishing Company, 1976.
[6] Federal Highway Administration (FHWA), US Department of Transportation (2000). Highway Capacity Manual. Washington, DC.
[7] D. Taylor, Analysis of Traffic Signal Clearance Interval Requirements for Bicycle-Automobile Mixed Traffic. In Transportation Research Record 1405, TRB, National Research Council, Washington, DC, 1993, pp. 13–20.

Chapter 9
Confirmation of validity of bicycle microscopic behavior model

Whether a bicycle microscopic behavior model at the mixed traffic flow un-signalized intersections can be truly applied in mixed traffic flow simulation models, traffic safety, traffic management and other fields, reflecting its practical value, depends on the validity and reliability of the model.

Especially when this model is used as the core behavior model of a microscopic computer simulation platform for mixed traffic flow.

With the continuous development of simulation technology, traffic flow simulation is more and more widely used in various fields of traffic engineering.

In order to draw the correct conclusion when applying the simulation model for evaluation, we must ensure that the simulation model can accurately reflect the actual traffic system.

Therefore, confirming the validity of the behavior model is very important and necessary work in behavior modeling and simulation research.

In this chapter, we first introduce the commonly used methods to confirm the validity of the model, and then use the actual field data to analyze the validity of the microscopic cyclist behavior model at the un-signalized intersections built in the previous chapter, and finally draw conclusions.

9.1 Model validation method

9.1.1 Basic concepts

The concept of simulation model validation first appeared in the 1960s. With the widespread application of computer simulation technology in various disciplines and engineering fields, the effectiveness of simulation models has drawn more and more attention.

The model validity has gradually become an important part of simulation research.

In 1967, Fishman and Kiviat of the American Rand Corporation clearly pointed out that the effectiveness of simulation models can be divided into two parts—**model validation** and **model verification** [1]. This view is widely adopted by the international simulation community.

Model validation has a specific meaning, which is to evaluate the reliability or usability of the model by comparing the consistency between the model outputs and the actual system outputs under the same input conditions and operating environment.

Model verification is to determine whether the computer implementation of the model is correct.

Model validation includes **theoretical model validation**, **data validation** and **operational validation**.

In this book, we only confirm the validity of the data and the validity of the theoretical model. The validity of the sample data has been confirmed in Chapter 8. In this chapter, we only confirm the validity of the theoretical model.

The theoretical model validation is the process of confirming the theoretical basis and assumptions used in the theoretical model and the rationality of the description of the problem entity by the theoretical model. Model validity is a measure of model availability and is related to experimental conditions, application purposes, accuracy requirements and user requirements.

The validity of the model can be expressed by the probability that the user accepts the model as valid—*Pa* (probability of accepting model as valid).

Model accuracy is a measure of the accuracy of the model results. The final method for judging whether the model results are accurate is to "test by practice," that is, to evaluate the accuracy of the model by comparing the deviation of the model from the operating results of the actual system.

9.1.2 Common methods for model validation

There are a variety of theoretical model validation methods. Commonly used model validation methods can be divided into two categories: the *subjective validation methods* and the *statistical test methods*.

The statistical test method mainly relies on various mathematical methods for quantitative analysis and comparison. The subjective method is sometimes aided by mathematical methods, sometimes relies entirely on qualitative analysis.

The representative methods are listed as follows:

- **The subjective validation evaluation:** The model is evaluated by experts familiar with the actual system whether the model and its output are reasonable.
- **Theoretical comparison method:** The model results are compared with the theoretical calculation results to judge the **validation** of the model.
- **The manual calculation method:** All input values and internal variables of the model adopt certain specified values to see if the model results are the same as those calculated manually.
- **The model comparison method:** The results of the model are compared with those models that are generally considered to be validated, and the effectiveness of the model is evaluated on the basis of its bias from these validated models.
- **The historical data method:** This method uses a part of the historical data to build the model, and another part of the data to verify the model's correctness.

- **The parameter validity test:** By changing the model's internal parameter values and input values, observe the effect on the model output results and judge whether this influence relationship is consistent with the real system.
- **The tracking test method:** Track the behavior of different types of entities in the model to determine if the model is logically correct.
- **The sensitivity analysis method:** This method uses sensitivity analysis techniques to determine the input/output relationship of the model and test its rationality.
- **The multistage validation method:** The method includes three stages: first, the model hypothesis is proposed; then the validity of the hypothesis is tested; finally, the correctness of the model input/output relationship is tested.

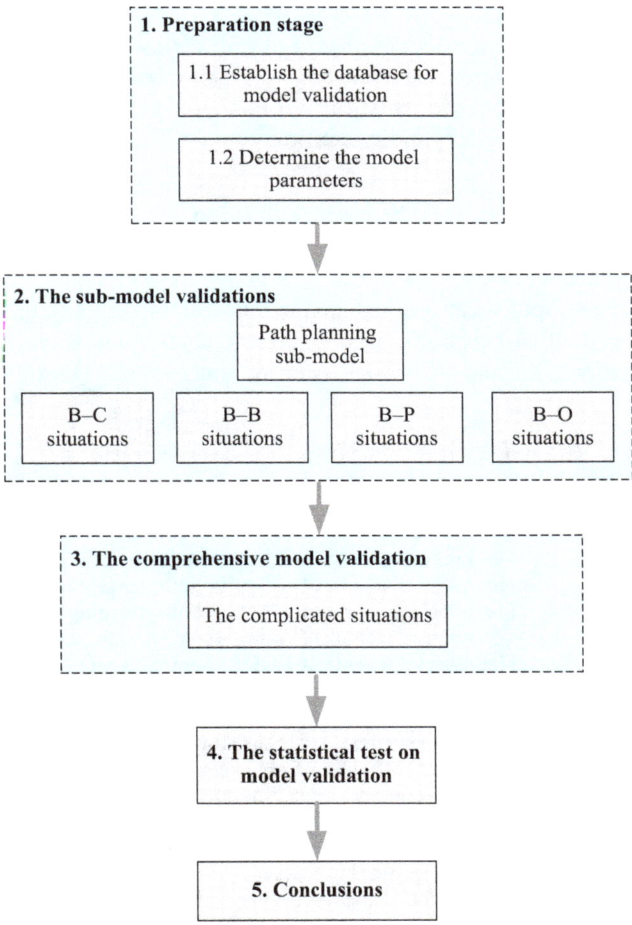

Figure 9.1 Basic process of model validity confirmation

- **The sub-model validity analysis method:** The model is decomposed into several sub-models according to certain principles, and the validity of the overall model is obtained by confirming the sub-models.
- **The prediction validity method:** Compare the model's predicted values with the actual system output to see if they are the same.
- **The statistical test method:** Use mathematical statistics methods to test and evaluate the validity of the model.
- **The curve method:** The model is verified by comparing the agreement between the model result curve and the actual system result curve.
- **The animation method:** Animate the output of the model and judge the model's correctness based on the intuition.

9.1.3 Validation of bicycle micro-behavior model

Our methods for confirming the validity of the bicycles' microscopic behavior model at un-signalized intersections in mixed traffic flows mainly adopt the previous methods: the historical data method, the sub-model validity analysis, the prediction validity method, the statistical test method and the curve method.

The basic flow of model validity confirmation is shown in Figure 9.1.

1. Preparation stage
   (i) Establish the database for model validation

   The database includes a bicycle's behavior data of simple and complicated cases.

   The **simple** situations include the bike–motor vehicle conflict situation (B–C conflict situation), bicycle–bicycle conflict situation (B–B conflict situation), bicycle–pedestrian conflict situation (B–P conflict situation)

Table 9.1 NBC model parameter values

Parameters	Values	Notes
c_L	1	The relative weight of riding distance in path planning
c_{safe} (Male)	3	The relative weight of safety in path planning for male cyclists
c_{safe} (Female)	4.5	The relative weight of safety in path planning for female cyclists
c_V	0.8	The relative weight of riding speed in path planning
c_{LOS}	0.7	The relative weight of LOS in path planning
M_C	5.5	The type coefficient of conflicting motor cars
M_B	1	The type coefficient of conflicting bicycle
M_P	0.5	The type coefficient of conflicting pedestrians
M_O	1	The type coefficient of obstacles
Δt (Male)	0.8	The subjective reaction time of male cyclists
Δt (Female)	1.2	The subjective reaction time of female cyclists
γ_0	1	The goal driving coefficient
2φ	120°	The perspective of cyclists
c	0.3	The weight of the conflicting vehicle outside the cyclist's perspective

and bike–obstacle conflict situation (B–O conflict situation). Each simple situation selects about ten sample cases that are not used for model parameter identification.

The **complicated** situation, that is, when crossing a non-signal intersection, a bicycle not only encounters a single conflicting vehicle (such as a motor vehicle, bicycle and pedestrian) but also encounters multiple conflicting vehicles, such as a bicycle traveling in a direction, and a lateral conflict direction motor vehicles or situations of conflicting motor vehicle traffic, that is, gap acceptance behavior. Each complicated situation selects about five sample cases that are not used for model parameter identification.

(ii) Determine the model parameters

The parameters of the model are determined in Chapter 8. Table 9.1 shows the parameter values in the microscopic behavior model of bicycles at mixed traffic flow un-signalized intersections.

The model inputs include basic information of the un-signalized intersection Ω; dynamic and static information of other relevant conflicting vehicles V_j at the intersection; information of the subject bicycle i, such as the initial position $\vec{x}_0$, the time arriving at the intersection t_0, the destination D_f and the cyclist's basic information such as gender.

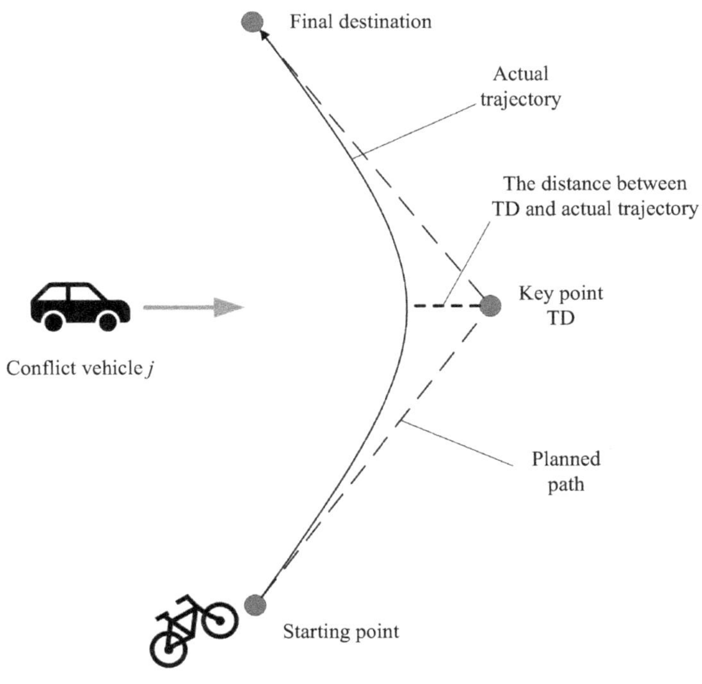

Figure 9.2 TDs in path planning sub-model verification

Table 9.2 Summary of model validation results (PP sub-model)

Case ID	The position of the inflection point of the actual trajectory			TD positions calculated by PP sub-model			Model results-actual data				Note
	Arriving T (s)	X (m)	Y (m)	Expected T_1 (s)	X (m)	Y (m)	ΔT	ΔX	ΔY	ΔL	
BC1	98	−7.39	29.37	100	−7	29	0.4	0.39	−0.37	0.53	
BC3	140	−8.76	25.76	140	−9	26	0.0	0.26	0.24	0.35	F
BC4	90	−15.38	25.06	90	−15	25	0.0	0.38	−0.06	0.38	
BC8	47	−14.07	20.00	50	−14	20	0.6	0.06	−0.50	0.50	
BC9	141	−10.61	23.71	140	−11	23	−0.2	0.11	−0.71	0.71	
BC10	97	−7.08	26.52	100	−7	26	0.6	0.08	−0.52	0.53	F
COM2	111	−20.76	23.45	110	−21	24	−0.2	0.26	0.05	0.27	
COM3	121	−16.65	17.56	120	−17	18	−0.2	0.15	−0.06	0.17	F
COM5	115	−15.70	22.57	110	−16	23	−0.4	0.20	0.43	0.47	
COM8	112	−11.06	23.89	110	−12	24	−1.0	−0.45	0.11	0.46	F
COM9	95	−5.83	25.17	100	−6	24	1.0	−0.17	−1.17	1.18	F
Absolute maximum error							1.00	0.45	1.17	1.18	
Mean error							0.05	0.12	−0.23	0.51	

F stands for the female cyclist situation.

The outputs of the model are the microscopic dynamic behaviors of the bicycle at the un-signalized intersection Ω, including acceleration $\vec{a}$, speed $\vec{v}$ and trajectory $\vec{x}$.

2. The sub-model validations

   We use the statistical analysis and curve comparison methods to validate the cyclist path planning (PP) sub-model and the riding sub-models on four simple conflict situations (B–C, B–B, B–P and B–O situations).

   Simply put, we compare the calculation results of the sub-models with field data and compare the results with actual bicycle dynamic data.

   If the validity of the sub-models does not meet the requirements, we should return to adjusting the model parameter values or the sub-model structure.
3. The comprehensive model validation

   After confirming the validation of the sub-models and the simple case in step 2, we use statistical analysis and curve comparison methods to validate the model comprehensively.

   To put it simply, it is to calculate the comprehensive model in the complex situation of bicycle crossing the un-signalized intersections and compare and analyze the model results with the actual bicycle dynamic data.

   The model validation in steps 2 and 3 focuses on the analysis of the individual bicycle behaviors.
4. The statistical test on model validation

   In this stage, we use mathematical statistics methods to test and evaluate the overall behavior model.

Table 9.3 The model validation results of simple situations (B–C conflict situations)

Case ID	**Mean acceleration error** $(\vec{a}_{\text{mod}} - \vec{a}_{em})$ **(m/s/s)**		**Mean speed error** $(\vec{v}_{\text{mod}} - \vec{v}_{em})$ **(m/s)**		**Mean position error** $(\vec{x}_{\text{mod}} - \vec{x}_{em})$ **(m)**		**Sample size**	**Note**
	X	*Y*	*X*	*Y*	*X*	*Y*		
BC1	−0.294	−0.536	0.269	−0.209	0.243	0.428	22	
BC2	−0.342	−0.454	0.221	−0.174	0.236	0.404	23	
BC3	−0.025	−0.373	0.359	−0.148	0.322	0.321	24	F
BC4	0.231	0.045	−0.039	0.052	−0.369	0.070	16	
BC5	−0.035	−0.656	0.347	−0.390	0.319	0.322	25	F
BC6	0.167	−0.781	0.088	−0.059	−0.362	0.341	27	
BC7	−0.413	−0.360	0.162	−0.143	0.234	0.372	25	
BC8	−0.163	−0.118	0.109	0.082	−0.062	0.167	35	
BC9	−0.467	−0.316	0.036	−0.183	0.355	0.137	17	
BC10	−0.264	0.081	−0.056	0.024	0.066	0.017	31	F
Absolute maximum error	0.467	0.781	0.359	0.390	0.369	0.428		
Mean error	−0.160	−0.347	0.150	−0.115	0.098	0.258		

For example, we will analyze the statistical value of the model results, such as the mean speed, whether are statistically consistent with the actual data.

5. Conclusions

Finally, we conclude the validity of the model by combining the previous validation analysis results.

9.2 Model validity analysis

9.2.1 The sub-model validations

9.2.1.1 Path planning sub-model

Here, a total of 11 examples of simple and complex situations of B–C conflicts are selected for microscopic cyclist's PP sub-model validation. When verifying the *r*

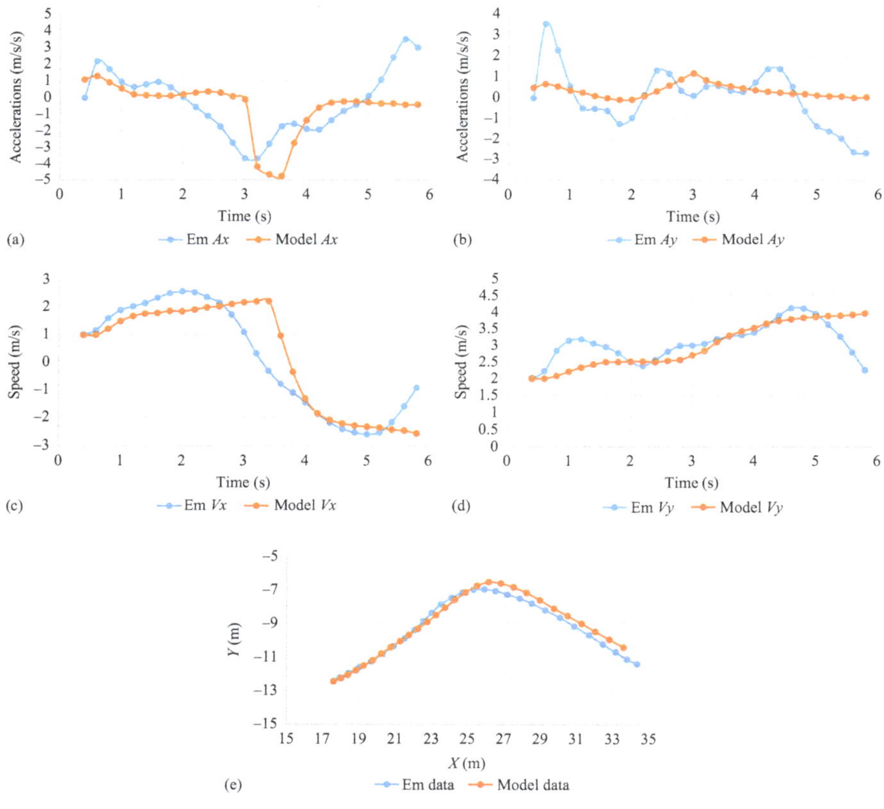

Figure 9.3 Comparisons of model outputs and the empirical data (Case ID: BC10): (a) the accelerations in the X direction; (b) the accelerations in the Y direction; (c) the speeds in the X direction; (d) the speeds in the Y direction; (e) the motion trajectories

PP model, the space and time of the intersection are discretized first, and the minimum space h=0.5 m is adopted.

Because the conflict situation of the verification sample is relatively simple, and there is only one conflicting motor vehicle, we consider that there is only one TD in the cyclist's planned path crossing the un-signalized intersection, which is located near the possible conflict area (see Figure 9.2).

In addition, for simplicity, we do not consider the cyclist's dynamic PP behaviors when performing the PP sub-model verification. That is, the cyclist determines the route when entering the intersection, and then he/she does not change it anymore. The general situation of model calculation and verification is listed in Table 9.2.

It can be seen from the data in Table 9.2 that the PP sub-model data fit well with the actual bicycle motion trajectory data. The maximum error (or distance) between the TD position and the inflection point of the actual trajectory line is 1.2 m, and the average is 0.5 m.

In general, if the error threshold is set to 0.5 m, 3 TDs in the 11 cases and the actual trajectory inflection point are slightly different (they are BC1, BC9 and COM9, respectively), and the PP sub-model accuracy is 73%. If we set the error threshold to 1 m, there is only one TD in 11 cases and the actual trajectory inflection point is slightly different (COM9), and the model accuracy rate can reach 91%.

However, considering that the output of the PP sub-model is the intermediate result of the entire bicycle crossing behavior model at un-signalized intersection, from the perspective of the model validation, the final result of the model, that is, the comparison of the bicycle's motion trajectory and actual data, is the most effective way to validate the model issue concerned. Therefore, the larger purpose

Table 9.4 Model validation results of simple situations (B–B conflict situations)

Case ID	**Mean acceleration error** $(\vec{a}_{\text{mod}}-\vec{a}_{em})$ **(m/s/s)**		**Mean speed error** $(\vec{v}_{\text{mod}}-\vec{v}_{em})$ **(m/s)**		**Mean position error** $(\vec{x}_{\text{mod}}-\vec{x}_{em})$ **(m)**		**Sample size**	**Note**
	X	*Y*	*X*	*Y*	*X*	*Y*		
BB11	0.487	0.405	0.027	0.012	−0.125	0.051	22	
BB11	−0.312	−0.398	−0.095	−0.075	−0.119	0.032	23	
BB29	0.298	0.359	0.058	0.194	0.053	−0.077	25	
BB29	−0.264	0.149	−0.253	0.214	−0.098	0.347	25	F
BB44	0.183	0.009	0.022	−0.026	−0.090	0.017	37	
BB44	0.197	0.256	−0.023	−0.050	0.028	−0.098	38	
BB45	0.288	0.552	−0.014	−0.044	0.010	−0.133	35	F
BB45	0.184	0.018	0.023	−0.016	−0.091	0.011	30	
BB46	−0.097	0.166	0.065	−0.018	−0.064	−0.005	33	
BB46	0.105	0.092	−0.022	−0.076	0.049	−0.069	30	
Absolute maximum error	0.487	0.552	0.253	0.214	0.125	0.347		
Mean error	0.107	0.161	−0.021	0.012	−0.045	0.008		

of the verification of the PP sub-model is to verify the logical correctness and reliability of the NCB hierarchical behavior theory model.

From this perspective, no matter the accuracy of the PP sub-model is taken as 0.5 or 1 m, it is consistent with the actual situations. Therefore, the verification results of the PP sub-model are satisfactory.

9.2.1.2 B–C conflict situations

In this section, we randomly picked ten simple B–C conflict situations to determine the validation of the NCB bicycle model. The general results are shown in Table 9.3.

As can be seen from the data in the table, the model results fit well with the empirical bicycle dynamic behavior data.

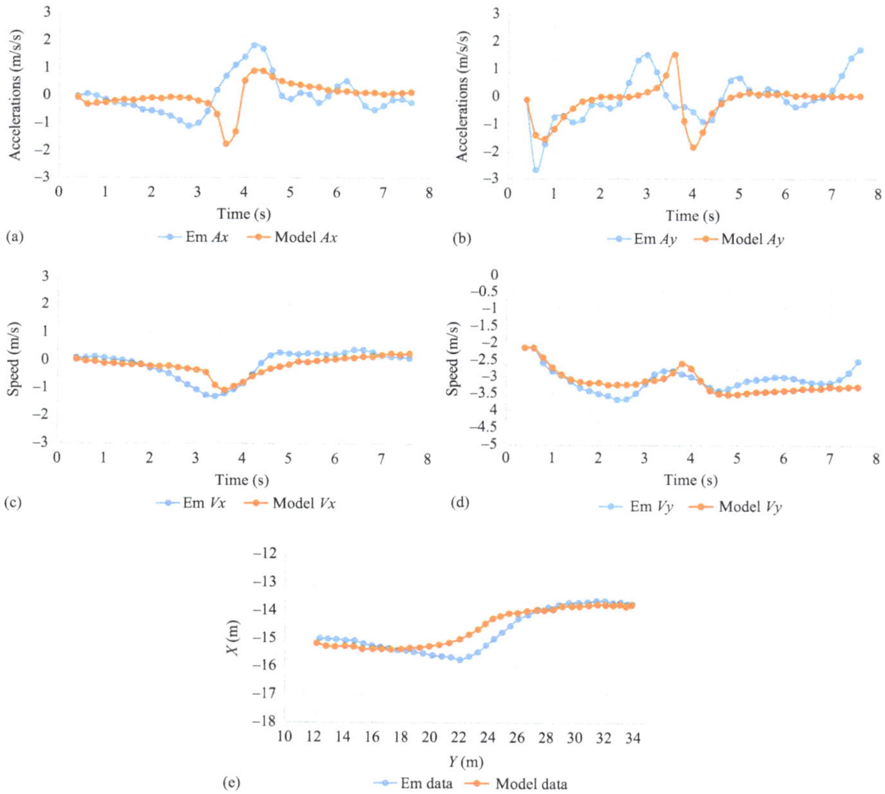

Figure 9.4 Comparisons of model outputs and the empirical data (Case ID: BB44): (a) the accelerations in the X direction; (b) the accelerations in the Y direction; (c) the speeds in the X direction; (d) the speeds in the Y direction; (e) the motion trajectories

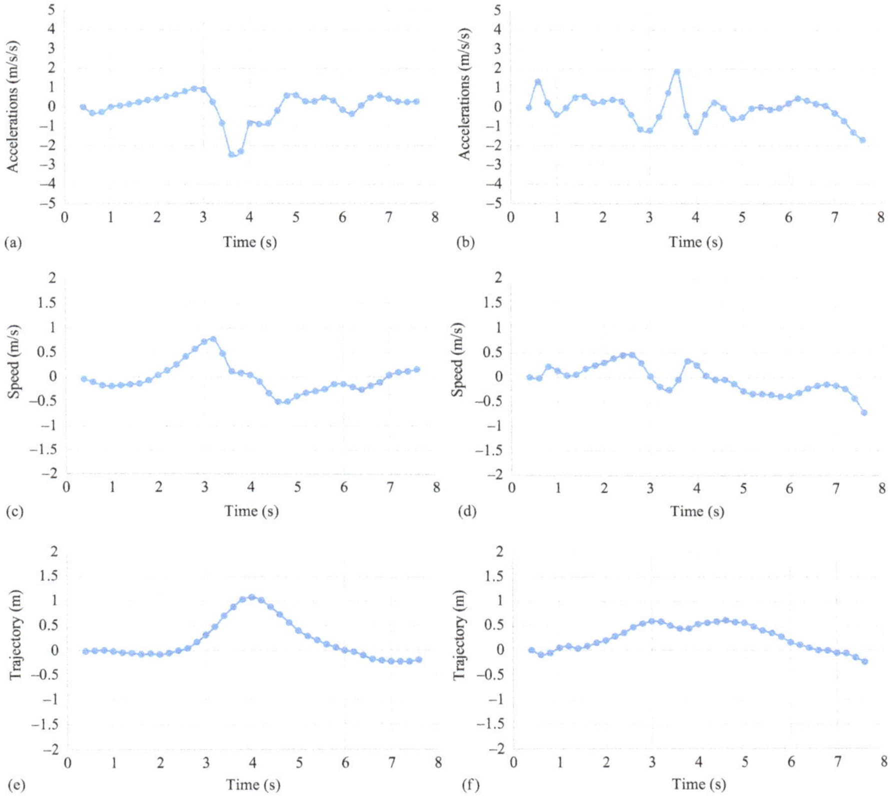

Figure 9.5 The errors between the model outputs and the empirical data (Case ID: BB44): (a) the acceleration errors in the X direction; (b) the acceleration errors in the Y direction; (c) the speed errors in the X direction; (d) the speed errors in the Y direction; (e) the trajectory errors in the X direction; (f) the trajectory errors in the Y direction

The absolute maximum acceleration errors in X and Y directions are 0.46 and 0.78 m/s/s, respectively; the absolute maximum speed errors in X and Y directions are 0.36 and 0.39 m/s, respectively; and the absolute maximum position error in X and Y directions are 0.37 and 0.43 m, respectively.

We use the curve method to compare the bicycle model with the actual acceleration, speed and position data every 0.2 s, and have a more intuitive understanding (see Figure 9.3).

9.2.1.3 B–B conflict situations

In this section, we randomly picked ten simple B–B conflict situations to determine the validation of the NCB bicycle model. The general results are shown in

Table 9.4. As can be seen from the data in the table, the model results fit well with the empirical bicycle dynamic behavior data.

The absolute maximum acceleration errors in X and Y directions are 0.49 and 0.55 m/s/s, respectively; the absolute maximum speed errors in X and Y directions are 0.25 and 0.21 m/s, respectively; and the absolute maximum position error in X and Y directions are 0.13 and 0.35 m, respectively.

We use the curve method to compare the bicycle model with the actual acceleration, speed and position data every 0.2 s, and have a more intuitive understanding (see Figures 9.4 and 9.5).

9.2.1.4 B–P conflict situations

In this section, we randomly picked ten simple B–P conflict situations to determine the validation of the NCB bicycle model. The general results are shown in Table 9.5. As can be seen from the data in the table, the model results fit well with the empirical bicycle dynamic behavior data.

The absolute maximum acceleration errors in X and Y directions are 0.27 and 0.50 m/s/s, respectively; the absolute maximum speed errors in X and Y directions are 0.09 and 0.39 m/s, respectively, and the absolute maximum position error in X and Y directions are 0.09 and 0.32 m, respectively.

We use the curve method to compare the bicycle model with the actual acceleration, speed and position data every 0.2 s and have a more intuitive understanding (see Figures 9.6 and 9.7).

Table 9.5 Model validation results of simple situations (B–P conflict situations)

Case ID	Mean acceleration error $(\vec{a}_{\text{mod}} - \vec{a}_{em})$ (m/s/s)		Mean speed error $(\vec{v}_{\text{mod}} - \vec{v}_{em})$ (m/s)		Mean position error $(\vec{x}_{\text{mod}} - \vec{x}_{em})$ (m)		Sample size	Note
	X	Y	X	Y	X	Y		
BP13	0.038	0.305	0.034	0.048	−0.002	0.039	31	
BP14	0.158	0.201	−0.013	0.386	0.040	0.195	26	
BP15	0.157	−0.256	−0.021	0.371	0.046	0.324	27	
BP20	−0.083	0.500	−0.016	0.058	0.060	0.084	41	
BP24	−0.273	−0.238	−0.086	0.065	0.094	−0.039	40	
BP25	0.110	−0.336	−0.052	0.067	0.072	−0.037	39	
BP26	0.225	0.326	0.040	0.264	−0.008	0.074	31	F
BP26	0.098	0.327	0.025	0.295	0.058	0.041	33	
BP35	0.010	−0.007	0.012	0.036	0.003	0.075	25	
BP36	0.038	0.305	0.034	0.048	−0.002	0.039	30	F
Absolute maximum error	0.273	0.500	0.086	0.386	0.094	0.324		
Mean error	0.048	0.113	−0.004	0.164	0.036	0.080		

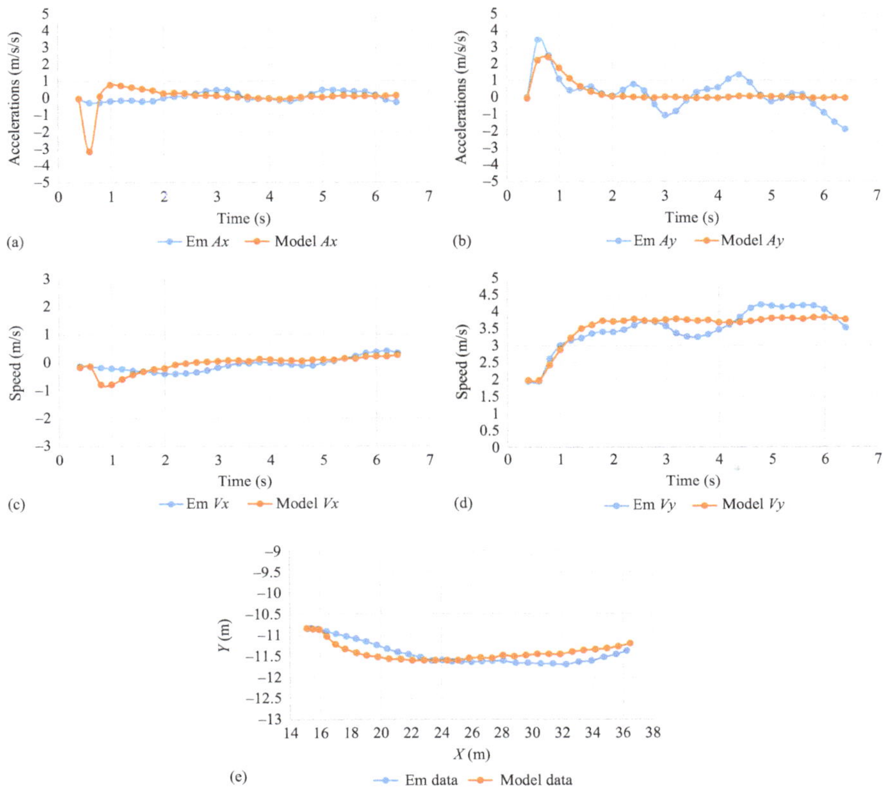

Figure 9.6 Comparisons of model outputs and the empirical data (Case ID: BP24): (a) the accelerations in the X direction; (b) the accelerations in the Y direction; (c) the speeds in the X direction; (d) the speeds in the Y direction; (e) the motion trajectories

9.2.1.5 B–O situations

In this section, we randomly picked ten simple B–O bypass situations to determine the validation of the NCB bicycle model. The general results are shown in Table 9.6. As can be seen from the data in the table, the model results fit well with the empirical bicycle dynamic behavior data.

The absolute maximum acceleration errors in *X* and *Y* directions are 0.89 and 1.04 m/s/s, respectively; the absolute maximum speed errors in *X* and *Y* directions are 0.19 and 0.33 m/s, respectively; and the absolute maximum position error in *X* and *Y* directions are 0.22 and 0.42 m, respectively.

We use the curve method to compare the bicycle model with the actual acceleration, speed and position data every 0.2 s and have a more intuitive understanding (see Figures 9.8 and 9.9).

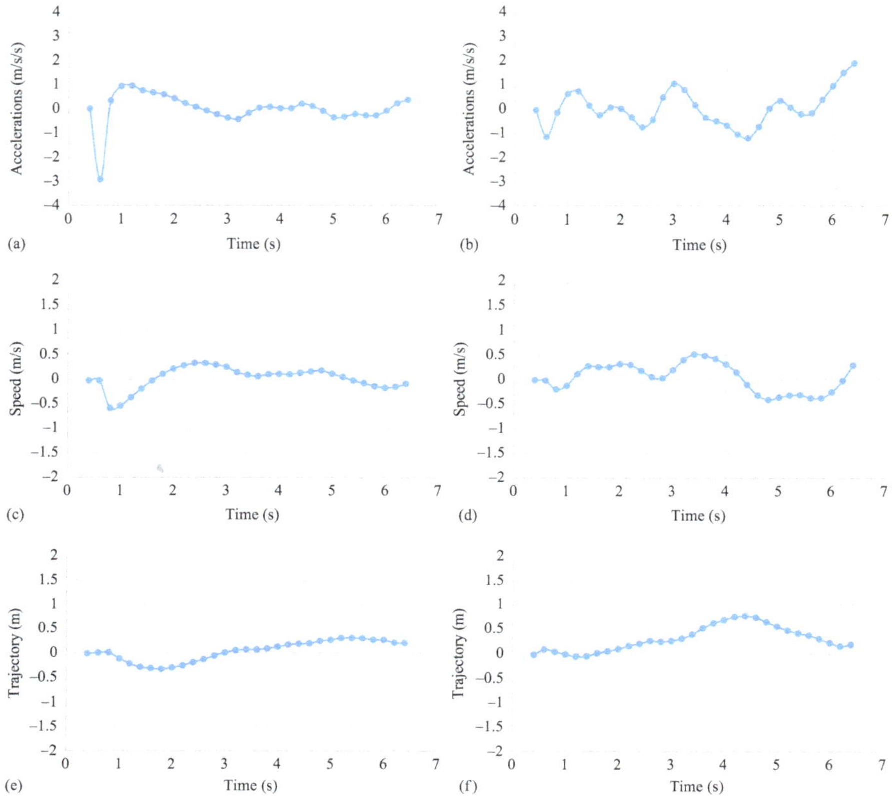

Figure 9.7 The errors between the model outputs and the empirical data (Case ID: BP24): (a) the acceleration errors in the X direction; (b) the acceleration errors in the Y direction; (c) the speed errors in the X direction; (d) the speed errors in the Y direction; (e) the trajectory errors in the X direction; (f) the trajectory errors in the Y direction

9.2.2 Comprehensive model validation

In this section, we randomly picked eight complicated samples of bicycles crossing un-signalized intersections to determine the model validation.

Complex situations include the gap acceptance of bicycles between two consecutive conflicting vehicles when crossing the un-signalized intersection; including conflicting both motorcars and conflicting bicycles (or conflicting pedestrians, see Figure 9.10); including conflicting motorcars and fellow bicycles, etc.

The general situation is shown in Table 9.7. It can be seen from the data in the table that even under complicated and changeable conditions, the model results and

Table 9.6 Model validation results of simple situations (B–O situations)

Case ID	**Mean acceleration error** $(\vec{a}_{\text{mod}}-\vec{a}_{em})$ **(m/s/s)**		**Mean speed error** $(\vec{v}_{\text{mod}}-\vec{v}_{em})$ **(m/s)**		**Mean position error** $(\vec{x}_{\text{mod}}-\vec{x}_{em})$ **(m)**		**Sample size**	**Note**
	X	*Y*	*X*	*Y*	*X*	*Y*		
BO9	−0.131	−0.031	0.130	−0.100	0.034	−0.163	27	
BO10	−0.172	0.584	0.006	0.030	−0.048	−0.087	40	
BO11	0.050	−0.868	−0.022	0.129	0.000	0.278	35	
BO12	−0.886	0.703	0.088	−0.301	0.101	−0.087	32	
BO12	−0.872	0.879	0.102	−0.167	0.125	−0.418	28	
BO12	−0.371	0.398	0.144	−0.331	0.010	−0.109	30	F
BO13	−0.682	0.583	0.091	−0.253	0.152	−0.135	33	
BO15	−0.657	0.643	0.075	−0.122	0.144	−0.258	35	
BO15	−0.558	0.806	0.187	−0.061	0.219	−0.406	21	
BO15	−0.538	1.042	0.035	0.109	0.131	−0.177	29	
Absolute maximum error	0.886	1.042	0.187	0.331	0.219	0.418		
Mean error	−0.482	0.474	0.084	−0.107	0.087	−0.156		

the actual dynamic behavior data of the bicycle are satisfactory, though there are not as good as the simple situations of B–C, B–B, B–P and B–O.

The absolute maximum acceleration errors in *X* and *Y* directions are 1.09 and 0.96 m/s/s, respectively; the absolute maximum speed errors in *X* and *Y* directions are 1.27 and 0.48 m/s, respectively; and the absolute maximum position error in *X* and *Y* directions are 0.12 and 0.66 m, respectively.

We use the curve method to compare the bicycle model with the actual acceleration, speed and position data every 0.2 s and have a more intuitive understanding (see Figures 9.11 and 9.12).

9.2.3 Macroscopic model validity analysis

From the above verification results of the bicycle behavior model, we can see that the verification results of both the sub-model and the comprehensive model are quite satisfactory. The following is a macro verification of the bicycle behavior model. Table 9.8 summarizes the calculation results of all sub-models and comprehensive verification. The error maximums and averages verified by each sub-model and comprehensive example are listed.

Figures 9.13 and 9.14 are the corresponding figures.

9.2.4 Conclusions of model validity analysis

By comparing the model results with actual data, we can conclude that the bicycle model verification results are as follows.

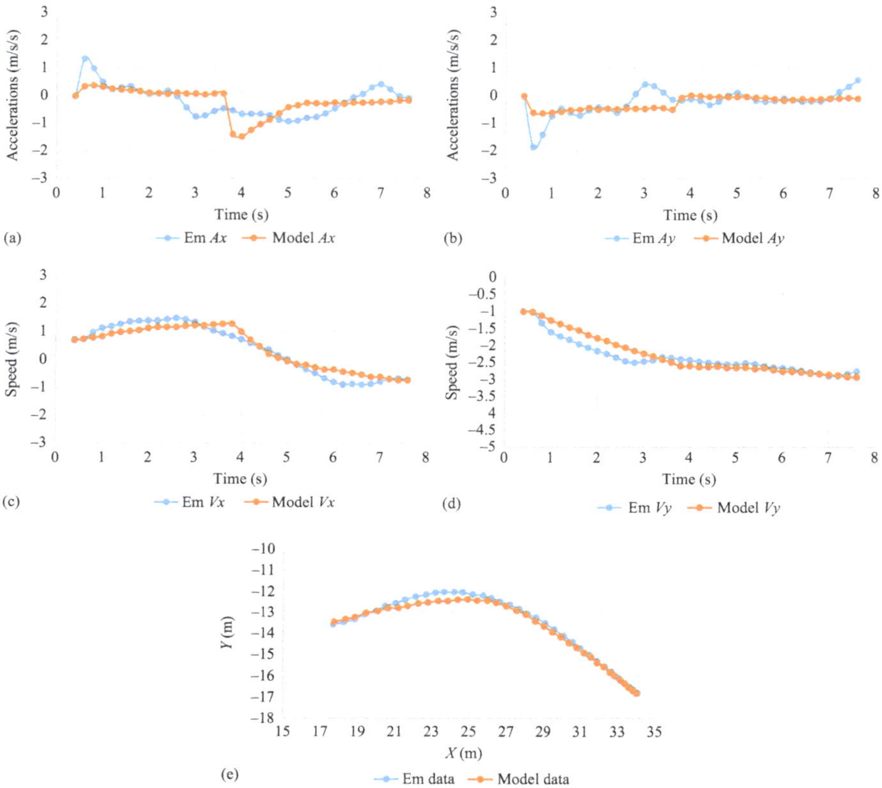

Figure 9.8 Comparisons of model outputs and the empirical data (Case ID: BO9): (a) the accelerations in the X direction; (b) the accelerations in the Y direction; (c) the speeds in the X direction; (d) the speeds in the Y direction; (e) the motion trajectories

9.2.4.1 Sub-models for each simple collision avoidance situation

The verification results of each sub-model are satisfactory. For each simple collision avoidance situation, including B–C, B–B, B–P and B–O, we found that the maximum value of the motion trajectory error is basically within 0.5 m (sampling at 0.2 s), and it is generally around 0.2 m.

9.2.4.2 Path planning model

Compared with other sub-models, the output result of the PP model is not so close to the actual value. In the model validation, the final result of the model (i.e. the trajectory of the bicycle) is the most concerned issue. The PP sub-model output is the intermediate result of the model, and the main purpose of its verification is to verify the logical correctness and reliability of the two-layer behavior model. Therefore, from this perspective, the verification results of the PP sub-model are also satisfactory.

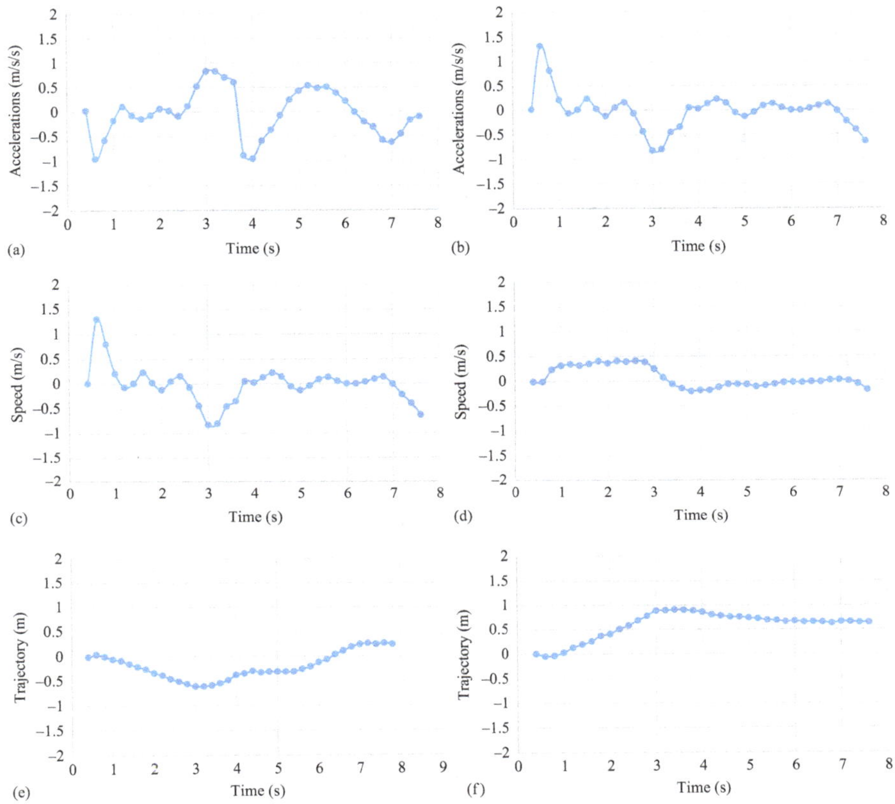

Figure 9.9 *The errors between the model outputs and the empirical data (Case ID: BO9): (a) the acceleration errors in the X direction; (b) the acceleration errors in the Y direction; (c) the speed errors in the X direction; (d) the speed errors in the Y direction; (e) the trajectory errors in the X direction; (f) the trajectory errors in the Y direction*

We believe that the larger purpose of this sub-model verification is to verify the logical correctness and reliability of the two-level behavioral theoretical model. From this perspective, the verification results of the PF model are also satisfactory.

9.2.4.3 Complex situations

Though the comprehensive verification results of complex situations are not as good as those of simple conflict situations, the verification results of this model are quite satisfactory, considering the complexity of people's behavior.

Figure 9.10 An example of a typical complex situation (the subject bicycle is inside the circle)

Table 9.7 The model validation results of complicated situations

Case ID	Mean acceleration error $(\vec{a}_{\text{mod}} - \vec{a}_{em})$ (m/s/s)		Mean speed error $(\vec{v}_{\text{mod}} - \vec{v}_{em})$ (m/s)		Mean position error $(\vec{x}_{\text{mod}} - \vec{x}_{em})$ (m)		Sample size	Note
	X	*Y*	*X*	*Y*	*X*	*Y*		
COM2	0.113	−0.752	0.709	0.313	−0.041	0.045	37	
COM3	0.221	−0.550	−0.037	−0.014	0.010	0.225	37	F
COM4	−0.274	−0.857	−0.080	0.014	0.090	0.299	36	F
COM5	0.123	0.070	0.061	−0.016	−0.062	0.011	41	
COM6	−0.171	0.227	1.270	0.475	−0.072	−0.658	40	
COM7	−0.402	−0.960	−0.155	−0.327	−0.072	−0.099	40	
COM8	1.087	0.390	0.473	0.459	−0.105	0.025	35	
COM 9	−0.320	−0.950	0.121	0.139	0.116	0.303	35	F
Absolute maximum error	1.087	0.960	1.270	0.475	0.116	0.658		
Mean error	0.047	−0.423	0.295	0.130	−0.017	0.019		

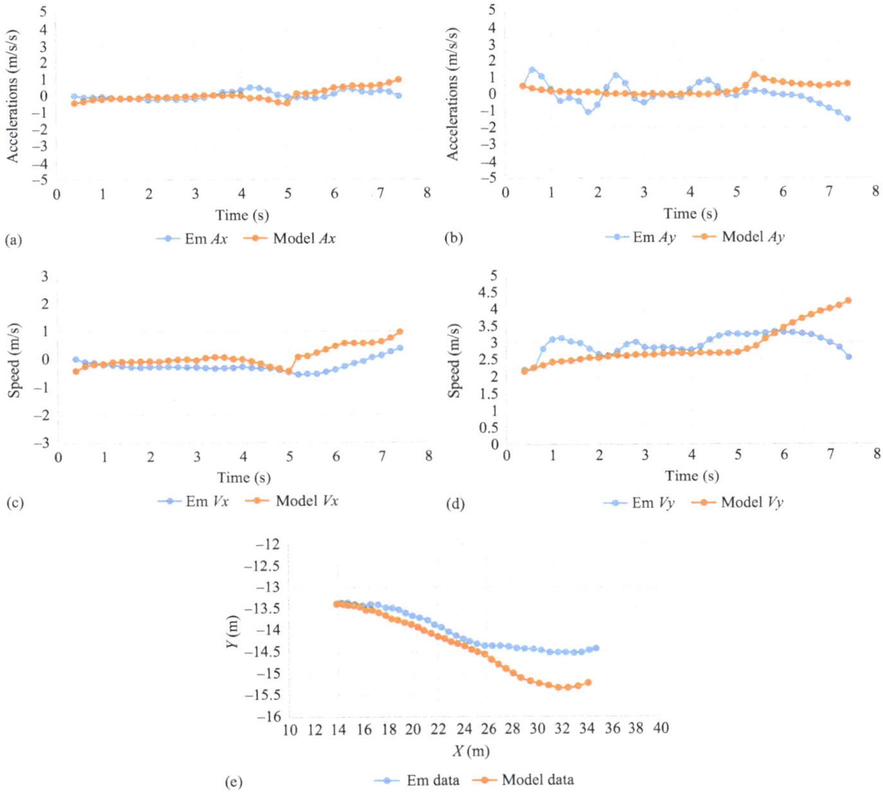

Figure 9.11 Comparisons of model outputs and the empirical data (Case ID: COM2): (a) the accelerations in the X direction; (b) the accelerations in the Y direction; (c) the speeds in the X direction; (d) the speeds in the Y direction; (e) the movement trajectories

In microscopic verification, the maximum average trajectory error in complex cases is 1.1 m, and the average error is within 0.5 m (especially the average error in the X direction is 0.05 m). The average error in speed is 0.3 m/s; the maximum error in acceleration is 0.7 m/s^2, while the average value is 0.02 m/s^2. This is a generally accepted error range in microscopic traffic flow simulation models.

In summary, it can be considered that the NCB behavior model at unsignalized intersections proposed in the book is a relatively reliable, accurate and flexible behavior model.

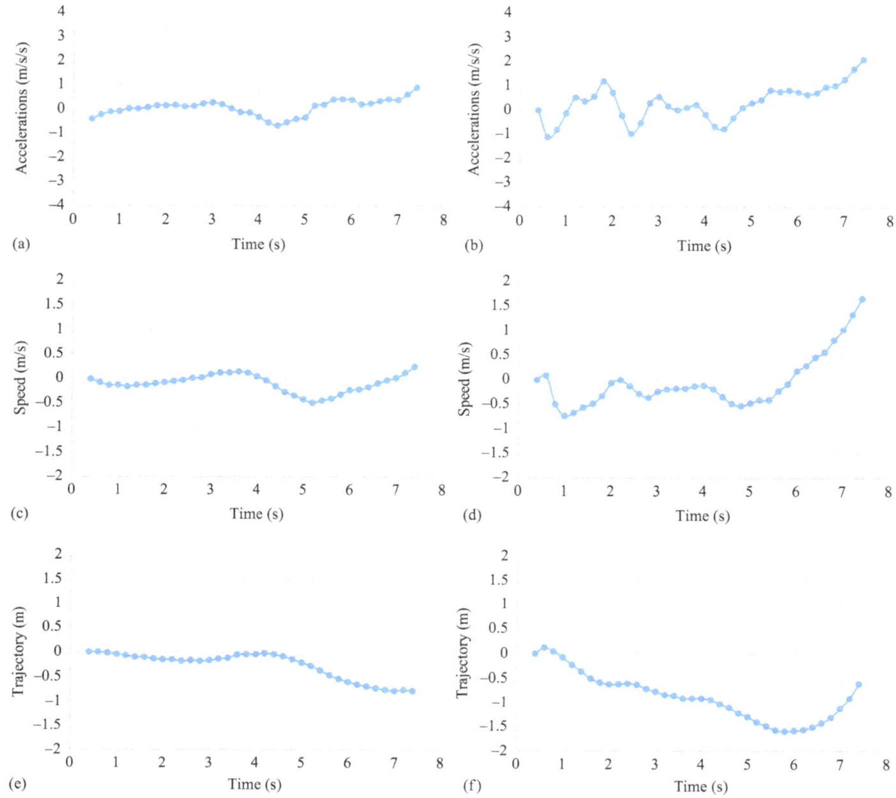

Figure 9.12 The errors between the model outputs and the empirical data (Case ID: COM2): (a) the acceleration errors in the X direction; (b) the acceleration errors in the Y direction; (c) the speed errors in the X direction; (d) the speed errors in the Y direction; (e) the trajectory errors in the X direction; (f) the trajectory errors in the Y direction

9.3 Summary

In this chapter, we use field data to verify the validity of the microscopic behavior model of the bicycle crossing un-signalized intersection established in Chapters 7 and 8. The validity of the model includes data validation (Chapter 6), model logic verification (**PP** sub-model), conflict sub-model verification (four simple conflict situations), comprehensive verification (complex conflict situations in mixed traffic flow), and microscopic and macroscopic verification analysis.

In the end, we determined the validity and reliability of the microscopic NCB model at un-signalized intersection proposed in the book.

Table 9.8 Summary of the bicycle model validation results

Model types	Item	ΔX (m)	ΔY (m)	ΔVx (m/s)	ΔVy (m/s)	ΔAx (m/s²)	ΔAy (m/s²)
The PP sub-model	Absolute maximum error	0.445	1.171	–	–	–	–
	Mean error	0.116	−0.231	–	–	–	–
The B–C conflict situation	Absolute maximum error	0.467	0.781	0.359	0.390	0.369	0.428
	Mean error	−0.160	−0.347	0.150	−0.115	0.098	0.258
The B–B conflict situation	Absolute maximum error	0.487	0.552	0.253	0.214	0.125	0.347
	Mean error	0.107	0.161	−0.021	0.012	−0.045	0.008
The B–P conflict situation	Absolute maximum error	0.273	0.500	0.086	0.386	0.094	0.324
	Mean error	0.048	0.113	−0.004	0.164	0.036	0.080
The B–O situation	Absolute maximum error	0.886	1.042	0.187	0.331	0.219	0.418
	Mean error	−0.482	0.474	0.084	−0.107	0.087	−0.156
The COM conflict situation	Absolute maximum error	1.087	0.960	1.270	0.475	0.116	0.658
	Mean error	0.047	−0.423	0.295	0.130	−0.017	0.019

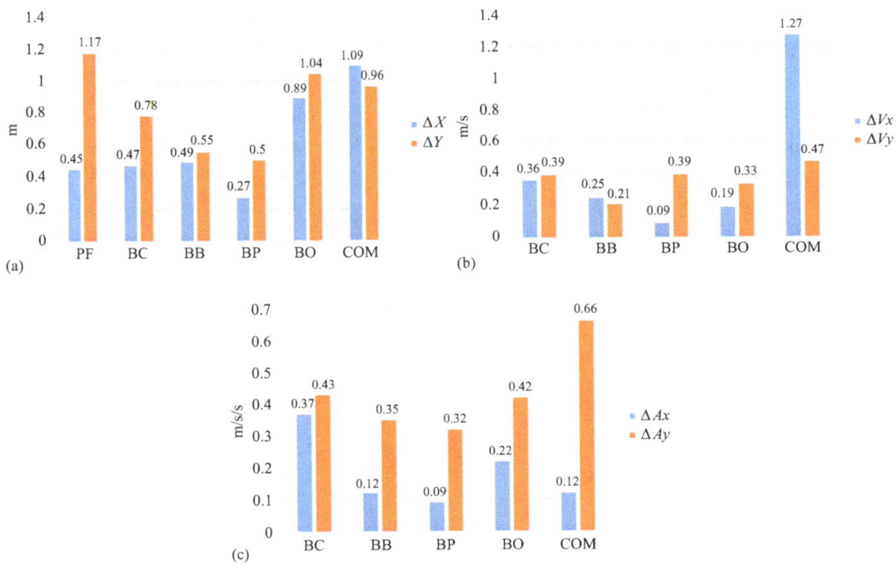

Figure 9.13 The summary of the bicycle model validation results (absolute maximum error): (a) the absolute maximum trajectory errors; (b) the absolute maximum speed errors; (c) the absolute maximum acceleration errors

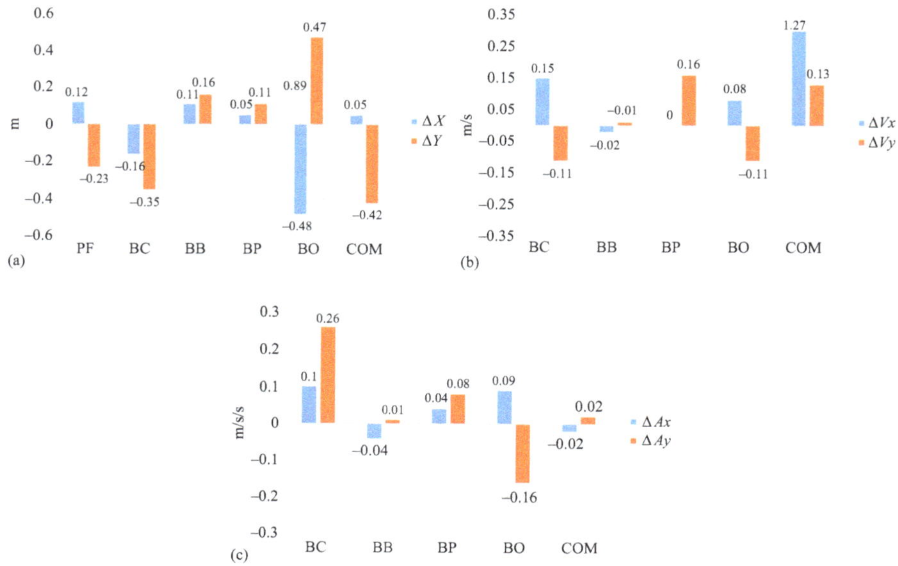

Figure 9.14 *The summary of the bicycle model validation results (mean error): (a) the mean trajectory errors; (b) the mean speed errors; (c) the mean acceleration errors*

Reference

[1] W. Wang (Ed.). Confirmation and Verification of Simulation Model Validity. Changsha, National University of Defense Technology Press, 1998.

Chapter 10

Neural network-based bicycle collision avoidance behavioral model at un-signalized intersections

Most of the existing microscopic behavior simulation models of bicycles are constructed by formulating corresponding behavior rules or by establishing complex behavior models. This results in complex rules of bicycle behavior models, difficult calibration of model parameters and difficulty in generalization and application.

Artificial neural networks (ANNs) have good nonlinear mapping and pattern recognition capabilities and are often used to model the complex behavior of traffic flows.

In this chapter, we use the neural network (NN) to study the bicycle's conflict avoidance behaviors when crossing the un-signalized intersections under mixed traffic flow conditions.

In this chapter, we first introduce the basic concepts and learning rules of NNs in the first section; in Section 10.2, we will introduce the structure and algorithm principle of back propagation (BP) NN; in Section 10.3, we will introduce the modeling ideas and steps of applying NN to construct the bicycle conflict-avoidance (BCA) behavior model. In Section 10.4, we will construct three types of BPNN-based BCA behavior models at un-signalized intersections. Finally, in Section 10.5, we will verify the NN-based bicycle's conflict avoidance behavior model at un-signalized intersections by field data.

10.1 Introduction to artificial neural networks (ANNs)

10.1.1 Artificial neuron model

ANN is a type of information processing system that consists of a large number of simple working artificial neuron connections to simulate the structure and function of the brain. Similar to biological neurons, each artificial processing unit can process multiple external input signals and convert them into output signals according to certain rules.

Because there are multiple connection relationships and signal processing methods between processing units, it is possible to construct ANN with a variety of complex relationships between the inputs and outputs, and obtain good results in associative memory, nonlinear mapping, pattern recognition and pattern recognition.

Figure 10.1 shows how a single artificial neuron works.

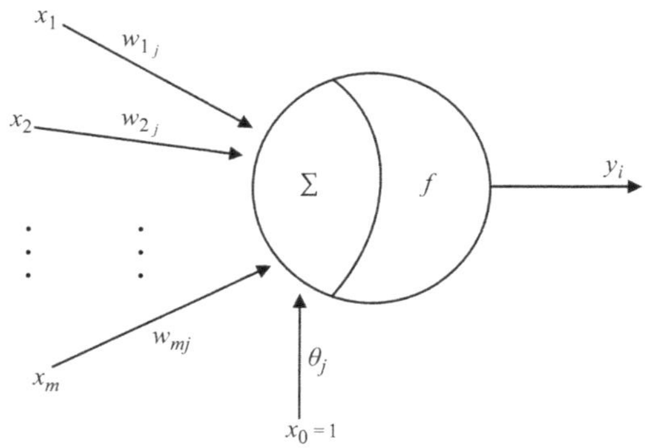

Figure 10.1 A single artificial neuron works

In this figure, x_i ($i = 1, 2, \ldots, m$) represents multiple input signals of neuron j; y_i is the output signal of neuron j; w_{ij} is the connection weight of input neuron node and neuron node j; θ_j is the threshold of neuron j, and $f(\cdot)$ is the transfer function, which expresses the nonlinear relationship between the input and output of the neuron; the relationship between the input and output of the neuron is

$$\begin{cases} s_j = \sum_{i=1}^{m} w_{ij}x_i - \theta_j \\ y_j = f(s_j) \end{cases} \tag{10.1}$$

10.1.2 Transfer function

The mathematical model of artificial neurons includes three functions: weighting, summing and transmitting. Different transfer functions make each processing unit have different information processing characteristics. The transfer function $f(\cdot)$ can transform the possible infinite domain into a specified limited range of output to reflect the relationship between the neuron input signal and its activation state.

There are four types of commonly used transfer functions: linear, threshold, S-type and probability, as shown in Figure 10.2.

Among them, the S-type function is a very important and most commonly used transfer function. Among them, the sigmoid function and the hyperbolic tangent function are common.

The sigmoid function is usually defined as

$$y = f(s) = \frac{1}{1 + e^{-ks}} \tag{10.2}$$

The sigmoid function defines a nonlinear gain for each artificial processing unit. When the input s changes between $(-\infty, 0)$, the gain increases from 0 to the maximum value; then s is $(0, +\infty)$, and the gain recovers from the maximum value to 0. This is the same situation as biological neurons.

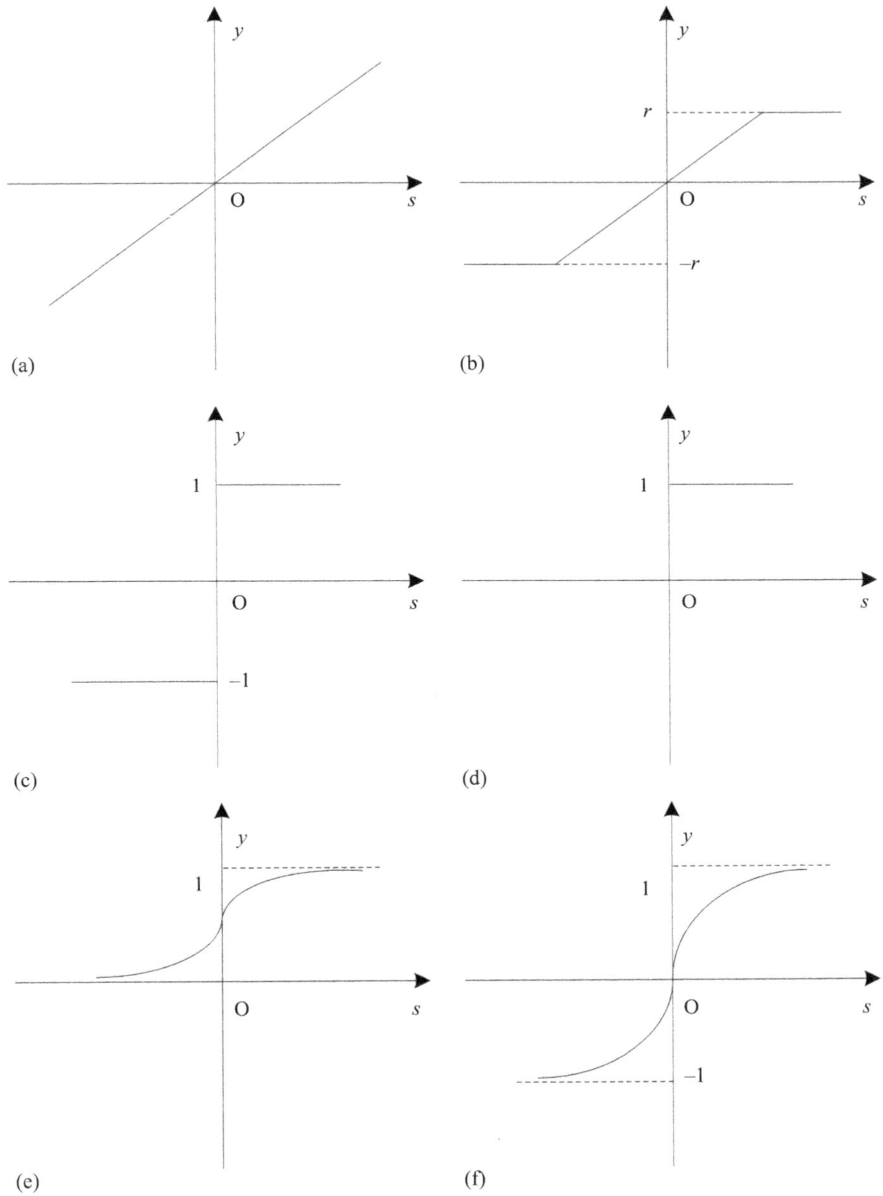

Figure 10.2 Commonly used transfer functions of artificial neurons: (a) linear function; (b) nonlinear piecewise function; (c) Sign function; (d) step function; (e) sigmoid function; (f) hyperbolic tangent function

The hyperbolic tangent function is similar to the sigmoid function, and biologists often use it to build mathematical models of nerve cell excitability. The difference from the sigmoid function is that it is symmetrical with respect to the origin. This function is usually used when the input signal is in the range of ± 1.

10.1.3 Learning method and learning rules

10.1.3.1 Learning methods

Training and learning is an important part of ANNs. In the training and learning process, external information needs to be converted into the input and output of the NN, and the connection weights between the various processing units are continuously adjusted according to some rules until the given input information can obtain a standard output.

According to different network evaluation standards, the learning method of ANNs can be reduced to supervised learning and unsupervised learning.

In supervised learning, the output evaluation standards of the guided learning network are provided to the network from outside, which is equivalent to supervised learning where teachers demonstrate the correct results, while unsupervised learning only provides input data without corresponding output data. According to its unique network structure and learning rules, the network autonomously searches for possible patterns and rules while continuously adjusting network weights.

10.1.3.2 Learning rules

In the NN model, the methods and rules for changing the weight of the network are the core content of network training and learning. The weights determined through learning and training determine the performance of the network.

The learning rules and algorithms are different in different ANN models.

Common learning rules include Hebb learning rules, perceptron learning rule, Widrow–Hoff learning rule, error BP learning rule, winner-take-all learning rule and Grossberg learning rule.

The Hebb learning rule is the earliest proposed ANN learning rule. Its basic principle can be described as when two adjacent neurons are in an excited state at the same time, their connection should be strengthened.

The Widrow–Hoff learning rule can achieve the smallest square difference between the expected output of the neuron and the actual output, also known as the minimum mean square error (MSE).

The winner-take-all learning rule is a competitive learning rule, which is often applied to NN models that use unsupervised learning.

At present, the error BP learning rule is widely used. BP learning rule mainly includes two processes of forward calculation output and error BP. The network calculates from the input layer through the hidden layer to the output layer, in turn, and compares the output. Layer output and expected output calculate the error, and then according to the direction of reducing the error, successively modify each connection weight from the output layer through the hidden layer to the input layer and repeat the previous process until the error accuracy requirements are met.

10.2 How BP neural network works

10.2.1 Structure of BP neural network

The BPNN model is an NN with no feedback and no interconnected multilayer structure within layers based on the BP learning algorithm. BPNN consists of the input, hidden and output layers. Each layer is composed of many simple and parallel processing units. There is no connection between processing units in the same layer. The processing units between layers are fully interconnected. Its structure is shown in Figure 10.3.

In an NN structure, the number of input and output nodes is determined by the problem itself.

The key to the NN model is to determine the number of hidden layers and the number of nodes in each layer.

The theory proves that as long as there are enough hidden layer nodes, an NN containing a hidden layer can approximate any nonlinear function.

The number of hidden nodes in an NN is difficult to determine. Generally, the appropriate number of nodes is selected by experience and trial and error.

10.2.2 BPNN standard learning process

Next we take a three-layer network structure as an example. The mathematical variables and specific learning process in the BPNN standard learning algorithm are as follows.

10.2.2.1 Mathematical variables

Input vector: $X^k = (x_1^k, x_2^k, \ldots, x_n^k)^{\mathrm{T}} (k = 1, 2, \ldots, m)$

Expected output vector corresponding to the input: $Y^k = (y_1^k, y_2^k, \ldots, y_q^k)^{\mathrm{T}}$

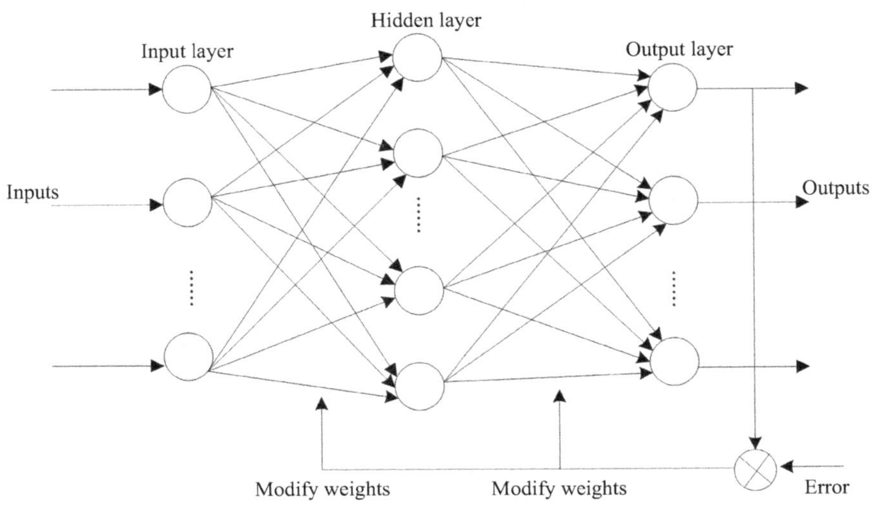

Figure 10.3 Three-layer BP neural network structure

Net input vector for the hidden layer: $S^k = (s_1^k, s_2^k, \ldots, s_p^k)^T$

Hidden layer output vector: $B^k = (b_1^k, b_2^k, \ldots, b_p^k)^T$

Net input vector for the output layer: $L^k = (l_1^k, l_2^k, \ldots, l_q^k)^T$

Output vector of the output layer: $C^k = (c_1^k, c_2^k, \ldots, c_q^k)^T$

Weight vector from the input layer to the hidden layer: $W = \{w_{ij}\}(i = 1, 2, \ldots, n, j = 1, 2, \ldots, p)$

Weight vector from hidden layer to output layer: $W = \{v_{jt}\}(t = 1, 2, \ldots, q, j = 1, 2, \ldots, p)$

The threshold of each processing unit of the hidden layer: $\theta = \{\theta_j\}$ $(j = 1, 2, \ldots, p)$

The threshold of each processing unit in the output layer: $\gamma = \{\gamma_t\}$ $(t = 1, 2, \ldots, q)$

And m is the number of learning mode pairs, n is the number of nodes in the input layer, q is the number of nodes in the output layer and p is the number of nodes in the hidden layer.

10.2.2.2 Learning process

The specific learning process of the BP algorithm is described as follows:

1. Initialization. The initial values of the interval [−1,1] are given to the connection weights W and V and the thresholds θ and γ.
2. Randomly select a learning mode pair (X^k, Y^k) and provide it to the network. No processing is performed on the pair at the input layer, that is, the output vector of the input layer is the same as the input mode vector; then the net inputs and outputs of the processing units of the hidden layer are calculated by the following formula:

$$s_j^k = \sum_{i=1}^{n} w_{ij} x_i^k - \theta_j \tag{10.3}$$

$$b_j^k = f(s_j^k) \quad j = 1, 2, \ldots, p \tag{10.4}$$

3. Calculate the net input and actual output of each processing unit in the output layer:

$$l_t^k = \sum_{i=1}^{p} v_{jt} b_j^k - \gamma_t \tag{10.5}$$

$$c_t^k = f(l_t^k) \tag{10.6}$$

4. Calculate the correction error d_t^k of each processing unit of the output layer according to the following formula:

$$d_t^k = (y_t^k - c_t^k) f'(l_t^k) \quad t = 1, 2, \ldots, q \tag{10.7}$$

5. Calculate the correction error e_t^k of each processing unit of the hidden layer:

$$e_j^k = \left[\sum_{t=1}^{q} v_{jt} d_t^k \right] f'\left(s_j^k\right) \quad j = 1, 2, \ldots, p \tag{10.8}$$

6. Adjust the connection weight V from the hidden layer to the output layer and the threshold γ of the neurons in the output layer, where α is the learning rate (lr) and $0<\alpha<1$:

$$\Delta v_{jt} = \alpha d_t^k b_j^k \quad \Delta \gamma_t = \alpha d_t^k \tag{10.9}$$

7. Adjust the connection weight W from the input layer to the hidden layer and the threshold value θ of the hidden layer neurons, where β is the lr, $0<\beta<1$:

$$\Delta w_{ij} = \beta e_j^k x_i^k \quad \Delta \theta_j = \beta e_j^k \tag{10.10}$$

8. Randomly select the next pattern pair and provide it to the network. After training m learning mode pairs, it is judged whether the global error E meets the accuracy requirement, if it is satisfied, it ends, otherwise the training is continued.

10.2.3 Problems and improvement of BP algorithm

Compared with earlier ANN models, BPNNs are more mature in theory and performance, highlighting their powerful nonlinear mapping capabilities, but there are still problems such as slow convergence and local minimum points. And the choice of lr, the number of hidden layers and the number of nodes still lack theoretical guidance.

Aiming at these problems of BPNN, researchers have done a lot of research on its performance improvement and put forward some improvement schemes. Common examples include cumulative error correction algorithms, BP algorithms with increasing momentum terms, lr adaptive adjustment algorithms, conjugate gradient methods and Gauss–Newton methods.

The cumulative error correction algorithm adjusts the weights and thresholds of the NN by calculating the cumulative error of the m learning mode pairs, so that the errors in the global sense fall in the direction of the gradient. Compared with the standard BP algorithm, the network parameters are corrected, the number of times is reduced, the learning time is shortened, and the learning efficiency of the network is improved.

The BP algorithm with dynamic momentum term—when adjusting the connection weight, the algorithm increases the previously learned adjustment amount (i.e., the dynamic momentum term) by a certain percentage, thereby reducing the oscillations during the training process and speeding up the network's convergence speed. The disadvantage is that it is difficult for us to determine the appropriate dynamic momentum factor. When the dynamic momentum coefficient is too large, this error correction term may not work, but it will prolong the convergence time and even cause oscillation. Generally, the maximum dynamic momentum coefficient does not exceed 0.9.

In BP learning algorithms, the lr is a key factor affecting the efficiency of network learning.

If the lr is too large, the network may not converge; if the lr is too small, the network convergence time may be too long. In practice, it is difficult to determine an optimal lr that is always appropriate.

In the lr adaptive adjustment algorithm, the lr can adaptively adjust the lr based on the trend of the error. When the error keeps decreasing and approaches the target, the lr can be appropriately increased. If the error keeps increasing and exceeds the set value, the lr should be appropriately reduced.

In addition, many scholars have improved the error function, transfer function and initial weight selection of BPNNs and have achieved good results in improving network convergence speed and network performance [1].

10.3 Modeling of bicycle conflict-avoidance behavior based on NN

10.3.1 Determine inputs and outputs

To use the BPNN to simulate the microscopic traffic behavior of bicycles, we first need to determine the inputs and outputs of the NN-based behavior model.

Based on the analysis in the previous chapters, we believe that the dynamic behavior of bicycles crossing un-signalized intersections is essentially the cyclist's process of dynamically identifying the traffic environment at intersections and making judgments and decisions.

For bicycle individuals, in the absence of conflict interference, they will choose to drive along the shortest straight line. In the case of conflict interference, they will choose a reasonable conflict avoidance path to prevent collision according to their spatial position relationship with the conflicting objects and their motion characteristics.

For a bicycle, its own position and speed, the relative positions and relative speeds of potentially conflicting objects around it, and the cyclist's goals may all be factors that influence bicycle behavior. Hence, we can take these factors as input variables to the NN model.

In addition, we assume that if the cyclist's behavior does not change, the default outputs of the bicycle behavior NN model should be zero. From this point of view, it is reasonable to take the change in bicycle motion state, that is, acceleration as the outputs of the NN model.

Therefore, based on the collected field data of the bicycle, we take the bicycle's position as the origin and take the relative distance and relative speed between the conflicting object and the bicycle, and the deviation between the cyclist's expected and the actual speed as the inputs of the NN model. We take the acceleration of the bicycle as an output to reflect changes in the state of the bicycle's motion.

The inputs and outputs of the model are both vectors with magnitude and direction. The specific variables are explained as follows:

b_x, b_y are the coordinates of the bicycle in the *X* and *Y* directions, in m;
c_x, c_y are the coordinates of potential conflicting objects in *X* and *Y* directions, in m;
v_{bx}, v_{by} are the components of the bicycle speed in the *X* and *Y* directions, in m/s;
v_{cx}, v_{cy} are the components of the potential conflicting objects speed in the *X* and *Y* directions, in m/s;
Δw_x is the *X*-direction component of the distance between the potential conflict object and bicycle;
Δw_y is the *Y*-direction component of the distance between the potential conflict object and bicycle;
Δv_x is the *X*-direction component of the relative speed of the potential conflict object to bicycle;
Δv_y is the *Y*-direction component of the relative speed of the potential conflict object to bicycle;
$\Delta v'_x$ is the *X*-direction component of deviation of the cyclist's expected and the actual speeds;
$\Delta v'_y$ is the *Y*-direction component of deviation of the cyclist's expected and the actual speeds;
d_x, d_y are the *X*- and *Y*-direction coordinates of the bicycle's destination across the intersection, respectively;
dt is the cyclist's expected arriving time at the destination;
t is the current time step;
λ is the simulation time step (0.04 s).

The input variables and calculation formulas of the BCA NN model are as follows:

$$\Delta w_x = c_x - b_x \tag{10.11}$$

$$\Delta w_y = c_y - b_y \tag{10.12}$$

$$\Delta v_x = v_{cx} - v_{bx} \tag{10.13}$$

$$\Delta v_y = v_{cy} - v_{by} \tag{10.14}$$

$$\Delta v'_x = \frac{d_x - b_x}{(dt - t) \times \lambda} - v_{bx} \tag{10.15}$$

$$\Delta v'_y = \frac{d_y - b_y}{(dt - t) \times \lambda} - v_{by} \tag{10.16}$$

The outputs of the model are the components of the bicycle's acceleration in the *X* and *Y* directions: ma_x and ma_y in m/s^2.

The specific bicycle behavior NN modeling is as follows.

First, we will obtain the trajectory and motion data of bicycles and surrounding potential conflicting road users (motor vehicles, bicycles, pedestrians) at un-signalized

Table 10.1 The inputs and outputs for the motorcar–bicycle conflict situations

Case ID	Time step	Δw_x	Δw_y	Δv_x	Δv_y	$\Delta v'_x$	$\Delta v'_y$	ma_x	ma_y
BC11	236	−6.4	8.0	0.17	−2.06	−0.14	1.75	0.46	0.29
BC11	241	−6.4	7.5	0.13	−2.73	−0.18	1.11	0.56	0.16
BC11	246	−6.3	6.9	0.05	−3.25	−0.25	0.63	0.66	0.64
BC11	251	−6.3	6.3	−0.07	−3.50	−0.34	0.43	0.78	0.09
BC11	256	−6.4	5.5	−0.18	−3.55	−0.46	0.41	0.62	−0.34
BC11	261	−6.4	4.8	−0.28	−3.59	−0.60	0.34	0.20	0.13
BC11	266	−6.5	4.1	−0.41	−3.68	−0.77	0.26	−0.20	0.74
...	...	...	...	...	...	...	...	...	...
...	...	...	...	...	...	...	...	...	...
BC51	62	−4.7	−5.1	6.39	1.59	0.78	−0.21	0.71	−0.21
BC51	67	−3.5	−4.8	6.14	1.80	0.79	−0.18	1.25	0.27
BC51	72	−2.2	−4.4	6.34	2.39	0.82	−0.09	1.49	0.38
BC51	77	−1.0	−3.9	7.05	2.65	0.81	0.01	1.16	0.16
BC51	82	0.5	−3.4	7.53	2.61	0.70	0.02	0.62	−0.05
BC51	87	2.0	−2.9	6.94	2.52	0.49	−0.06	0.20	−0.15
...	...	...	...	...	...	...	...	...	...

intersections and establish a microscopic behavioral database of car, motorcycle, bicycle and pedestrian conflicts.

The movement data of each traffic individual in the database is extracted at an interval of 0.2 s; then according to the previous calculation formula, the inputs of each sample are calculated using VBA programming and used as inputs for each learning mode pair for the NN.

It is worth noting that because the cyclist needs a certain perceived response time, this is specifically reflected in the fact that the acceleration and driving force of the bicycle has a certain delay, and the perceived response time is generally between 0.4 and 1 s. In each sample, the bicycle's accelerations after 0.5 s are used as the outputs of the corresponding learning mode pair.

Table 10.1 shows some examples of the inputs and outputs for the motorcar–bicycle conflict situations.

10.3.2 Data normalization

In order to reduce the impact of singular samples on network training, we usually need to normalize the training sample data before performing NN training.

Singular samples are sample vectors that are particularly large or small compared to other input samples. The existence of singular sample data will cause the network training time to increase and may cause the network to fail to converge. Therefore, it is best to normalize the training samples in advance.

Because the transfer function of the BPNN generally uses the S-type function, in order to improve the training speed and sensitivity of the network, and effectively avoid the saturation region of the S-type function, the input and output data are generally required to be between [−1,1].

For example, the output data of the common *sigmoid* function and *logsig* function are in the interval [0,1]. The *tansig* function requires the output data to be in the interval [−1,1], so the data need to be normalized.

The commonly used normalization processing methods are as follows:

1. $x'_i = \frac{2(x_i - x_{\min})}{(x_{\max} - x_{\min})} - 1$, and the normalization interval is [−1,1]; the sample restoration process is

$$x_i = 0.5 \times (x'_i + 1)(x_{\max} - x_{\min}) + x_{\min} \tag{10.17}$$

2. $x'_i = \frac{(x_i - x_{\min})}{(x_{\max} - x_{\min})}$, and the normalization interval is [0,1]; the sample restoration process is

$$x_i = x'_i(x_{\max} - x_{\min}) + x_{\min} \tag{10.18}$$

3. $x'_i = (x_i - x_{\text{mean}})/x_{std}$, where x_{mean} and x_{std} are the sample mean and variance of x. That is, the input data are normalized to standard variance and 0 mean; the sample restoration process is

$$x_i = x'_i x_{std} + x_{\text{mean}} \tag{10.19}$$

When constructing a BPNN model, a corresponding normalization method can be selected according to the type of the transfer function. When the transfer function is a linear function, normalization can also be omitted.

10.3.3 Learning sample division

When using BPNN to map complex functional relationships, a large number of samples are required.

Here, the learning samples are divided into the training set, cross-validation (CV) set and test set.

The training set mainly provides learning mode pairs for the NN to build a suitable network; the CV set is to calculate the error of the test sample by using the obtained NN and find the network structure corresponding to the minimum MSE of the CV set; the test set is mainly used to verify the validity of the NN model.

In addition, when selecting training set from the learning samples, we should include some boundary samples as much as possible so that the training set is in the same interval after normalization processing, reducing the impact of data fluctuations on NN performance.

10.3.4 Neural network structure

10.3.4.1 The number of hidden layers

In 1989, Robert–Nielson proved that a BP network with ONE hidden layer can complete any *N*-dimensional to *M*-dimensional mapping [2]. A continuous function of any closed interval can be approximated by a three-layer BP network.

In practical applications, the number of hidden layers of the BPNN is generally selected to be one or two. It has a direct relationship with the complexity of the research problem and the model accuracy requirements. It needs to be comprehensively determined according to the actual situation.

It is generally believed that increasing the number of hidden layers can reduce network errors and improve the accuracy of the NN model. However, the more hidden layers, the more complicated the calculation process, which greatly increases the training time.

At the same time, increasing the number of hidden layers makes it easy for the network to fall into a local minimum during the training process, and it is difficult to adjust the network weight to the global minimum error.

In addition, research has found that for more complex problems, increasing the number of hidden layers can effectively reduce the number of neuron nodes and sometimes can achieve better prediction results.

Hence, when we create an NN model for BCA behavior, we chose two hidden layers after some experiments and comparisons.

10.3.4.2 The number of hidden layer nodes

In the structure design of BPNN, the decision of the number of neuron nodes in the hidden layer is a very important and complex problem, because the number of hidden layers and the number of nodes in the puppet layer determine the structure of the NN, which determines the learning ability and adaptability of the NN model.

If there are too few hidden layer nodes, it may lead to poor network training effect, reduce the network's sample recognition ability and poor fault tolerance.

However, if there are too many hidden nodes, the network will be more "fragile," and it will easily fall into the local minimum point and fail to obtain the global best advantage. As a result, overfitting will occur, reducing the applicability of the network.

And if there are too many nodes in the hidden layers, the calculation is relatively complicated, which increases the training time of the network and reduces the learning efficiency of the network.

In order to avoid overfitting as much as possible when training the NN, and to improve the performance and generalization ability of the network, we usually choose as few hidden layer nodes as possible to meet the model accuracy requirements.

Research shows that the number of hidden layer nodes is not only related to the number of input and output units, but also to the degree of nonlinearity between the input and output, the form of the transfer function and the characteristics of the samples.

Therefore, when selecting the number of nodes in the hidden layer, we should comprehensively consider the complexity of the network structure and the size of the error and determine it by testing the adaptive capacity of the model.

At present, there is no accurate and scientific theoretical method for the selection of the number of nodes in the hidden layer. Most of them are determined by empirical formulas.

Some commonly used empirical formulas are as follows:

1. $$K < \sum_{i=1}^{N} C_i^h \tag{10.20}$$

where K is training sample, h is the number of nodes in the hidden layer and N is the number of input units.

2. $$h = \sqrt{N + M} + \varepsilon \tag{10.21}$$

where M is the number of output units, N is the number of input units and ε is a constant between 1 and 10.

3. $$h = \log_2 N \tag{10.22}$$

where N is the number of input units.

In constructing the bicycle behavior NN model, we first determine the approximate range of the number of nodes in the hidden layer according to the formulas (10.20) and (10.21).

Then in the MATLAB® program, a cyclic-self-structuring method is used to dynamically save the network corresponding to the number of nodes in different hidden layers, and then to call the NN to calculate the overall MSE of the CV set, and select the network structure corresponding to the smallest overall MSE of the test samples.

The specific process is shown in Figure 10.4.

10.3.5 Training algorithm and parameter selection

10.3.5.1 Training algorithm selection

Earlier, we introduced that the learning rules of traditional BPNNs are based on the steepest descent method to modify network weights and thresholds. It has the disadvantages of low learning efficiency, slow convergence speed and easy to fall into a local minimum.

The improved BP algorithm mostly uses heuristic learning algorithms or learning algorithms based on numerical optimization theory to train the network, which greatly improves the learning efficiency and applicability of BPNN [3].

Heuristic learning algorithms generally include the gradient descent BP algorithm with momentum (the corresponding function in MATLAB is *traingdm*), gradient descent with adaptive lr BP (the corresponding function in MATLAB is *traingda*), gradient descent with momentum and adaptive lr BP (the corresponding function in MATLAB is *traingdx*), etc.; based on numerical optimization, the theoretical training algorithms mainly include conjugate gradient methods (such as *traincgf*, *traincgb*, *trainscg*), Gauss–Newton method (*trainbfg*, *trainoss*), Levenberg–Marquardt algorithm (*trainlm*), etc.

Next we select several widely used learning algorithms for brief introduction and comparison.

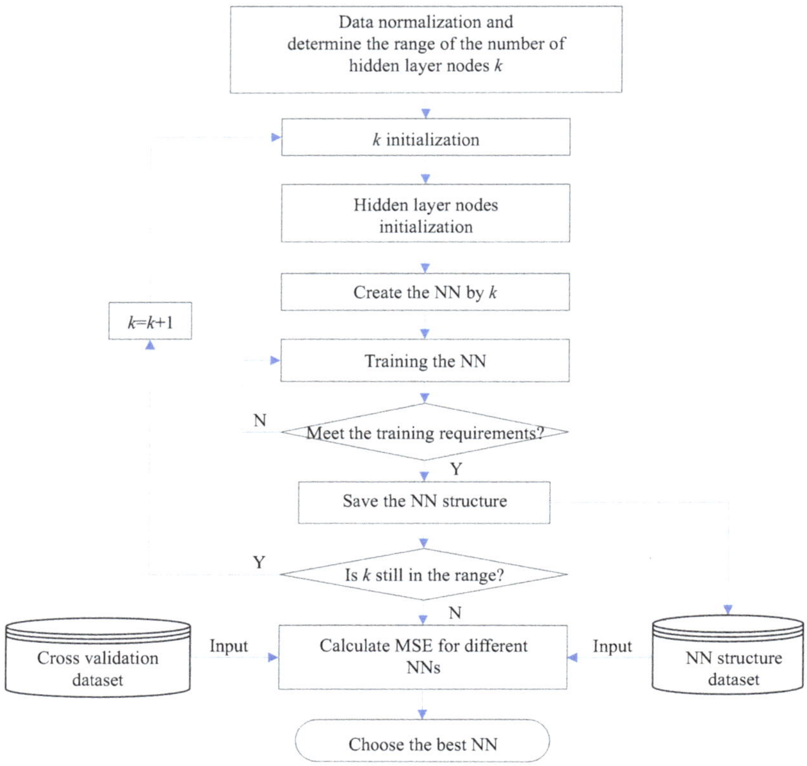

Figure 10.4 The process of choosing the reasonable number of hidden layer nodes

Gradient descent BP with momentum

When adjusting the network weights, the gradient descent BP with momentum takes into account the two error gradient descent directions of the current and last adjustments and calculates the change of the connection weights according to a certain scale factor:

$$\Delta w_{ji}(n+1) - \Delta w_{ji}(n) = \eta d(n) + \beta d(n-1) \tag{10.23}$$

where η is lr, β is momentum factor and $d(n)$ is the negative error gradient of the nth adjustment. Practical application shows that the BP algorithm with momentum term can speed up network convergence and improve network learning efficiency, but the determination of the momentum factor lacks theoretical guidance and requires multiple experiments to determine.

Gradient decent with momentum and adaptive lr BP

In the gradient descent with momentum and adaptive lr BP algorithm, when the error keeps decreasing and approaches the goal, the lr will be appropriately increased and the adjustment range of the weights will be increased, while the error continuously increases and exceeds the set value, the lr will be appropriately reduced and the adjustment range of the weights will be reduced, that is

$$\eta(n+1) = \begin{cases} k_1\eta(n) & E(n+1) < E(n) \\ k_2\eta(n) & E(n+1) \geq E(n) \end{cases} \tag{10.24}$$

where $\eta(n+1)$ is lr at the $(n+1)$th adjustment, k_1 is lr increment factor and k_2 is lr decrement factor; the adaptive lr can shorten the training time and improve the learning efficiency while increasing the learning algorithm reliability.

Conjugate gradient method

Conjugate gradient method is an improved algorithm based on gradient method, which can reduce the defects of gradient method oscillation and slow convergence. The basic idea is to find the direction in which the negative gradient direction of the error and the previous search direction are conjugated as the new search direction. The specific process is as follows:

1. Determine the initial search direction:

$$p_0 = -g_0 \tag{10.25}$$

2. One-dimensional search in this direction:

$$w_{n+1} = w_n + \eta_n p_n \tag{10.26}$$

3. Then use the conjugate direction to do a new search, usually adding the previous search direction to the current negative gradient direction:

$$p_n = d(n) + \gamma_n p_{n+1} \tag{10.27}$$

According to the selection of γ_n, a variety of conjugate gradient methods are derived, such as *Fletcher–Reeves*, *Polak–Ribiere* and other modified methods.

In addition, the Gauss–Newton method, nonlinear least square method (Levenberg–Marquardt), etc. are commonly used. The training method based on numerical optimization theory usually reduces the error gradient by one iteration. Therefore, the training and learning time required is shorter; but at the same time, this method occupies a large amount of system memory and has a slightly poorer performance. The momentum gradient descent method and adaptive lr method occupy less system memory. Under certain data size and error accuracy requirements, the network has a better learning ability and adaptability.

The training algorithm in this chapter uses the algorithm with dynamic momentum and lr adaptive adjustment.

10.3.5.2 Training parameters selection

Commonly used training parameters in BPNN are *transfer function*, lr (η), *momentum factor (β), margin of error* $E_{\min}$, *epochs*, etc. These parameters have a significant impact on the learning efficiency and the reliability of the network.

Transfer function selection

Transfer function selection: In BPNNs with double hidden layers, the most commonly used transfer functions are sigmoid function and linear function (purelin). The sigmoid function has the characteristics of flat ends and sharp changes in the middle. In a certain interval near the middle, it can play a good transfer function. Outside this interval, the output value of the transfer function is close to 1. The transfer effect is not obvious. This is just as it is often difficult for humans to produce strong thinking responses to problems that are far below and above their own intelligence and knowledge. S-type functions have strong nonlinear mapping capabilities and are often used as transfer functions for hidden layers (such as logsig and tansig functions); linear functions are rarely used alone, sometimes as the transfer of the output layer in a multilayer BPNN functions can play a good learning effect.

This book uses a four-layer BPNN structure with double hidden layers. The transfer functions of the hidden layer and the output layer are tansig (hyperbolic tangent function), tansig and purelin. The first normalization method is selected in data preprocessing.

Learning rate and momentum factor

Since the *triangdx* training algorithm with adaptive adjustment of the lr is selected in this paper, the lr will not be discussed here. The purpose of increasing the momentum term is to prevent network training from falling into a local minimum. In theory, the value of the momentum factor β is related to the magnitude of the network weight correction amount. In practice, a constant value between 0.3 and 0.9 is generally used. The value depends on the specific training situation.

Margin of error and epochs

In the learning process of the BPNN, the target accuracy and epochs of the system need to be determined in advance according to the actual needs and the convergence ability of the network. The system target accuracy is usually measured by the system MSE, that is, the MSE of the actual output and the target output E_{mse}:

$$E_{mse} = \sqrt{\frac{1}{N}\sum_{i=1}^{N} e_i^2} \tag{10.28}$$

where N denotes the number of learning mode pairs, e_i denotes the residual of predicted and actual values. E_{mse} reflects the average deviation between predicted and actual values. According to the experimental results, the target accuracy of the system is determined to be 0.01 and the maximum number of training is 5,000–8,000.

10.4 The bicycle conflict avoidance model based on BPNN

Based on the previous analysis, we will use the neural network toolbox in MATLAB to build a BCA model. First, according to different conflict situations, we built a BPNN-based BCA model in B–C conflict situations (BCA-C), a BPNN-based BCA model B–B conflict situations (BCA-B), and a BPNN-based BCA model B–P conflict situations (BCA-P), respectively.

By adding an input variable of the NN, the conflict situation type, we can integrate the three NN models (BCA-C, BCA-B and BCA-P) into one model (BCA model). Based on this, we investigate the effect of the cyclist's reaction time on the performance of the BCA model.

10.4.1 BPNN-based bicycle conflict avoidance model in B–C conflict situations

In our dataset, there are 60 effective sample cases (1,244 records) in the B–C conflict situation. We randomly pick 30 cases as the training set (567 records), 15 cases as CV set (326 records) and 25 cases as the test set (351 records).

According to the previous BPNN modeling process, the main parameter settings of the BPNN-based BCA-C model are shown in Table 10.2.

According to the parameters in Table 10.2, we choose different numbers of hidden layer nodes to train the NN. The specific training process is shown in Figure 10.5.

The MSE surface of the NN models of different numbers of hidden layer nodes by the CV set is shown in Figure 10.6.

From the MSE matrix, we select the network structure corresponding to the minimum MSE and determine that the final structure of the network is 6-8-12-2 (i.e., the first hidden layer node number $N1$ is 8 and the second hidden layer node number $N2$ is 12), and the corresponding NN is called BCA-C.

The weight matrix $w1$ between the input layer and the first hidden layer is

$$w1 = \begin{bmatrix} 1.3048 & -0.2488 & 0.2822 & 1.2882 & -1.0536 & 0.2144 \\ -0.6404 & -0.8252 & 0.1719 & 1.2129 & 0.1861 & -1.5434 \\ 0.6396 & 0.6503 & -0.7578 & 1.0919 & -0.2516 & -0.6304 \\ -0.8152 & 1.6947 & 0.8411 & 0.2111 & 0.0832 & -0.4329 \\ -0.6425 & 1.0672 & 1.0048 & 0.7676 & -0.9567 & -0.4810 \\ -0.0553 & -0.9906 & -0.0282 & 0.7838 & 0.7770 & -1.6035 \\ -0.0193 & -1.4975 & -0.4379 & -1.4711 & -0.8621 & 0.6468 \\ -0.1026 & 0.0309 & -0.5417 & 0.4174 & 1.7321 & 0.7057 \end{bmatrix}$$

Table 10.2 Main parameters of the BPNN model of BCA-C

Parameters	**Value**
Number of hidden neuron nodes $N1$, $N2$	7–20, 7–20
Data normalization interval	[−1,1]
Transfer function	*tansig, tansig, purelin*
Network training function	*traingdx*
Momentum factor (β)	0.9
Goal accuracy (goal)	0.01
Maximum training epochs	8,000

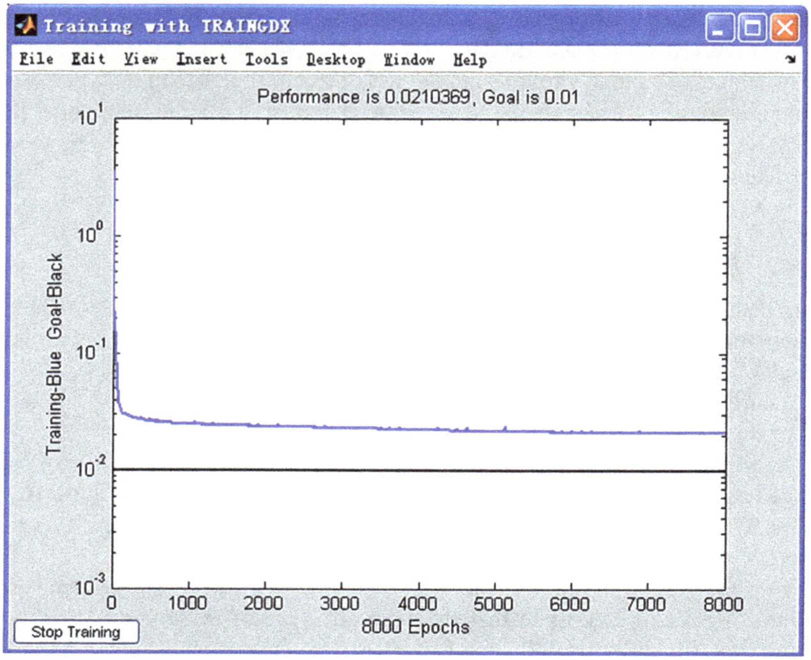

Figure 10.5 BP neural network training diagram

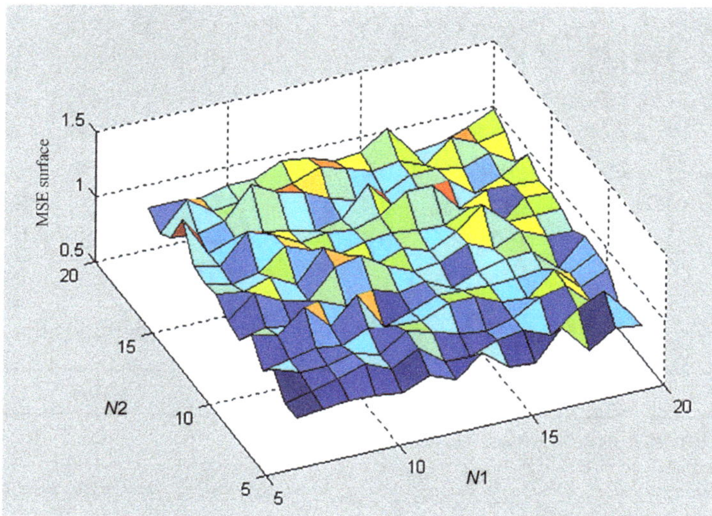

Figure 10.6 MSE surface of different numbers of hidden layer nodes

The weight matrix $w2$ between the first and second hidden layers is

$$w2=\begin{bmatrix} -0.1543 & -0.6962 & -0.4339 & 0.1874 & 1.8585 & -0.7884 & -0.0505 & -1.5432 \\ 0.9871 & -0.9571 & -0.9324 & 0.0762 & -0.3605 & 0.1895 & 0.0879 & 0.8283 \\ -0.6712 & 1.0878 & 0.4053 & -0.3331 & -0.4174 & -0.9715 & 0.1931 & 0.7720 \\ -0.6674 & 0.9667 & -0.1994 & -1.6855 & 0.0649 & -0.8355 & -0.1372 & -0.6642 \\ 0.5194 & 1.0117 & 0.4107 & -0.8306 & -1.2189 & -0.4515 & 0.5736 & -0.3736 \\ -0.3561 & -0.5613 & -1.2405 & -0.6371 & 0.19218 & -1.0005 & 0.3631 & 0.6115 \\ 0.7480 & -0.4714 & 0.9640 & -0.9049 & 0.15778 & 0.5529 & 0.2608 & -0.6319 \\ -0.7855 & 1.1956 & 0.3128 & -0.9159 & 0.27722 & -0.1233 & 0.2331 & 0.6584 \\ -1.0362 & 0.4601 & -0.0063 & -0.5398 & -0.7635 & 0.8048 & -1.8017 & -0.7240 \\ 0.7555 & -0.8637 & 0.2994 & 0.8847 & 0.5298 & 0.4871 & 1.4138 & -0.1793 \\ 0.7985 & -0.1535 & 0.6338 & 1.1287 & -0.5297 & -0.2698 & 0.1807 & 0.7823 \\ -0.3501 & -0.8951 & -1.0684 & 0.1882 & -0.7199 & 0.7344 & 1.0219 & -0.7756 \end{bmatrix}$$

The weight matrix $w3$ between the hidden layer and the output layer is

$$w3=\begin{bmatrix} -0.3549 & 0.0510 \\ -0.0525 & 1.3630 \\ -0.0493 & 0.3686 \\ -0.5857 & 0.5426 \\ -0.4770 & -0.3530 \\ 0.5065 & 0.0069 \\ 0.5418 & 0.1423 \\ 0.5897 & -0.2734 \\ 0.3572 & -0.6806 \\ -0.1372 & -0.8960 \\ 1.0471 & 0.0633 \\ 0.3421 & -0.9600 \end{bmatrix}$$

10.4.2 BPNN-based bicycle conflict avoidance model in B–B conflict situations

In our dataset, there are 41 effective sample cases (786 records) of the B–C conflict situation. We randomly pick 20 cases as the training set (398 records), 6 cases as CV set (126 records) and 15 cases as the test set (262 records).

Similar to the previous BCA-C NN training process, the main trained parameter results of the BPNN-based BCA-B model are shown in Table 10.3.

Table 10.3 Main parameter results of the BPNN model of BCA-B

Parameters	Value
Number of hidden neuron nodes *N*1, *N*2	8, 7
Data normalization interval	[−1,1]
Transfer function	*tansig, tansig, purelin*
Network training function	*traingdx*
Momentum factor (β)	0.8
Goal accuracy (goal)	0.01
Maximum training epochs	8,000

Table 10.4 Main parameter results of the BPNN model of BCA-P

Parameters	Value
Number of hidden neuron nodes *N*1, *N*2	8, 8
Data normalization interval	[−1,1]
Transfer function	*tansig, tansig, purelin*
Network training function	*traingdx*
Momentum factor (β)	0.9
Goal accuracy (goal)	0.01
Maximum training epochs	5,000

10.4.3 BPNN-based bicycle conflict avoidance model in B–P conflict situations

In our dataset, there are 50 effective sample cases (1,249 records) of the B–C conflict situation. We randomly pick 25 cases as the training set (508 records), 10 cases as CV set (276 records) and 15 cases as the test set (465 records).

Similar to the previous BCA-C NN training process, the main trained parameter results of the BPNN-based BCA-B model are shown in Table 10.4.

10.4.4 NN-based bicycle conflict avoidance model considering gender and type of conflict object

In this section, we try to add an impact coefficient δ of different types of conflicting objects on the bicycle and integrate the three types of models (BCA-C, BCA-B and BCA-P), that is, add a variable δ to the input layer of the NN.

Here, we consider that the impact coefficient δ of conflicting motor vehicles, bicycles and pedestrians is somewhat similar to the impact of different conflicting vehicle types in the NCB model. Therefore, the values of δ in the B–C, B–B and B–P conflict situations refer to M_C, M_B and M_P in the NCB model, respectively. More information refers to Table 8.4.

At the same time, we also consider the influence of gender factors on bicycle behavior and then add a gender variable g in the input layer (for male cyclists g take 1, for female cyclists g take 0), so the input layer nodes of the network increase to 8.

At the same time, we get 151 effective sample cases (3,279 records) of all bicycle conflict situations. We combine the training sets, CV sets and test set of the three different conflict situations and get the integrated training set of 75 cases with 1,473 records, CV set of 31 cases with 728 records and a test set of 45 cases with 1,078 records. The integrated training dataset is shown in Table 10.5.

The training results of the integrated BCA model main network parameters are shown in Table 10.6.

Table 10.5 Neural network dataset considering gender and type of conflict object

Case ID	Frame	δ	g	Δw_x	Δw_y	Δv_x	Δv_y	$\Delta v'_x$	$\Delta v'_y$	ma_x	ma_y
BC11	236	5.5	1	−6.4	8.0	0.17	−2.06	−0.14	1.75	0.46	0.29
BC11	251	5.5	1	−6.3	6.3	−0.07	−3.50	−0.34	0.43	0.78	0.09
BC11	261	5.5	1	−6.4	4.8	−0.28	−3.59	−0.60	0.34	0.20	0.13
...	...			...	...	...	...	...	...	...	...
BC21	19	5.5	0	4.0	10.5	−3.15	−3.35	−0.31	1.26	0.97	−0.86
BC21	24	5.5	0	3.4	9.7	−3.20	−3.93	−0.57	0.62	1.23	−0.70
BC21	29	5.5	0	2.8	8.9	−3.33	−4.06	−0.80	0.41	1.39	−1.31
...	...			...	...	...	...	...	...	...	...
BB11	28	1	1	7.9	−10.8	−2.78	3.98	−0.43	−0.72	−1.07	0.62
BB11	33	1	1	7.4	−9.9	−2.79	5.00	−0.46	0.29	−1.19	0.16
BB11	38	1	1	6.8	−8.9	−2.70	5.42	−0.38	0.72	−1.47	0.77
...	...			...	...	...	...	...	...	...	...
BB34	11	1	0	1.2	−4.0	3.51	2.50	0.85	−0.65	0.94	−0.02
BB34	16	1	0	1.9	−3.5	4.31	3.02	0.97	−0.20	1.07	−0.20
BB34	21	1	0	2.9	−2.8	4.56	3.25	0.95	−0.02	1.41	0.18
...	...			...	...	...	...	...	...	...	...
BP10	48	0.9	1	2.5	1.0	1.40	−3.27	0.11	−0.04	−0.03	0.68
BP10	53	0.9	1	2.8	0.3	1.28	−3.11	0.04	0.03	0.03	−0.23
BP10	58	0.9	1	3.0	−0.3	1.24	−3.32	0.01	−0.01	0.06	−1.14
...	...			...	...	...	...	...	...	...	...
BP26	98	0.9	0	3.5	5.8	−0.2	−3.3	0.41	0.19	0.11	0.40
BP26	103	0.9	0	3.4	5.1	−0.3	−3.3	0.32	0.19	−0.14	0.81
...	...			...	...	...	...	...	...	...	...

Note: g represents gender, for male cyclists it is 1, for female cyclists it is 0; δ is the influence coefficient of different conflict objects on bicycles. In this model, the influence coefficients of conflict motor vehicles, bicycles and pedestrians on bicycles are 5.5, 1 and 0.9, respectively.

Table 10.6 Main NN parameter results of the integrated BCA model

Parameters	Value
Number of hidden neuron nodes $N1$, $N2$	8, 14
Data normalization interval	[−1,1]
Transfer function	*tansig, tansig, purelin*
Network training function	*traingdx*
Momentum factor (β)	0.9
Goal accuracy (goal)	0.01
Maximum training epochs	5,000

In the previous model, the reaction time of the cyclist is set to 0.5 s. In the next section, we verify the NN-based BCA model. We also calibrate the collision avoidance NN model when the cyclist's response time is taken as 0.8 s and compare with the bicycle collision avoidance model of Sections 10.4.1–10.4.4.

10.5 Model simulation and verification

In Chapter 9, we introduced that in microscopic traffic simulation, an accurate and reliable behavior model is really crucial. Especially when it is the core behavior model of the computer simulation platform, it determines the effectiveness and practicability of the simulation platform.

Therefore, in order to ensure that the BCA simulation model we constructed can better simulate the actual traffic behavior, we need to verify and evaluate the model through case analysis.

In this chapter, we will use the sample data in the test set to verify the BCA model built in Section 10.4.

In Chapter 9, we have introduced commonly used model verification methods, including statistical test methods, curve methods, theoretical comparison methods, model comparison methods and subjective validity evaluation.

Here, we choose the statistical test and trajectory curve to evaluate the validity of the NN-based BCA model.

The statistical test method mainly uses mathematical statistics to evaluate the simulation results of the model as a whole; the trajectory curve method analyzes and compares the actual trajectory and the calculated trajectory of the predicted sample, which can intuitively reflect the error between the actual data and the predicted data.

10.5.1 Model simulation

10.5.1.1 Model initialization state extraction

When using the established BPNN-based bicycle avoidance model for simulation experiments, the initial state of the test case samples of the model needs to be known first, and then the motion trajectory of the bicycle is predicted given the motion trajectory of the conflict object.

The initial state of the test case sample mainly includes the starting point information and destination point information of the bicycle. The starting point information is mainly the starting point coordinates, initial speed and initial time of the bicycle. The destination point information is mainly the coordinates of the destination point and the bicycle's expected arrival time at the destination.

10.5.1.2 Simulation of bicycles' conflict avoidance behavior

First, based on the initialization information of the test case sample, we use the previous input variable calculation formulas to calculate the input variables and pass them to the NN model to get the bicycle's acceleration at the next time step $t+1$ then use the predicted acceleration to calculate speed and position of the bicycle at the next time step $t+1$ by

$$v_x(t+1) = v_x(t) + a_x(t+1) \times 0.2 \tag{10.29}$$

$$w_x(t+1) = w_x(t) + (v_x(t) + v_x(t+1)) \times 0.1 \tag{10.30}$$

where $a_x(t+1)$ is the bicycle's X-direction acceleration component calculated from the input information at time step t; $v_x(t)$ is bicycle X-direction speed component at time step t; $w_x(t)$ is bicycle X-direction coordinate at time step t; the interval between each iteration is 0.2 s. The same method can be used to calculate the Y-direction speed component and coordinates of the bicycle.

Then, based on the updated bicycle status, we calculate the input information at time step $t+1$, then call the NN model to predict the bicycle's acceleration at time step $t+2$, repeat the previous steps until the last record of the test case sample data, and the model ends.

The specific flowchart is shown in Figure 10.7.

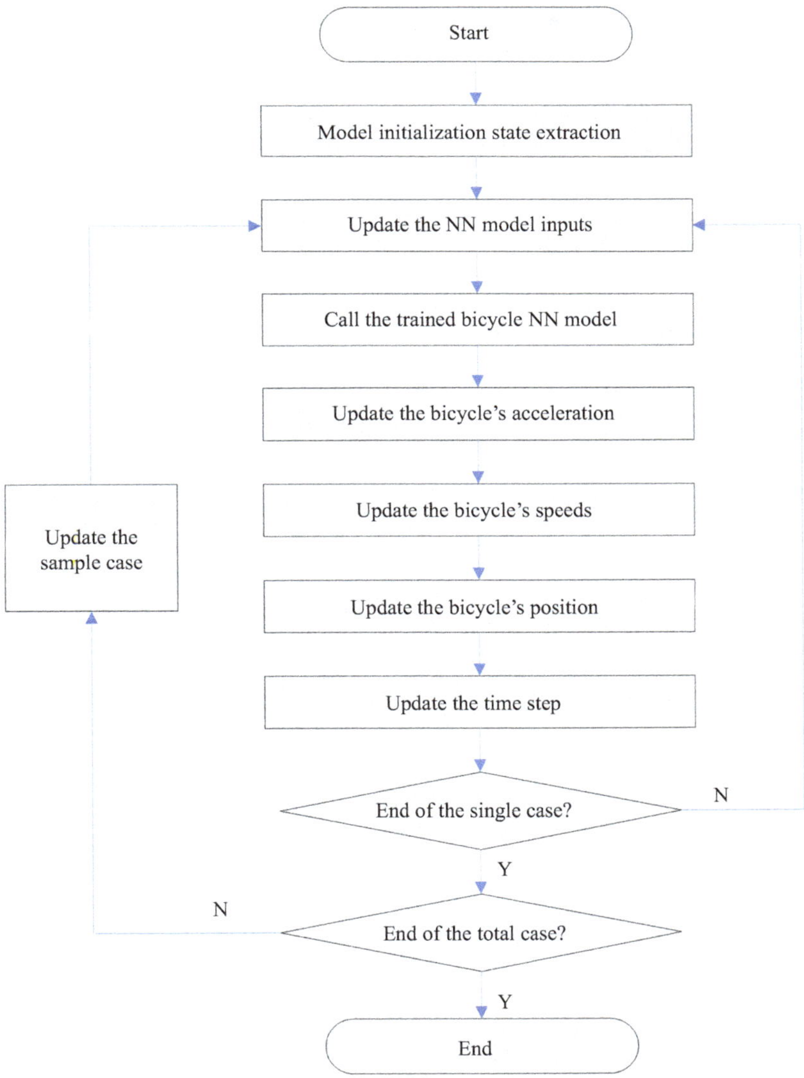

Figure 10.7 NN-based bicycle behavior flowchart

Here, we first use VBA function in EXCEL to extract the initialization information of all test case samples, and then import them into MATLAB, and call the bicycle NN model established earlier in MATLAB to calculate the motion parameters of each test case sample. On this basis, the effectiveness and reliability of the NN-based BCA models under different scenarios are analyzed in the next section.

10.5.2 Model verification

In this section, we will use the trajectory curve method and mathematical statistics method to verify the bicycle collision avoidance NN model.

10.5.2.1 Model curve comparisons

Here, we first perform trajectory curve comparisons on the BPNN-based BCA-C, BCA-B and BCA-P constructed in Sections 10.4.1–10.4.3.

Then, we compare the curves of the integrated BPNN-based BCA model constructed in Section 10.4.4.

We calculate and compare the trajectory, speed and acceleration curves of the bicycle.

The BCA-C model

Figures 10.8–10.10 are examples of the calculation results of a BCA-C model verification case sample.

The BCA-B model

Figures 10.11–10.13 are examples of the calculation results of a BCA-B model verification case sample.

The BCA-P model

Figures 10.14–10.16 are examples of the calculation results of a BCA-P model verification case sample.

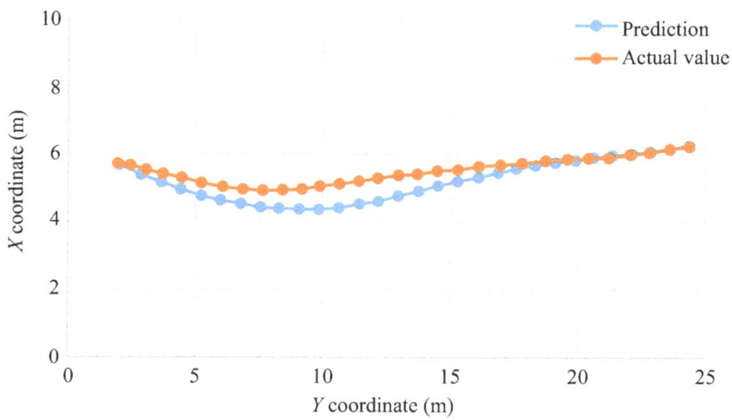

Figure 10.8 The comparisons of bicycle trajectories between field and BCA-C model

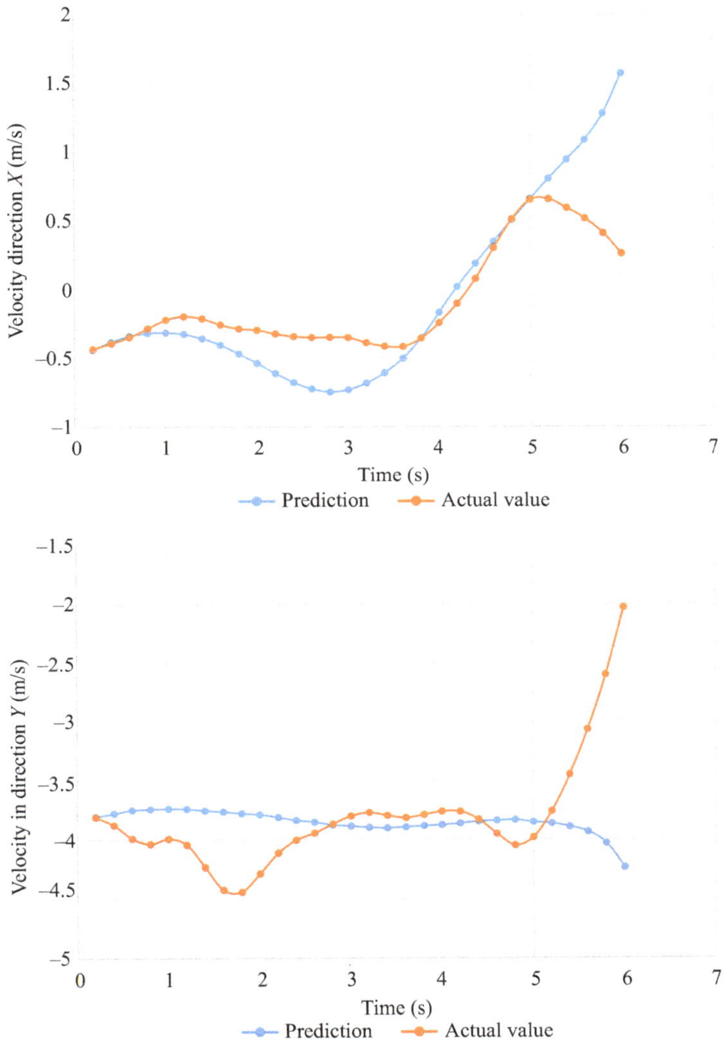

Figure 10.9 The comparisons of bicycle speed between field and BCA-C model

The integrated BCA model

We perform curve verification on the integrated BCA model. Here we pick a B–C conflict situation as an example, the specific comparison results are shown in Figures 10.17–10.19.

In the previous models, we set the cyclist's reaction time as 0.5 s. In addition, we also constructed and verified the corresponding BPNN BCA model when the cyclist's reaction time is 0.8 s. Here, we will not compare the curves one by one,

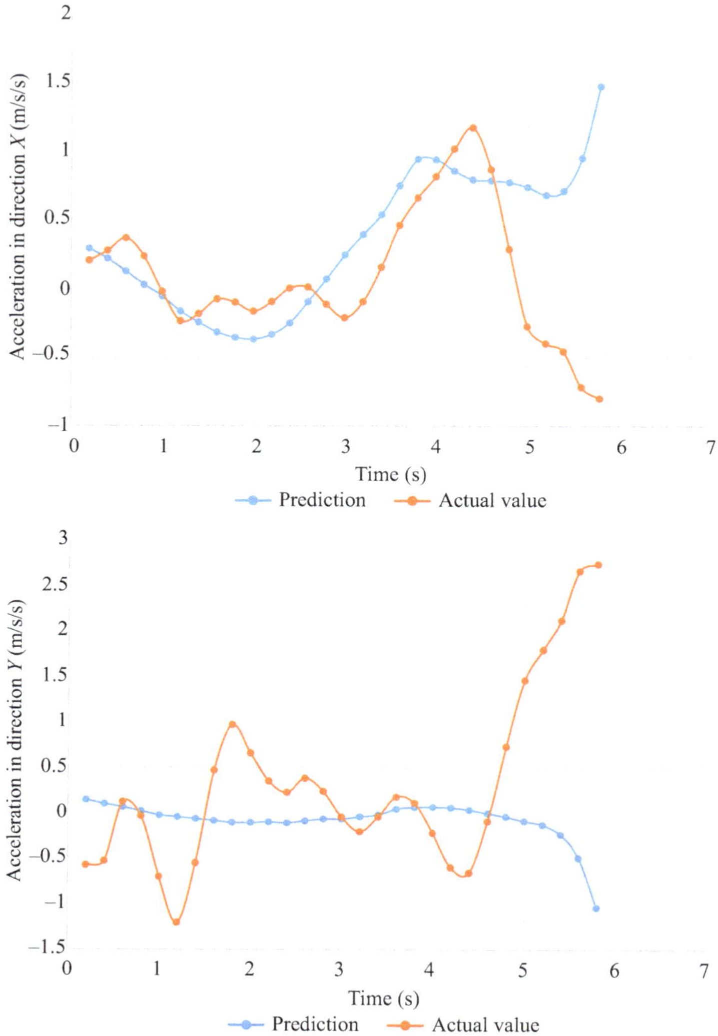

Figure 10.10 The comparisons of bicycle acceleration between field and BCA-C model

while in the next section, we will use mathematical statistics to verify the effectiveness of the various BCA NN models.

10.5.2.2 Comparison of mathematical statistics

In mathematical statistics method, we mainly choose two measurement parameters: the average value of absolute deviation of prediction results ($\bar{d}$), and the MSE of absolute deviation of prediction results (S_e).

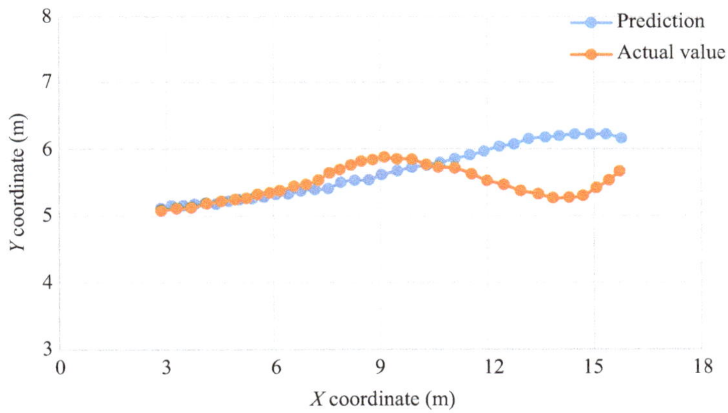

Figure 10.11 The comparisons of bicycle trajectories between field and BCA-B model

Among them, the $\bar{d}$ mainly reflects the average deviation between the predicted value and the actual value. Here, we use $\bar{d}$ to evaluate the effectiveness and reliability of the bicycle NN model.

The S_e mainly represents the fluctuation of absolute deviation to evaluate the accuracy of the model; the smaller S_e, the higher the accuracy of the bicycle NN model.

The specific calculation formulas for $\bar{d}$ and S_e are as follows:

$$\bar{d} = \frac{\sum_{i=1}^{n} d_i}{N} = \frac{\sum_{i=1}^{n} |x_i - \widehat{x}_i|}{N} \tag{10.31}$$

$$S_e = \sqrt{\frac{\sum_{i=1}^{n} (d_i - \bar{d})^2}{N}} \tag{10.32}$$

In the bicycle microscopic simulation, the accuracy and reliability of the bicycle simulation model are mainly reflected in the degree of fitness between the model simulation trajectory and the actual trajectory.

In this section, we will calculate the $\bar{d}$ and S_e of the bicycle's trajectory in the X-direction and Y-direction in the test case sample, thus evaluating the validity and accuracy of the BCA model.

The specific results are shown in Table 10.7.

As can be seen from Table 10.7, the $\bar{d}$ of the X-direction trajectory error of each type of the BCA model in the statistical comparison results is between 0.1 and 0.7 m, and $\bar{d}$ of the Y-direction trajectory error is between 0.4 and 0.9 m. The average simulation error of the model in the bicycle's trajectory is small, which indicates that the various types of BCA model based on BPNN can represent the

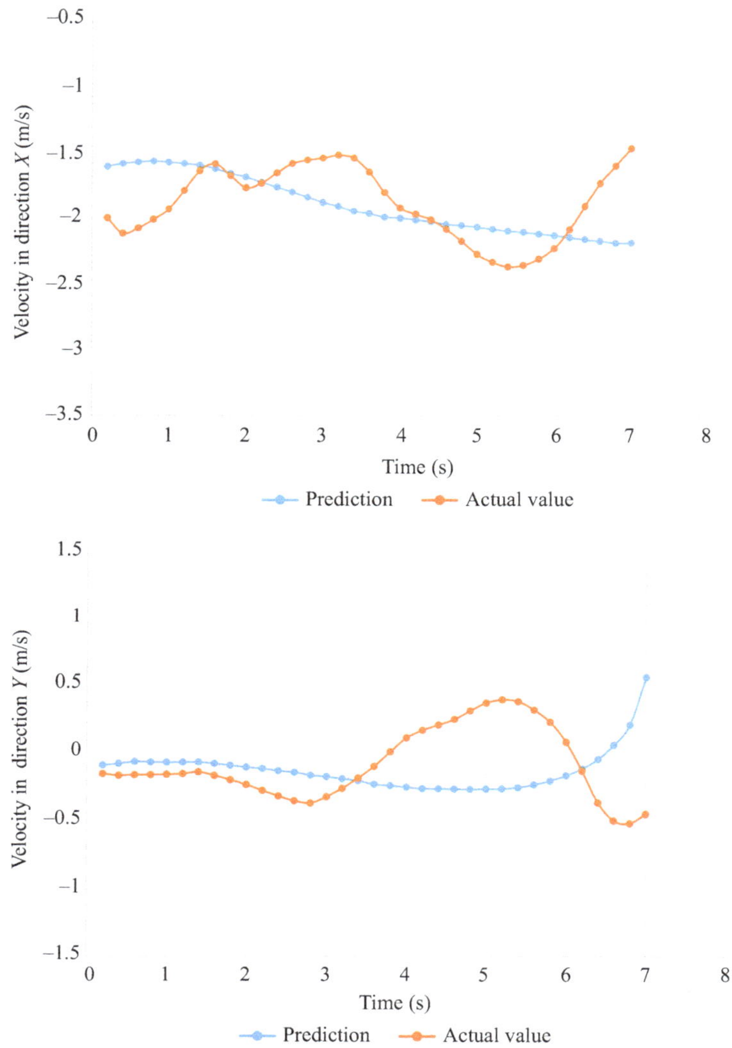

Figure 10.12 The comparisons of bicycle speeds between field and BCA-B model

avoidance rule of bicycles in conflict situations. At the same time, the S_e of each BCA model is less than 0.9, indicating that the model has higher accuracy and stability.

10.5.2.3 Discussion of model validation results

Through the previous simulation experiments and error statistical analysis, we can find that the BPNN model used can fit the trajectory of the bicycle well, which proves the effectiveness of the BPNN-based BCA simulation model. At the same

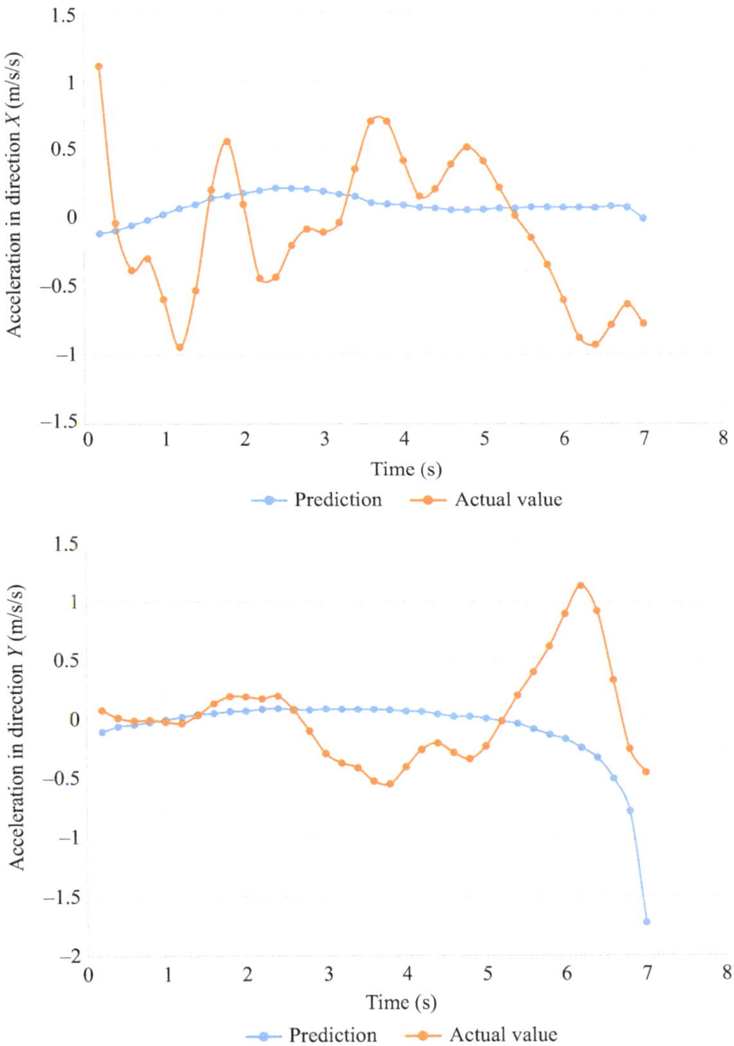

Figure 10.13 The comparisons of bicycle accelerations between field and BCA-B model

time, the effectiveness of the BCA influencing factors we have selected is also verified, which has certain reference significance for the microscopic traffic simulation modeling of bicycles.

By comparing the test error statistics of different BPNN models (Type 1 vs. Type 2 in Table 10.7), we find that the integrated BPNN model, which integrates gender and conflict influence coefficients, is still effective in different conflict situations. This shows that using NN to construct and improve simulation models is

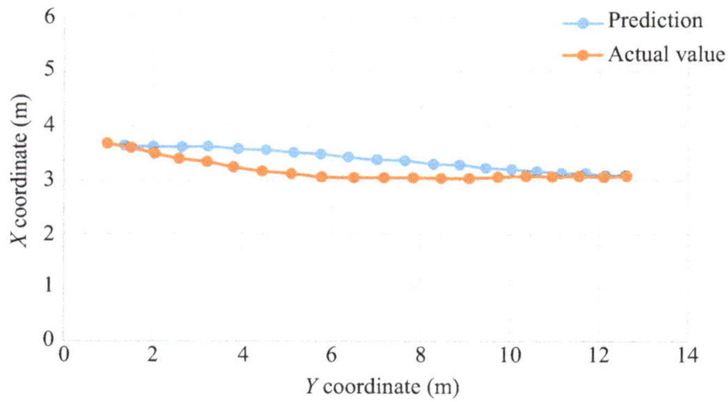

Figure 10.14 The comparisons of bicycle trajectories between field and BCA-P model

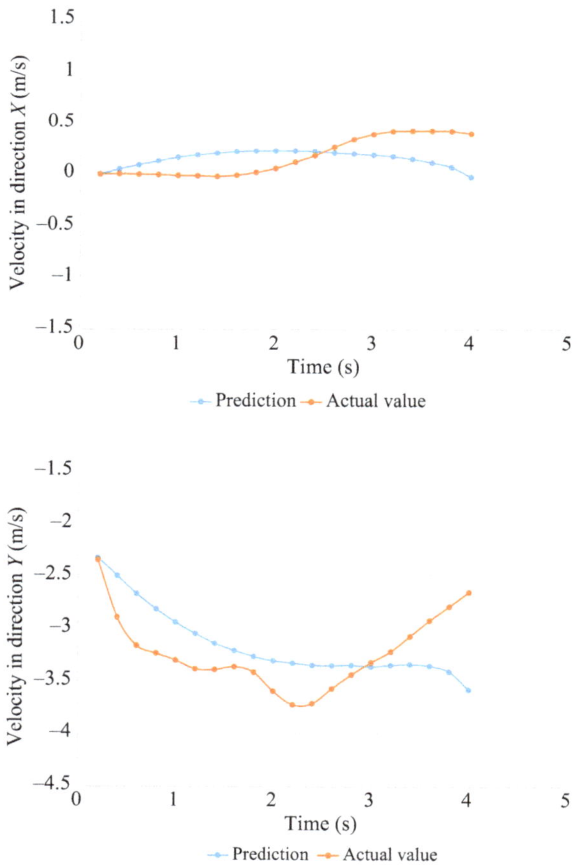

Figure 10.15 The comparisons of bicycle speed between field and BCA-P model

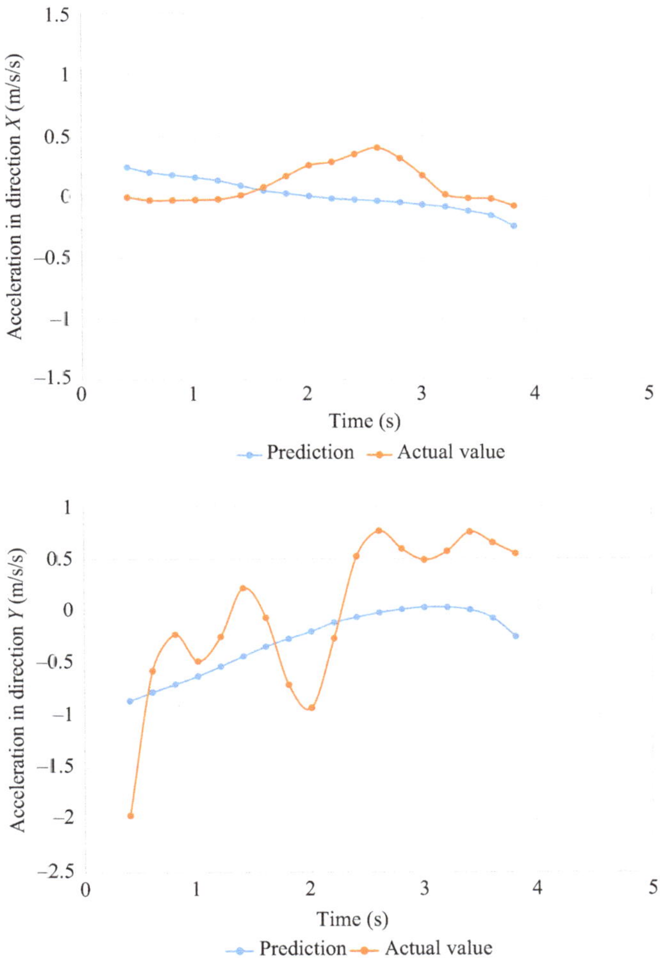

Figure 10.16 The comparisons of bicycle acceleration between field and BCA-P model

more convenient than traditional mathematical models. After adding suitable input influence variables, we only need to retrain the NN by adjusting the number of nodes in the hidden layer, so that the model's parameter calibration workload smaller.

By comparing the test error statistics of different bicycle BPNN models with different reaction times (Type 2 vs. Type 3 in Table 10.7), we can see that when the cyclist's reaction time is taken as 0.8 s, the error of the model is larger.

This shows that the value of cyclist's reaction time has an impact on model performance. In the actual traffic environment, the reaction time of 0.5 s is more in line with the actual situation.

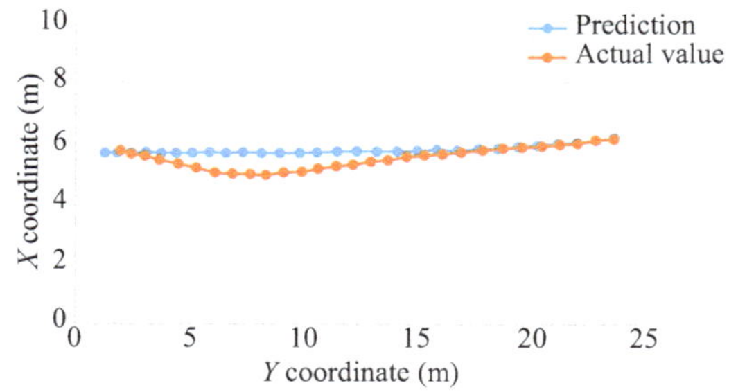

Figure 10.17 The comparisons of bicycle trajectories between field and integrated BCA model

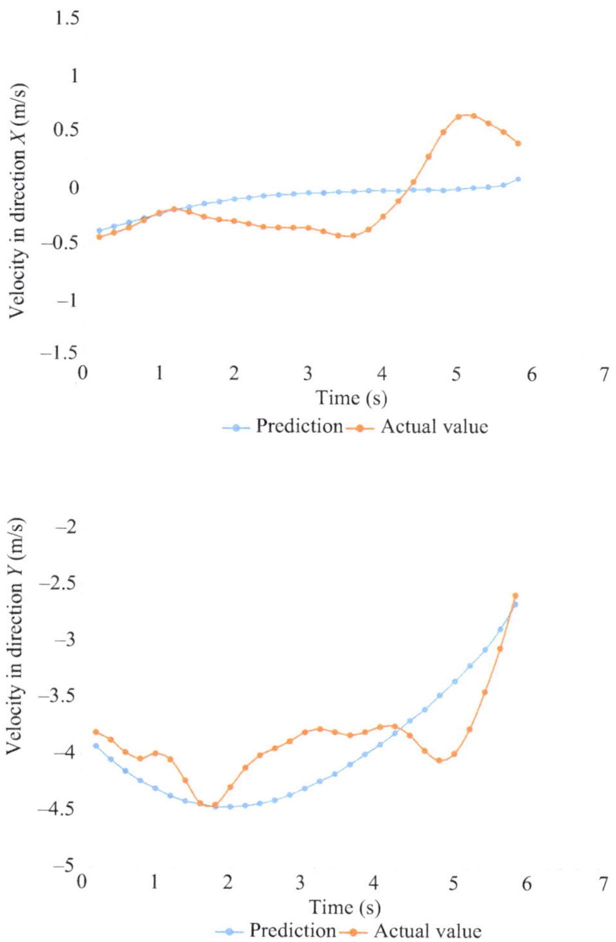

Figure 10.18 The comparisons of bicycle speed between field and integrated BCA model

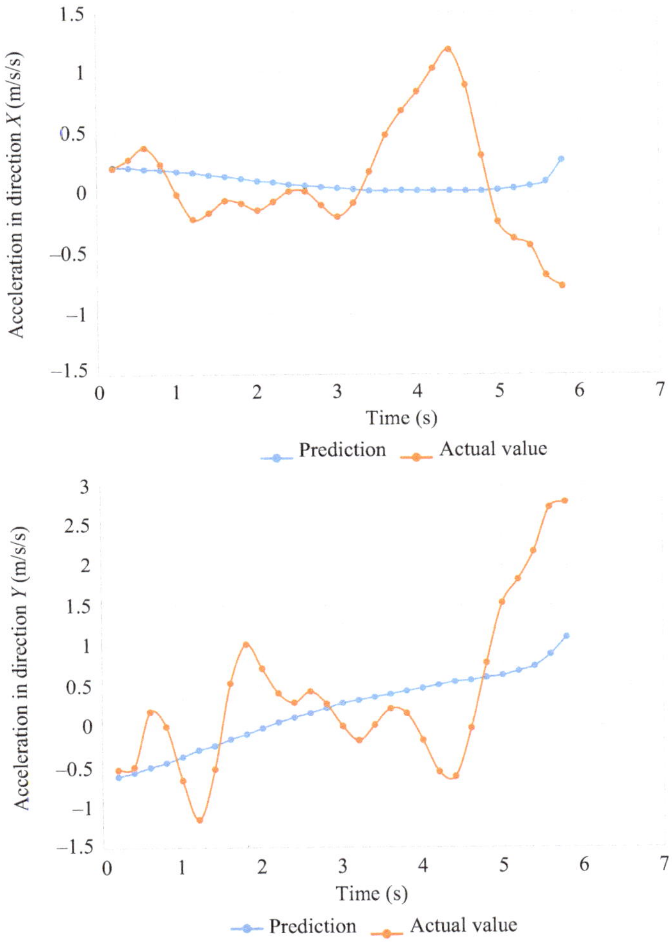

Figure 10.19 The comparisons of bicycle acceleration between field and integrated BCA model

10.6 Summary

In this chapter, we proposed an NN-based bicycle collision avoidance behavioral model at un-signalized intersections. First, we make some brief introduction to the ANNs and how BPNNs work. Then we described the modeling process of the BCA behavior model based on NN, mainly including the determination of model inputs and outputs, data normalization, learning sample division, NN structure selection, training algorithm and parameter selection. Last, we compare three different types of the BPNN-based BCA models and verify the effectiveness of our BPNN-based BCA simulation model.

Table 10.7 Verification results of bicycle trajectory statistics on various BPNN-based BCA models

Type 1 (models built for different conflict situations)				**The trajectory error in the *X*-direction**		**The trajectory error in the *Y*-direction**	
Model	NN structure	Goal accuracy	Max. training epochs	$\bar{d}$	S_e	$\bar{d}$	S_e
BCA-C	6-8-12-2	0.01	8,000	0.62	0.76	0.46	0.35
BCA-B	6-8-7-2	0.01	8,000	0.34	0.29	0.85	0.88
BCA-P	6-8-8-2	0.01	5,000	0.30	0.35	0.41	0.30
Type 2 (integrated BCA model with 0.5 s reaction time)							
Conflict situation	NN structure	Goal accuracy	Max. training epochs	$\bar{d}$	S_e	$\bar{d}$	S_e
B–C	8-8-14-2	0.01	8,000	0.58	0.61	0.90	0.67
B–B				0.38	0.41	0.53	0.64
B–P				0.17	0.18	0.54	0.53
Total				0.34	0.43	0.69	0.64
Type 3 (integrated BCA model with 0.8 s reaction time)							
Conflict situation	NN structure	Goal accuracy	Max. training epochs	$\bar{d}$	S_e	$\bar{d}$	S_e
Total	8-8-14-2	0.01	8,000	0.43	0.44	0.80	0.88

References

[1] X.G. Wang, Z. Tang, and H. Tamura. A Modified Error Function for the Back-Propagation Algorithm. Neurocomputing, 2004, 57:477–484.

[2] R. Hecht-Nielsen. Neurocomputer Applications. In Neural Computers, Berlin, Heidelberg, Springer, 1989, pp. 445–453.

[3] W. Jin, Z.J. Li, L.S. Wei, and Z. Han. The improvements of BP neural network learning algorithm. In WCC 2000-ICSP 2000. 2000 5th International Conference on Signal Processing Proceedings. 16th World Computer Congress 2000. IEEE, 2000, 3: 1647–1649.

Index

www.ingramcontent.com/pod-product-compliance
Lightning Source LLC
LaVergne TN
LVHW011839200926
848484LV00002B/378

* 9 7 8 1 7 8 5 6 1 9 5 1 9 *